TREATISE ON INVERTEBRATE PALEONTOLOGY

Prepared under the Guidance of the
Joint Committee on Invertebrate Paleontology

Paleontological Society | *Society of Economic Paleontologists and Mineralogists* | *Palaeontographical Society*

Directed and Edited by

RAYMOND C. MOORE

Part P
ARTHROPODA 2
CHELICERATA with sections on PYCNOGONIDA and PALAEOISOPUS

BY LEIF STØRMER, ALEXANDER PETRUNKEVITCH, AND JOEL W. HEDGPETH

GEOLOGICAL SOCIETY OF AMERICA
and
UNIVERSITY OF KANSAS PRESS

1955

Second Printing 1963
Third Printing 1973
Fourth Printing 1978
Fifth Printing 1992*

Library of Congress Catalogue Card Number 53-12913
ISBN 0-8137-3016-3

*Distributed by the Geological Society of America, Inc., P.O. Box 9140, Boulder, Colorado 80301, from which current price lists of Parts in print may be obtained and to which all orders and related correspondence should be directed. Editorial office for the *Treatise*: Paleontological Institute, 121 Lindley Hall, The University of Kansas, Lawrence, Kansas 66045.

The *Treatise on Invertebrate Paleontology* has been made possible by (1) a grant of funds from The Geological Society of America through the bequest of Richard Alexander Fullerton Penrose, Jr., for preparation of illustrations and partial defrayment of organizational expense; (2) contribution of the knowledge and labor of specialists throughout the world, working in co-operation under sponsorship of The Palaeontographical Society, The Paleontological Society, and The Society of Economic Paleontologists and Mineralogists; and (3) acceptance by the University of Kansas Press of publication without cost to the Societies concerned and without any financial gain to the Press.

TREATISE ON INVERTEBRATE PALEONTOLOGY

Directed and Edited by
RAYMOND C. MOORE

PARTS

The indicated Parts (excepting the first and last) are to be published at whatever time each is ready. All may be assembled ultimately in bound volumes. In the following list, already published Parts are marked with a double asterisk (**) and those in press or nearing readiness for press are marked with a single asterisk (*). Each is cloth bound with title in gold on the cover. Copies are available on orders sent to the Geological Society of America at 3300 Penrose Place, Boulder, Colorado 80301, at prices quoted, which very incompletely cover costs of producing and distributing them but on receipt of payment the Society will ship copies without additional charge to any address in the world.

The list of contributing authors is subject to change.

(A)—INTRODUCTION.
B—PROTISTA 1 (chrysomonads, coccolithophorids, diatoms, etc.).
****C**—PROTISTA 2 (foraminifers).
****D**—PROTISTA 3 (radiolarians, tintinnines) (xii+195 p., 1050 figs.). $3.00.
****E**—ARCHAEOCYATHA, PORIFERA (xviii + 122 p., 728 figs.). $3.00.
****F**—COELENTERATA *(in press, 1955)*.
****G**—BRYOZOA (xii + 253 p., 2,000 figs.). $3.00.
H—BRACHIOPODA.
****I**—MOLLUSCA 1 (chitons, scaphopods, gastropods).
J—MOLLUSCA 2 (gastropods).
****K**—MOLLUSCA 3 (nautiloid cephalopods).
****L**—MOLLUSCA 4 (ammonoid cephalopods).
M—MOLLUSCA 5 (dibranchiate cephalopods).
****N**—MOLLUSCA 6 (pelecypods).
****O**—ARTHROPODA 1 (tribolitomorphs).
****P**—ARTHROPODA 2 (chelicerates, pycnogonids) (this volume).
****Q**—ARTHROPODA 3 (ostracodes).
****R**—ARTHROPODA 4 (branchiopods, cirripeds, malacostracans, myriapods, insects).
****S**—ECHINODERMATA 1 (cystoids, blastoids, edrioasteroids, etc.).
****T**—ECHINODERMATA 2 (crinoids).
****U**—ECHINODERMATA 3 (echinozoans, asterozoans).
****V**—GRAPTOLITHINA (xvii + 101 p., 358 figs.). $3.00.
****W**—MISCELLANEA (worms, conodonts, problematical fossils.
(X)—ADDENDA (index).

CONTRIBUTING AUTHORS

AMSDEN, T. W., Oklahoma Geological Survey, Norman, Okla.
ARKELL, W. J., Sedgwick Museum, Cambridge University, Cambridge, Eng.
BAIRSTOW, LESLIE, British Museum (Natural History), London, Eng.
BARKER, R. WRIGHT, Shell Development Co., Houston, Tex.
BASSLER, R. S., U.S. National Museum, Washington, D.C.
BATTEN, ROGER L., U.S. Geological Survey, Washington, D.C.
BAYER, FREDERICK M., U.S. National Museum, Washington, D.C.
BEAVER, HAROLD H., Baylor University, Waco, Tex.
BELL, W. CHARLES, University of Texas, Austin, Tex.
BERDAN, JEAN M., U.S. Geological Survey, Washington, D.C.
BOSCHMA, H., Rijksmuseum van Natuurlijke Historie, Leiden, Netherlands.
BOWSHER, ARTHUR L., U.S. Geological Survey, Fairbanks, Alaska.

Branson, Carl C., University of Oklahoma, Norman, Okla.
Bulman, O. M. B., Sedgwick Museum, Cambridge University, Cambridge, Eng.
Campbell, Arthur S., St. Mary's College, St. Mary's College, Calif.
Carpenter, Frank M., Museum of Comparative Zoology, Harvard University, Cambridge, Mass.
Casey, Raymond, Geological Survey of Great Britain, London, Eng.
Chavan, André, Thoiry (Ain), France.
Clark, Austin H., U.S. National Museum, Washington, D.C.
Cline, Lewis M., University of Wisconsin, Madison, Wis.
Cole, W. Storrs, Cornell University, Ithaca, N.Y.
Cooper, G. Arthur, U.S. National Museum, Washington, D.C.
Cox, L. R., British Museum (Natural History), London, Eng.
Curry, Dennis, Pinner, Middlesex, Eng.
Davies, L. M., Edinburgh, Scot.
Dechaseaux, C., Laboratoire de Paléontologie à la Sorbonne, Paris, France.
Durham, J. Wyatt, Museum of Palaeontology, University of California, Berkeley, Calif.
Eames, F. E., Anglo-Iranian Oil Company, London, Eng.
Elliott, Graham F., Kensington, Eng.
Emerson, W. K., University of California, Berkeley, Calif.
Exline, Harriet, Rolla, Mo.
Fell, H. Barraclough, Victoria University College, Wellington, N.Z.
Fisher, D. W., New York State Museum, Albany, N.Y.
Flower, Rousseau H., New Mexico Institute of Mining and Technology, Socorro, N.Mex.
Frizzell, Donald L., Missouri School of Mines, Rolla, Mo.
Furnish, William M., State University of Iowa, Iowa City, Iowa.
Gardner, Julia, U.S. Geological Survey, Washington, D.C.
George, T. Neville, Glasgow University, Glasgow, Scot.
Glaessner, M. F., University of Adelaide, Adelaide, S.Austral.
Haas, Fritz, Chicago Natural History Museum, Chicago, Ill.
Hanna, G. Dallas, California Academy of Sciences, San Francisco, Calif.
Häntzschel, Walter, Geologisches Staatsinstitut, Hamburg, Ger.
Harrington, H. J., University of Kansas, Lawrence, Kans.
Hass, Wilbert H., U.S. Geological Survey, Washington, D.C.
Hatai, Kotora, Tohoku University, Sendai, Japan.
Hawkins, H. L., Reading University, Reading, Eng.
Hedgpeth, Joel, Scripps Institution of Oceanography, University of California, La Jolla, Calif.
Henningsmoen, Gunnar, Paleontologisk Institutt, University of Oslo, Oslo, Norway.
Hertlein, L. G., California Academy of Sciences, San Francisco, Calif.
Hessland, Ivar, Geologiska Institutet, University of Stockholm, Stockholm, Swed.
Hill, Dorothy, University of Queensland, Brisbane, Queensl.
Howe, Henry V., Louisiana State University, Baton Rouge, La.
Howell, B. F., Princeton University, Princeton, N.J.
Hyman, Libbie H., American Museum of Natural History, New York, N.Y.
Jaanusson, Valdar, Paleontological Inst., Uppsala, Sweden.
Kamptner, Erwin, Naturhistorisches Museum, Wien, Aus.
Keen, Myra, Stanford University, Stanford, Calif.
Kesling, Robert V., Paleontological Museum, University of Michigan, Ann Arbor, Mich.
King, Ralph H., University of Kansas, Lawrence, Kans.
Knight, J. Brookes, U.S. National Museum, Washington, D.C.
Kummel, Bernhard, Museum of Comparative Zoology, Harvard University, Cambridge, Mass.
La Rocque, Aurèle, Ohio State University, Columbus, Ohio.
Laubenfels, M. W. de, Oregon State College, Corvallis, Ore.
Lecompte, Marius, Institut Royal des Sciences Naturelles, Bruxelles, Belg.
Leonard, A. Byron, University of Kansas, Lawrence, Kans.

Levinson, S. A., Humble Oil & Refining Company, Houston, Tex.

Lochman-Balk, Christina, New Mexico Institute of Mining and Geology, Socorro, N.Mex.

Loeblich, A. R., Jr., U.S. National Museum, Washington, D.C.

Loeblich, Helen Tappan, U.S. Geological Survey, Washington, D.C.

Lohman, Kenneth E., U.S. Geological Survey, Washington, D.C.

Lowenstamm, Heinz A., California Institute of Technology, Pasadena, Calif.

Ludbrook, N. H., Department of Mines, Adelaide, S.Austral.

Marwick, J., New Zealand Geological Survey, Wellington, N.Z.

Melville, R. V., Geological Survey and Museum, London, Eng.

Miller, Arthur K., State University of Iowa, Iowa City, Iowa.

Montanaro-Gallitelli, Eugenia, Istituto di Geología e Paleontología, Università, Modena, Italy.

Moore, Raymond C., University of Kansas, Lawrence, Kans.

Morrison, J. P. E., U.S. National Museum, Washington, D.C.

Muir-Wood, H. Margaret, British Museum (National History), London, Eng.

Newell, Norman D., American Museum of Natural History (and Columbia University), New York, N.Y.

Okulitch, Vladimir J., University of British Columbia, Vancouver, B.C.

Olsson, Axel A., Coral Gables, Florida.

Palmer, Katherine Van Winkle, Paleontological Research Institution, Ithaca, N.Y.

Peck, Raymond E., University of Missouri, Columbia, Mo.

Petrunkevitch, Alexander, Osborn Zoological Laboratory, Yale University, New Haven, Conn.

Poulsen, Chr., Universitetets Mineralogisk-Geologiske Institut, København, Denm.

Powell, A. W. B., Auckland Institute and Museum, Auckland, N.Z.

Puri, Harbans, Florida State Geological Survey, Tallahassee, Fla.

Rasetti, Franco, Johns Hopkins University, Baltimore, Md.

Regnéll, Gerhard, Paleontologiska Institution, Lunds Universitets, Lund, Swed.

Reichel, M., Bernouillanum, Basel University, Basel, Switz.

Richter, Emma, Senckenberg Natur-Museum, Frankfurt-a.-M., Ger.

Richter, Rudolf, Universität Frankfurt-a.-M., Frankfurt-a.-M., Ger.

Robertson, Robert, Stanford University, Stanford, Calif.

Schindewolf, O. H. Geologisch-paläontologisches Institut der Universität Tübingen, Tübingen, Ger.

Schmidt, Herta, Senckenbergische Naturforschende Gesellschaft, Frankfurt-a.-M., Ger.

Scott, Harold W., University of Illinois, Urbana, Ill.

Shaver, Robert, University of Mississippi, University, Miss.

Sieverts-Doreck, Hertha, Stuttgart-Degerloch, Ger.

Smith, Allyn G., Berkeley, Calif.

Sohn, I. G., U.S. Geological Survey, Washington, D.C.

Spencer, W. K., Crane Hill, Ipswich, Suffolk, Eng.

Stainbrook, Merrill A., Brandon, Iowa.

Stehli, Frank, California Institute of Technology, Pasadena, Calif.

Stenzel, H. B., University of Houston, Houston, Tex.

Stephenson, Lloyd W., U.S. Geological Survey, Washington, D.C.

Størmer, Leif, Paleontologisk Institutt, University of Oslo, Oslo, Nor.

Stubblefield, C. J., Geological Survey and Museum, London, Eng.

Stumm, Erwin C., Museum of Paleontology, University of Michigan, Ann Arbor, Mich.

Swain, Frederick M., University of Minnesota, Minneapolis, Minn.

Sylvester-Bradley, P. C., University of Sheffield, Sheffield, Eng.

Tasch, Paul, University of Wichita, Wichita, Kans.

Teichert, Curt, U.S. Geological Survey, Federal Center, Denver, Colo.

Thompson, M. L., University of Kansas, Lawrence, Kans.

Thompson, R. H., University of Kansas, Lawrence, Kans.

Tiegs, O. W., University of Melbourne, Melbourne, Victoria, Austral.

Tripp, Ronald P., Glasgow, Scot.

Ubaghs, G., Université de Liège, Liège, Belg.

Vokes, H. E., Johns Hopkins University, Baltimore, Md.

Wanner, J., Scheidegg (Allgäu), Bayern, Ger.

Weir, John, Glasgow University, Glasgow, Scot.

Weller, J. Marvin, University of Chicago, Chicago, Ill.

Wells, John W., Cornell University, Ithaca, N.Y.

Whittington, H. B., Museum of Comparative Zoology, Harvard University, Cambridge, Mass.

Williams, Alwyn, Queens University of Belfast, Belfast, N.Ire.

Wills, L. J., Romsley, Eng.

†Withers, T. H., Bournemouth, Eng.

Wright, C. W., Kensington, Eng.

†Wrigley, Arthur, Norbury, London, Eng.

Yochelson, Ellis L., U.S. Geological Survey, Washington, D.C.

Yonge, C. M., University of Glasgow, Glasgow, Scot.

Youngquist, Walter L., University of Kansas, Lawrence, Kans.

---

†—Deceased.

EDITORIAL PREFACE

The aim of the *Treatise on Invertebrate Paleontology,* as originally conceived and consistently pursued, is to present the most comprehensive and authoritative, yet compact statement of knowledge concerning invertebrate fossil groups that can be formulated by collaboration of competent specialists in seeking to organize what has been learned of this subject up to the mid-point of the present century. Such work has value in providing a most useful summary of the collective results of multitudinous investigations and thus should constitute an indispensable text and reference book for all persons who wish to know about remains of invertebrate organisms preserved in rocks of the earth's crust. This applies to neozoologists as well as paleozoologists and to beginners in study of fossils as well as to thoroughly trained, long-experienced professional workers, including teachers, stratigraphical geologists, and individuals engaged in research on fossil invertebrates. The making of a reasonably complete inventory of present knowledge of invertebrate paleontology may be expected to yield needed foundation for future research and it is hoped that the *Treatise* will serve this end.

The *Treatise* is divided into parts which bear index letters, each except the initial and concluding ones being defined to include designated groups of invertebrates. The chief purpose of this arrangement is to provide for independence of the several parts as regards date of publication, because it is judged desirable to print and distribute each segment as soon as possible after it is ready for press. Pages in each part will bear the assigned index letter joined with numbers beginning with 1 and running consecutively to the end of the part. When the parts ultimately are assembled into volumes, no renumbering of pages and figures is required.

The outline of subjects to be treated in connection with each large group of invertebrates includes (1) description of morphological features, with special reference to hard parts, (2) ontogeny, (3) classification, (4) geological distribution, (5) evolutionary trends and phylogeny, and (6) systematic description of genera, subgenera, and higher taxonomic units. In general, paleoecological aspects of study are omitted or little emphasized because comprehensive treatment of this subject is being undertaken in a separate work, prepared under auspices of a committee of the United States National Research Council. A selected list of references is furnished in each part of the *Treatise.*

Features of style in the taxonomic portions of this work have been fixed by the Editor with aid furnished by advice from the Joint Committee on Invertebrate Paleontology representing the societies which have undertaken to sponsor the *Treatise.* It is the Editor's responsibility to consult with authors and co-ordinate their work, seeing that manuscript properly incorporates features of adopted style. Especially he has been called on to formulate policies in respect to many questions of nomenclature and procedure.

The subject of family and subfamily names is reviewed briefly in a following section of this preface, and features of *Treatise* style in generic descriptions are explained.

A generous grant of $25,000 has been made by the Geological Society of America for the purpose of preparing *Treatise* illustrations. Administration of expenditures has been in charge of the Editor and most of the work by photographers and artists has been done under his direction at the University of Kansas, but sizable parts of this program have also been carried forward in Washington and London.

FORM OF ZOOLOGICAL NAMES

Many questions arise in connection with the form of zoological names. These include such matters as adherence to stipulations concerning Latin or Latinized nature of words accepted as zoological names, gender of generic and subgeneric names, nominative or adjectival form of specific names, required endings for some family-group names, and numerous others. Regulation extends to capitalization, treatment of particles belonging to modern patronymics, use of neo-Latin letters, and approved methods for converting diacritical marks. The magnitude and complexities of nomenclature problems surely are enough to warrant the complaint of those who hold that zoology is the study of animals rather than of names applied to them.

CLASSIFICATION OF ZOOLOGICAL NAMES

In accordance with the "Copenhagen Decisions on Zoological Nomenclature" (London, 135 p., 1953), zoological names may be classified usefully in various ways. The subject is summarized here with introduction of designations for some categories which the *Treatise* proposes to distinguish in systematic parts of the text for the purpose of giving readers comprehension of the nature of various names together with authorship and dates attributed to them.

CO-ORDINATE NAMES OF TAXA GROUPS

Five groups of different-rank taxonomic units (termed *taxa,* sing., *taxon*) are discriminated, within each of which names are treated as co-ordinate, being transferrable from one category to another without change of authorship or date. These are: (1) Species Group (subspecies, species); (2) Genus Group (subgenus, genus); (3) Family Group (tribe, subfamily, family, superfamily); (4) Order/Class Group (suborder, order, subclass, class); and (5) Phylum Group (subphylum, phylum). In the first 3 of these groups, but not others, the author of the first-published valid name for any taxon is held to be the author of all other taxa in the group which are based on the same nominate type and the date of publication for purposes of priority is that of the first-published name. Thus, if author A in 1800 introduces the family name X-idae to include 3 genera, one of which is *X-us;* and if author B in 1850 divides the 20 genera then included in X-idae into subfamilies called X-inae and Y-inae; and if author C in 1950 combines X-idae with other later-formed families to make a superfamily X-acea (or X-oidea, X-icae, etc.); the author of X-inae, X-idae and X-acea is A, 1800, under the Rules. Because taxonomic concepts introduced by authors B and C along with appropriate names surely are not attributable to author A, some means of recording responsibility of B and C are needed. This is discussed later in explaining proposed use of "*nom. transl.*"

The co-ordinate status of zoological names belonging to the species group is stipulated in Art. 11 of the present Rules; genus group in Art. 6 of the present Rules; family group in paragraph 46 of the Copenhagen Decisions; order/class group and phylum group in paragraphs 65 and 66 of the Copenhagen Decisions.

ORIGINAL AND SUBSEQUENT FORMS OF NAMES

Zoological names may be classified according to form (spelling) given in original publication and employed by subsequent authors. In one group are names which are entirely identical in original and subsequent usage. Another group comprises names which include with the original subsequently published variants of one sort or another. In this second group, it is important to distinguish names which are inadvertent changes from those constituting intentional emendations, for they have quite different status in nomenclature. Also, among intentional emendations, some are acceptable and some quite unacceptable under the Rules.

VALID AND INVALID NAMES

Valid names. A valid zoological name is one that conforms to all mandatory provisions of the Rules (Copenhagen Decisions, p. 43-57) but names of this group are divisible into subgroups as follows: (1) *"inviolate names,"* which as originally published not only meet all mandatory requirements of the Rules but are not subject to any sort of alteration (most generic and subgeneric names); (2) *"perfect names,"* which as they appear in original publication (with or without precise duplication by subsequent authors) meet all mandatory requirements and need no correction of any kind but which nevertheless are legally alterable under present Rules (as in changing the form of ending of a published class/order-group name); (3) *"imperfect names,"* which as originally published and with or without subsequent duplication meet mandatory requirements but contain defects such as incorrect gender of an adjectival specific name (for example, *Spironema recta* instead of *Spironema rectum*) or incorrect stem or form of ending of a family-group name (for example, Spironemidae instead of Spironematidae); (4) *"transferred names,"* which are derived by valid emendation from either of the 2nd or 3rd subgroups or from a pre-existing transferred name (as illustrated by change of a family-group name from -inae to -idae or making of a superfamily name); (5) *"improved names,"* which include necessary as well as somewhat arbitrarily made emendations allowable under the Rules for taxonomic categories not now covered by regulations as to name form and alterations that are distinct from changes that distinguish the 4th subgroup (including names derived from the 2nd and 3rd subgroups and possibly some alterations of 4th subgroup names). In addition, some zoological names included among those recognized as valid are classifiable in special categories, while at the same time belonging to one or more of the above-listed subgroups. These chiefly include (7) *"substitute names,"* introduced to replace invalid names such as junior homonyms; and (8) *"conserved names,"* which are names that would have to be rejected by application of the Rules except for saving them in their original or an altered spelling by action of the International Commission on Zoological Nomenclature in exercising its plenary powers to this end. Whenever a name requires replacement, any individual may publish a "new name" for it and the first one so introduced has priority over any others; since newness is temporary and relative, the replacement designation is better called substitute name rather than new name. Whenever it is considered desirable to save for usage an otherwise necessarily rejectable name, an individual cannot by himself accomplish the preservation, except by unchallenged action taken in accordance with certain provisions of the Copenhagen Decisions; otherwise he must seek validation through ICZN.

It is useful for convenience and brevity of distinction in recording these subgroups of valid zoological names to introduce Latin designations, following the pattern of *nomen nudum, nomen novum,* etc. Accordingly, the subgroups are (1) *nomina inviolata* (sing., *nomen inviolatum,* abbr., *nom. inviol.*); (2) *nomina perfecta* (sing., *nomen perfectum,* abbr., *nom. perf.*); (3) *nomina imperfecta* (sing., *nomen imperfectum,* abbr., *nom. imperf.*); (4) *nomina translata* (sing., *nomen translatum,* abbr., *nom. transl.*); (5) *nomina correcta* (sing., *nomen correctum,* abbr., *nom. correct.*); (6) *nomina substituta* (sing., *nomen substitutum,* abbr., *nom. subst.*); (7) *nomina conservata* (sing., *nomen conservatum,* abbr., *nom. conserv.*).

Invalid names. Invalid zoological names consisting of originally published names that fail to comply with mandatory provisions of the Rules and consisting of inadvertent changes in spelling of names have no status in nomenclature. They are not available as replacement names and they do not preoccupy for purposes of the Law of Homonomy. In addition to *nomen nudum,* invalid names may be distinguished as follows: (1) *"denied names,"* which consist of originally published names (with or without subsequent duplication) that do not meet mandatory requirements of the Rules; (2) *"null names,"* which comprise unintentional alterations of names; and (3) *"vain or void names,"* which consist of invalid emendations of previously published valid or invalid names. Void names do have status in nomenclature, being classified as junior synonyms of valid names.

Proposed Latin designations for the indicated kinds of invalid names are as follows: (1) *nomina negata* (sing., *nomen negatum*, abbr., *nom. neg.*); (2) *nomina nulla* (sing., *nomen nullum*, abbr., *nom. null.*); (3) *nomina vana* (sing., *nomen vanum*, abbr., *nom. van.*). It is desirable in the *Treatise* to identify invalid names, particularly in view of the fact that many of these names (*nom. neg., nom null.*) have been considered incorrectly to be junior objective synonyms (like *nom. van.*), which have status in nomenclature.

SUMMARY OF NAME CLASSES

Partly because only in such publications as the *Treatise* is special attention to classes of zoological names called for and partly because new designations are now introduced as means of recording distinctions explicitly as well as compactly, a summary may be useful. In the following tabulation valid classes of names are indicated in boldface type, whereas invalid ones are printed in italics.

Definitions of Name Classes

nomen conservatum (nom. conserv.). Name otherwise unacceptable under application of the Rules which is made valid, either with original or altered spelling, through procedures specified by the Copenhagen Decisions or by action of ICZN exercising its plenary powers.

nomen correctum (nom. correct.). Name with intentionally altered spelling of sort required or allowable under the Rules but not dependent on transfer from one taxonomic category to another ("improved name"). (*See* Copenhagen Decisions, paragraphs 50, 71-2-a-i, 74, 75, 79, 80, 87, 101; in addition, change of endings for categories not now fixed by Rules.)

nomen imperfectum (nom. imperf.). Name that as originally published (with or without subsequent identical spelling) meets all mandatory requirements of the Rules but contains defect needing correction ("imperfect name"). (*See* Copenhagen Decisions, paragraphs 50-1-b, 71-1-b-i, 71-1-b-ii, 79, 80, 87, 101.)

nomen inviolatum (nom. inviol.). Name that as originally published meets all mandatory requirements of the Rules and also is uncorrectable or alterable in any way ("inviolate name"). (*See* Copenhagen Decisions, paragraphs 152, 153, 155-157).

nomen negatum (nom. neg.). Name that as originally published (with or without subsequent identical spelling) constitutes invalid original spelling and although possibly meeting all other mandatory requirements of the Rules, is not correctable to establish original authorship and date ("denied name"). (*See* Copenhagen Decisions, paragraph 71-1-b-iii.)

nomen nudum (nom. nud.). Name that as originally published (with or without subsequent identical spelling) fails to meet mandatory requirements of the Rules and having no status in nomenclature, is not correctable to establish original authorship and date ("naked name"). (*See* Copenhagen Decisions, paragraph 122.)

nomen nullum (nom. null.). Name consisting of an unintentional alteration in form (spelling) of a previously published name (either valid name, as *nom. inviol., nom. perf., nom. imperf., nom. transl.;* or invalid name, as *nom. neg., nom. nud., nom. van.,* or another *nom. null.*) ("null name"). (*See* Copenhagen Decisions, paragraphs 71-2-b, 73-4.)

nomen perfectum (nom. perf.). Name that as originally published meets all mandatory requirements of the Rules and needs no correction of any kind but which nevertheless is validly alterable ("perfect name").

nomen substitutum (nom. subst.). Replacement name published as substitute for an invalid name, such as a junior homonym (equivalent to "new name").

nomen translatum (nom. transl.). Name that is derived by valid emendation of a previously published name as result of transfer from one taxonomic category to another within the group to which it belongs ("transferred name").

nomen vanum (nom. van.). Name consisting of an invalid intentional change in form (spelling) from a previously published name, such invalid emendations having status in nomenclature as junior objective synonyms ("vain or void name"). (*See* Copenhagen Decisions, paragraphs 71-2-a-ii, 73-3.)

Except as specified otherwise, zoological names accepted in the *Treatise* may be understood to be classifiable either as *nomina inviolata* or *nomina perfecta* (omitting from notice *nomina correcta* among specific names) and these are not discriminated. Names which are not accepted for one reason or another include junior homonyms, a few senior synonyms classifiable as *nomina negata* or *nomina nuda,* and numerous junior synonyms which include both objective *(nomina vana)* and subjective (all classes of valid names) types; effort to classify the invalid names as completely as possible is intended.

NAME CHANGES IN RELATION TO GROUP CATEGORIES

SPECIFIC AND SUBSPECIFIC NAMES

Detailed consideration of valid emendation of specific and subspecific names is unnecessary here because it is well understood and relatively inconsequential. When the form of adjectival specific names is changed to obtain agreement with the gender of a generic name in transferring a species from one genus to another, it is never needful to label the changed name as a *nom. transl.* Likewise, transliteration of a letter accompanied by a diacritical mark in manner now called for by the Rules (as in changing originally published *bröggeri* to *broeggeri*) or elimination of a hyphen (as in changing originally published *cornu-oryx* to *cornuoryx* does not require "*nom. correct.*" with it. Revised provisions for emending specific and subspecific names are stated in the report on Copenhagen Decisions (p. 43-46, 51-57).

GENERIC AND SUBGENERIC NAMES

So rare are conditions warranting change of the originally published valid form of generic and subgeneric names that lengthy discussion may be omitted. Only elimination of diacritical marks of some names in this category seems to furnish basis for valid emendation. It is true that many changes of generic and subgeneric names have been published, but virtually all of these are either *nomina vana* or *nomina nulla*. Various names which formerly were classed as homonyms are not now, for two names that differ only by a single letter (or in original publication by presence or absence of a diacritical mark) are construed to be entirely distinct. Revised provisions for emendation of generic and subgeneric names also are given in the report on Copenhagen Decisions (p. 43-47).

Examples in use of classificatory designations for generic names as previously given are the following, which also illustrate designation of type species, as explained later.

Kurnatiophyllum THOMSON, 1875 [**K. concentricum;* SD GREGORY, 1917] [=*Kumatiophyllum* THOMSON, 1876 *(nom. null.); Cymatophyllum* THOMSON, 1901 *(nom. van.); Cymatiophyllum* LANG, SMITH & THOMAS, 1940 *(nom. van.)*].

Stichophyma POMEL, 1872 [**Manon turbinatum* RÖMER, 1841; SD RAUFF, 1893] [=*Stychophyma* VOSMAER, 1885 *(nom. null.); Sticophyma* MORET, 1924 *(nom. null.)*].

Stratophyllum SMYTH, 1933 [**S. tenue*] [=*Ethmoplax* SMYTH, 1939 (*nom. van. pro Stratophyllum*); *Stratiphyllum* LANG, SMITH & THOMAS, 1940 (*nom. van. pro Stratophyllum* SMYTH) (*non Stratiphyllum* SCHEFFEN, 1933)].

Placotelia OPPLIGER, 1907 [**Porostoma marconi* FROMENTEL, 1859; SD DELAUBENFELS, herein] [=*Plakotelia* OPPLIGER, 1907 *(nom. neg.)*].

Walcottella DELAUB., *nom. subst.*, 1955 [*pro Rhopalicus* SCHRAMM., 1936 (*non* FÖRSTER, 1856)].

Cyrtograptus CARRUTHERS, 1867 [*nom. correct.* LAPWORTH, 1873 (*pro Cyrtograpsus* CARRUTHERS, 1867), *nom. conserv.* proposed BULMAN, 1955 (ICZN pend.)]

FAMILY-GROUP NAMES; USE OF "*NOM. TRANSL.*"

The Rules now specify the form of endings only for subfamily (-inae) and family (-idae) but decisions of the Copenhagen Congress direct classification of all family-group assemblages (taxa) as co-ordinate, signifying that for purposes of priority a name published for a unit in any category and based on a particular type genus shall date from its original publication for a unit in any category, retaining this priority (and authorship) when the unit is treated as belonging to a lower or higher category. By exclusion of -inae and -idae, respectively reserved for subfamily and family, the endings of names used for tribes and superfamilies must be unspecified different letter combinations. These, if introduced subsequent to designation of a subfamily or family based on the same nominate genus, are *nomina translata,* as is also a subfamily that is elevated to family rank or a family reduced to subfamily rank. In the *Treatise* it is desirable to distinguish the valid emendation comprised in the changed ending of each transferred family group name by the abbreviation "*nom. transl.*" and record of the author and date belonging to this emendation. This is particularly important in the case of superfamilies, for it is the author who introduced this taxon that one wishes to know about rather than the author of the superfamily as defined by the Rules, for the latter is merely the individual who first defined some lower-rank family-group taxon that contains the nominate genus of the superfamily. The publication of the author containing introduction of the superfamily *nomen translatum* is likely to furnish the information on taxonomic considerations that support definition of the unit.

Examples of the use of "*nom. transl.*" are the following.

Subfamily STYLININAE d'Orbigny, 1851

[*nom. transl.* EDWARDS & HAIME, 1857 (*ex* Stylinidae D'ORBIGNY, 1851]

Superfamily ARCHAEOCTONOIDEA Petrunkevitch, 1949

[*nom. transl.* PETRUNKEVITCH, herein (*ex* Archaeoctonidae PETRUNKEVITCH, 1949)]

Superfamily CRIOCERATITACEAE Hyatt, 1900

[*nom. transl.* WRIGHT, 1952 (*ex* Crioceratitidae HYATT, 1900)]

FAMILY-GROUP NAMES; USE OF "*NOM. CORRECT.*"

Valid emendations classed as *nomina correcta* do not depend on transfer from one category of family-group units to another but most commonly involve correction of the stem of the nominate genus; in addition, they include somewhat arbitrarily chosen modification of ending for names of tribe or superfamily. Examples of the use of "*nom. correct.*" are the following.

Family STREPTELASMATIDAE Nicholson, 1889

[*nom. correct.* WEDEKIND, 1927 (*ex* Streptelasmidae NICHOLSON, 1889, *nom. imperf.*)]

Family PALAEOSCORPIIDAE Lehmann, 1944

[*nom. correct.* PETRUNKEVITCH, herein (*ex* Palaeoscorpionidae LEHMANN, 1944, *nom. imperf.*)]

Family AGLASPIDIDAE Miller, 1877

[*nom. correct.* STØRMER, herein (*ex* Aglaspidae MILLER, 1877, *nom. imperf.*)]

Superfamily AGARICIICAE Gray, 1847

[*nom. correct.* WELLS, herein (*ex* Agaricioidae VAUGHAN & WELLS, 1943, *nom. transl. ex* Agariciidae GRAY, 1847)]

FAMILY-GROUP NAMES; USE OF "*NOM. CONSERV.*"

It may happen that long-used family-group names are invalid under strict application of the Rules. In order to retain the otherwise invalid name, appeal to ICZN is needful. Examples of use of *nom. conserv.* in this connection, as cited in the *Treatise,* are the following.

Family ARIETITIDAE Hyatt, 1874

[*nom. correct.* HAUG, 1885 (*pro* Arietidae HYATT, 1875), *nom. conserv.* proposed ARKELL, 1955 (ICZN pend.)]

Family STEPHANOCERATIDAE Neumayr, 1875

[*nom. correct.* FISCHER, 1882 (*pro* Stephanoceratinen NEUMAYR, 1875, invalid vernacular name), *nom conserv.* proposed ARKELL, 1955 (ICZN pend.)]

FAMILY-GROUP NAMES; REPLACEMENTS

Family-group names are formed by adding letter combinations (prescribed for family and subfamily but not now for others) to the stem of the name belonging to genus (nominate genus) first chosen as type of the assemblage. The type genus need not be the oldest in terms of receiving its name and definition, but it must be the first-published as name-giver to a family-group taxon among all those included. Once fixed, the family-group name remains tied to the nominate genus even if its name is changed by reason of status as a junior homonym or junior synonym, either objective or subjective. According to the Copenhagen Decisions, the family-group name requires replacement only in the event that the nominate genus is found to be a junior homonym, and then a substitute family-group name is accepted if it is formed from the oldest available substitute name for the nominate genus. Authorship and date attributed to the replacement family-group name are determined by first publication of the changed family-group name.

The aim of family-group nomenclature is greatest possible stability and uniformity, just as in case of other zoological names. Experience indicates the wisdom of sustaining family-group names based on junior subjective synonyms if they have priority of publication, for opinions of different workers as to the synonymy of generic names founded on different type species may not agree and opinions of the same worker may alter from time to time. The retention similarly of first-published family-group names which are found to be based on junior objective synonyms is less clearly desirable, especially if a replacement name derived from the senior objective synonym has been recognized very long and widely. To displace a much-used family-group name based on the senior objective synonym by disinterring a forgotten and virtually unused family-group name based on a junior objective synonym because the latter happens to have priority of publication is unsettling. Conversely, a long-used family-group name founded on a junior objective synonym and having priority of publication is better continued in nomenclature than a replacement name based on the senior objective synonym. The Copenhagen Decisions (paragraph 45) take account of these considerations by providing a relatively simple procedure for fixing the desired choice in stabilizing family-group names. In conformance with this, the *Treatise* assigns to contributing authors responsibility for adopting provisions of the Copenhagen Decisions.

Replacement of a family-group name may be needed if the former nominate genus is transferred to another family-group. Then the first-published name-giver of a family-group assemblage in the remnant taxon is to be recognized in forming a replacement name.

FAMILY-GROUP NAMES; AUTHORSHIP AND DATE

All family-group taxa having names based on the same type genus are attributed to the author who first published the name for any of these assemblages, whether tribe, subfamily, or family (superfamily being almost inevitably a later-conceived taxon). Accordingly, if a family is divided into subfamilies or a subfamily into tribes, the name of no such subfamily or tribe can antedate the family name. Also, every family containing differentiated subfamilies must have a nominate *(sensu stricto)* subfamily, which is based on the same type genus as that for the family, and the author and date set down for the nominate subfamily invariably are identical with those of the family, without reference to whether the author of the family or some subsequent author introduced subdivisions.

Changes in the form of family-group names of the sort constituting *nomina correcta,* as previously discussed, do not affect authorship and date of the taxon concerned, but in publications such as the *Treatise* it is desirable to record the authorship and date of the correction.

ORDER/CLASS-GROUP NAMES; USE OF "*NOM. CORRECT.*"

Because no stipulation concerning the form of order/class-group names is given yet by the Rules, emendation of all such names actually consists of arbitrarily devised changes in the form of endings. Nothing precludes substitution of a new name for an old one, but a change of this sort is not considered to be an emendation. Examples of the use of "*nom. correct.*" as applied to order/class-group names are the following.

Order DISPARIDA Moore & Laudon, 1943

[*nom. correct.* Moore, 1952 (*ex* Disparata Moore & Laudon, 1943)]

Suborder FAVIINA Vaughan & Wells, 1943

[*nom. correct.* Wells, herein (*ex* Faviida Vaughan & Wells, 1943)]

Suborder FUNGIINA Verrill, 1865

[*nom. correct.* Wells, herein (*ex* Fungiida Duncan, 1884, *ex* Fungacea Verrill, 1865)]

TAXONOMIC EMENDATION

Emendation has two measurably distinct aspects as regards zoological nomenclature. These embrace (1) alteration of a name itself in various ways for various reasons, as has been reviewed, and (2) alteration of taxonomic scope or concept in application of a given zoological name, whatever its hierarchical rank. The latter type of emendation primarily concerns classification and inherently is not associated with change of name, whereas the other type introduces change of name without necessary expansion, restriction, or other modification in applying the name. Little attention generally has been paid to this distinction in spite of its significance.

Most zoologists, including paleozoologists, who have signified emendation of zoological names refer to what they consider a material change in application of the name such as may be expressed by an importantly altered diagnosis of the assemblage covered by the name. The abbreviation "*emend.*" then may accompany the name, with statement of the author and date of the emendation. On the other hand, a multitude of workers concerned with systematic zoology think that publication of "*emend.*" with a zoological name is valueless because more or less alteration of taxonomic sort is introduced whenever a subspecies, species, genus, or other assemblage of animals is incorporated under or removed from the coverage of a given zoological name. Inevitably associated with such classificatory expansions and restrictions is some degree of emendation affecting diagnosis. Granting this, still it is true that now and then somewhat radical revisions are put forward, generally with published statement of reasons for changing the application of a name. To erect a signpost at such points of most significant change is worth while, both as aid to subsequent workers in taking account of the altered nomenclatural usage and as indication that not-to-be-overlooked discussion may be found at a particular place in the literature. Authors of contributions to the *Treatise* are encouraged to include records of all specially noteworthy emendations of this nature, using the abbreviation "*emend.*" with the name to which it refers and citing the author and date of the emendation.

In Part G (Bryozoa) and Part D (Protista 3) of the *Treatise,* the abbreviation "*emend.*" is employed to record various sorts of name emendations, thus conflicting with usage of "*emend.*" for change in taxonomic application of a name without alteration of the name itself. This is objectionable. In Part E (Archaeocyatha, Porifera) and later-issued divisions of the *Treatise,* use of "*emend.*" is restricted to its customary sense, that is, significant alteration in taxonomic scope of a name such as calls for noteworthy modifications of a diagnosis. Other means of designating emendations that relate to form of a name are introduced.

STYLE IN GENERIC DESCRIPTIONS

DEFINITION OF NAMES

Most generic names are distinct from all others and are indicated without ambiguity by citing their originally published spelling accompanied by name of the author and date of first publication. If the same generic name has been applied to 2 or more distinct taxonomic units, however, it is necessary to differentiate such homonyms, and this calls for distinction between junior homonyms and senior homonyms. Because a junior homonym is invalid, it must be replaced by some other name. For example, *Callopora* HALL, 1851, introduced for Paleozoic trepostome bryozoans, is invalid because GRAY in 1848 published the same name for Cretaceous-to-Recent cheilostome bryozoans, and BASSLER in 1911 introduced the new name *Hallopora* to replace HALL's homonym. The *Treatise* style of entry is:

Hallopora BASSLER, 1911 [*pro Callopora* HALL, 1851 (*non* GRAY, 1848)].

In like manner, a needed replacement generic name may be introduced in the *Treatise* (even though first publication of generic names otherwise in this work is avoided). The requirement that an exact bibliographic reference must be given for the replaced name commonly can be met in the *Treatise* by citing a publication recorded in the list of references, using its assigned index number, as shown in the following example.

Mysterium DELAUBENFELS, *nom. subst.* [*pro Mystrium* SCHRAMMEN, 1936 (ref. 40, p. 60) (*non* ROGER, 1862)] [**Mystrium porosum* SCHRAMMEN, 1936].

For some replaced homonyms, a footnote reference to the literature is necessary. A senior homonym is valid, and in so far as the *Treatise* is concerned, such names are handled according to whether the junior homonym belongs to the same major taxonomic division (class or phylum) as the senior homonym or to some other; in the former instance, the author and date of the junior homonym are cited as:

Diplophyllum HALL, 1851 [*non* SOSHKINA, 1939] [**D. caespitosum*].

Otherwise, no mention of the existence of a junior homonym is made.

CITATION OF TYPE SPECIES

The name of the type species of each genus and subgenus is given next following the generic name with its accompanying author and date, or after entries needed for definition of the name if it is involved in homonymy. The originally published combination of generic and trivial names for this species is cited, accompanied by an asterisk (*), with notation of the author and date of original publication. An exception in this procedure is made, however, if the species was first published in the same paper and by the same author as that containing definition of the genus which it serves as type; in such case, the initial letter of the generic name followed by the trivial name is given without repeating the name of the author and date, for this saves needed space. Examples of these 2 sorts of citations are as follows:

Diplotrypa NICHOLSON, 1879 [**Favosites petropolitanus* PANDER, 1830].

Chainodictyon FOERSTE, 1887 [**C. laxum*].

If the cited type species is a junior synonym of some other species, the name of this latter also is given, as follows:

Acervularia SCHWEIGGER, 1819 [**A. baltica* (=**Madrepora ananas* LINNÉ, 1758)].

It is judged desirable to record the manner of establishing the type species, whether by original designation or by subsequent designation, but various modes of original designation are not distinguished. According to convention adopted in the *Treatise,* absence of any indication as to manner of fixing the type species is to be understood as signifying original designation. If the type species has been fixed by subsequent designation, this is indicated by the letters "SD" followed by the name of the author and date of such subsequent designation, as follows:

Hexagonaria GÜRICH, 1896 [**Cyathophyllum hexagonum* GOLDFUSS, 1826; SD LANG, SMITH & THOMAS, 1940].

SYNONYMS

Citation of synonyms is given next following record of the type species and if 2 or more synonyms of differing date are recognized, these are arranged in chronological order. Objective synonyms are indicated by accompanying designation "(obj.)," others being understood to constitute subjective synonyms. Examples showing *Treatise* style in listing synonyms follow.

Calapoecia BILLINGS, 1865 [**C. anticostiensis;* SD LINDSTRÖM, 1833] [=*Columnopora* NICHOLSON, 1874; *Houghtonia* ROMINGER, 1876].

Staurocyclia HAECKEL, 1882 [**S. cruciata* HAECKEL, 1887] [=*Coccostaurus* HAECKEL, 1882 (obj.); *Phacostaurus* HAECKEL, 1887 (obj.)].

A synonym which also constitutes a homonym is recorded as follows:

Lyopora NICHOLSON & ETHERIDGE, 1878 [**Palaeopora? favosa* M'COY, 1850] [=*Liopora* LANG, SMITH & THOMAS, 1940 (*non* GIRTY, 1915)].

ABBREVIATIONS

Some authors' names and most stratigraphic and geographic names are abbreviated in order to save space. General principles for guidance in determining what names should be abbreviated are frequency of repetition, length of name, and avoidance of ambiguity. Abbreviations used in this division of the *Treatise* are explained in the following alphabetically arranged list.

Abbreviations

Abt., Abteilung, -en
Acad., Académie, Academy
Accad., Accademia
Afr., Africa
Akad., Akademie
Am., America, -n
Ann., Annal, -s, Annual
Ariz., Arizona
AsiaM., Asia Minor
Atl., Atlantic
auctt., auctores
Austral., Australia
Balt., Baltic
Bayer., Bayerischen
Bd., Band
Belg., Belgium
Blainv., Blainville, H. M. de
Bol., Boletim
Bull., Bulletin, -s
C., Central
Calif., California
Cam., Cambrian
Can., Canada, -ian
Carb., Carboniferous
Cien., Ciencias
Coll., Collection, -s; College
Colo., Colorado
Comp., Comparative
Conn., Connecticut
cosmop., comopolitan
Cret., Cretaceous
Czech., Czechoslovakia
Dev., Devonian
Downton., Downtonian
E., East
Ehr., Ehrenberg, C. G.
Encycl., Encyclopedia
Eng., England, English
Entomol., Entomologique, Entomologischen
Eu., Europe
Exped., Expedition
f., för, für
Fac., Faculdade
fig., figure, -s
Filos., Filosofia
Fr., France
Geol., Geological, -ie, -y
geol., geologischen
géol., géologique
Ger., Germany
Gesell., Gesellschaft
Grassi-C., Grassi & Calandruccio
Hist., Historie, -y
ICZN, International Commission on Zoological Nomenclature
Ill., Illinois
Imp., Imperial
Inst., Institute
Ire., Ireland
Jahrb., Jahrbuch
Jur., Jurassic
Kan., Kansas
Kgl., Königlich
Kl., Klasse
Koch-B., Koch & Berendt

L., Lower
Landesanst., Landesanstalt
Lind., Lindström, G.
M., Middle
Mag., Magazine
Math., Mathematische
Meek-W., Meek & Worthen
Mem., Memoir, -s
Mém., Mémoire, -s
Micros., Microscopical
Miller-G., Miller & Gurley
Mio., Miocene
Misc., Miscellaneous
Mon., Monograph, -ie
Mus., Musée, Museum
N., North
Nat., Natural
Natl., National
naturhist., naturhistorischen
naturwiss., naturwissenschaftlich
N.F., Neue Folge
no., number, -s
nom. correct., nomen correctum
nom. transl., nomen translatum
N.S.W., New South Wales
N.Y., New York
obj., objective
Okla., Oklahoma
ökonom., ökonomische
Oligo., Oligocene
Ord., Ordovician
p., page, -s
Palaeontogr., Palaeontographica, -l
Paläont., Paläontologie, -isch
Paleont., Paleontology, -ical
pend., pending
Penn., Pennsylvanian
Perm., Permian
Petr., Petrunkevitch, Alexander
phys., physische
pl., plate, -s
Pleisto., Pleistocene
Plio., Pliocene
Preuss., Preussischen
Quart., Quarterly
Rec., Recent; Recueil
reconstr., reconstructed, -ion
Roy., Royal, -e
Russ., Russia
S., Sea; South
Schr., Schrift, -en
Sci., Science, -s; Scientific
Scot., Scotland
SD., subsequent designation
ser., serial, series
Sil., Silurian
Sitzungsber., Sitzungsberichte
Skr., Skrifter
Soc., Société, Society
Spitz., Spitzbergen
Terr., Territory, -ies
Tert., Tertiary
Thorell-L., Thorell & Lindström
Trans., Transactions
Trempeal., Trempealeauan
Trias., Triassic
Tschern., Tschernyschew, B. F.
U., Upper
Univ., Université, -idade, -y
USA, United States of America
v., volume, -s
Ver., Verein
Verh., Verhandlung, -en
Vict., Victoria
Vid., Videnskap
W., West
Walc., Walcott, C. D.
Whitf., Whitfield, R. P.
Wis., Wisconsin
Wiss., Wissenschaft, -lich
Woodw., Woodward, H.
Wyo., Wyoming
Zeitschr., Zeitschrift
Zittel-E., Zittel & Eastman
Zool., Zoologie, -ical, -y

REFERENCES TO LITERATURE

Each part of the *Treatise* is accompanied by a selected list of references to paleontological literature consisting primarily of recent and comprehensive monographs available but also including some older works recognized as outstanding in importance. The purpose of giving these references is to aid users of the *Treatise* in finding detailed descriptions and illustrations of morphological features of fossil groups, discussions of classifications and distribution, and especially citations of more or less voluminous literature. Generally speaking, publications listed in the *Treatise* are not original sources of information concerning taxonomic units of various rank but they tell the student where he may find them; otherwise it is necessary to turn to such aids as the *Zoological Record* or NEAVE's *Nomenclator Zoologicus*. References given in the *Treatise* are arranged alphabetically by authors and accompanied by index numbers which serve the purpose of permitting citation most concisely in various parts of the text; these citations of listed papers are inclosed invariably in parentheses and are distinguishable from dates because the index numbers comprise no more than 3 digits. Ordinarily, index numbers for literature references are given at the end of generic or family diagnoses.

SOURCES OF ILLUSTRATIONS

At the end of figure captions an index number is given to supply record of the author of illustrations used in the *Treatise*, reference being made to an alphabetically arranged list of authors' names which follows. The names of authors, but generally not individual publications, are cited. Illustrations consisting of exact copies of previously published figures (except for possible change of scale) are distinguished by the use of an asterisk (*) with the index number, and previously unpublished illustrations are marked by the letter "n" (signifying "new") with the index number; all other indications of the sources of illustrations are construed to mean "after" the cited author or authors, that is, embodying some degree of change. Addition of the abbreviation "mod." denotes appreciable alteration of the source figure.

RAYMOND C. MOORE

PART P
ARTHROPODA 2

CHELICERATA
WITH SECTIONS ON PYCNOGONIDA AND PALAEOISOPUS

By Leif Størmer, Alexander Petrunkevitch, and Joel W. Hedgpeth

CONTENTS

CHELICERATA

By Leif Størmer

Subphylum CHELICERATA Heymons 1901

Terrestrial and aquatic arthropods with pair of preoral appendages developed as 2-, 3-, and ?4-jointed chelicerae or pincers; next following pair of appendages (pedipalps) primarily ambulatory, secondarily modified into prehensile, sensory, or masticatory organs. Body divided into a prosoma (cephalothorax) with 6 postoral segments in addition to the preoral portion, and an opisthosoma (abdomen) of primarily ?12 segments, the 1st (comprising 7th somite) commonly reduced. Genital ducts open in the 8th postoral somite. *Cam.-Rec.*

GROUPS AND THEIR OCCURRENCE

The subphylum comprises the classes Merostomata and Arachnida. The Pantopoda (Pycnogonida) have also been included in the Chelicerata but may belong to a separate group.

Whereas the merostomes are aquatic forms, the arachnids are confined to land, except for a few forms which are secondarily adapted to living in water. The different mode of life has influenced the development of several morphological structures, especially the appendages and respiratory organs (gills, lungs). In spite of this, other structures have been kept nearly unaltered in all members of the Chelicerata. Fig. 1 indicates the main ventral structure in a generalized chelicerate. The secondarily preoral position of the chelicerae is demonstrated in the ontogeny. The conservatism of the chelicerae is expressed in the eurypterid *Pterygotus,* where it has retained its general structure even though the pincers are enormously developed.

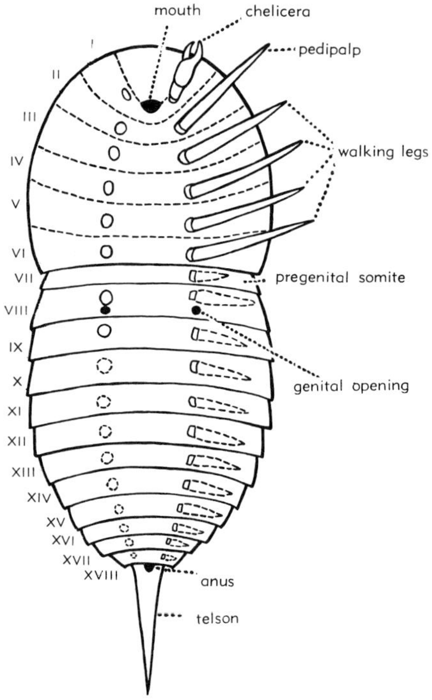

FIG. 1. Generalized chelicerate showing main morphological features of ventral side, diagrammatic.

Since fossil chelicerates are comparatively rare, our knowledge of the vertical and horizontal distribution of species is fairly limited. The vertical distribution of the orders is shown in Fig. 2. As for horizontal distribution, merostomes are known from North and South America, Europe, Africa, Asia and Australia, and are more frequent in the northern than in the southern hemisphere. Up to now fossil arachnids have been found almost exclusively in Europe and North America.

CLASSIFICATION

A tabular summary showing main divisions of the chelicerates as generally recognized and adopted in organizing the *Treatise* follows. The arrangement of merostomes is by Størmer, of arachnids by Petrunkevitch, and of pycnogonids by Hedgpeth.

Main Divisions of Chelicerata and Pycnogonida

Chelicerata *(subphylum), L.Cam.-Rec.*
- Merostomata *(class), L.Cam.-Rec.*
  - Xiphosura *(subclass), L.Cam.-Rec.*
    - Aglaspida *(order), L.Cam.-U.Ord.*
    - Xiphosurida, *Sil.-Rec.*
  - Eurypterida *(subclass), M.Ord.-Perm.*
- Arachnida *(class), Sil.-Rec.*
  - Latigastra *(subclass), Sil.-Rec.*
    - Scorpionida *(order), Sil.-Rec.*
    - Pseudoscorpionida, *Oligo.-Rec.*
    - Phalangida, *Penn.-Rec.*
    - Architarbida, *Carb.*
    - Acarida, *Dev.-Rec.*
  - Stethostomata *(subclass), Carb.*
    - Haptopodida *(order), Carb.*
    - Anthracomartida, *Carb.*
  - Soluta *(subclass), Dev.-Carb.*
    - Trigonotarbida *(order), Dev.-Carb.*
  - Caulogastra *(subclass), ?Dev., Carb.-Rec.*
    - Latisterna *(superorder), Jur.-Rec.*
      - Palpigradida *(order), Jur.-Rec.*
    - Camarostomata *(superorder), Carb.-Rec.*
      - Thelyphonida *(order), Carb.-Rec.*
      - Schizomida, *?Plio.-Rec.*
      - Kustarachnida, *Penn.*
    - Labellata *(superorder), ?Dev., Carb.-Rec.*
      - Phrynichida *(order), Carb.-Rec.*
      - Araneida, *?Dev., Carb-Rec.*
    - Rostrata *(superorder), Carb.-Rec.*
      - Solpugida *(order), Carb.-Rec.*
    - Podogona *(superorder), Carb.-Rec.*
      - Ricinuleida *(order), Carb.-Rec.*

Pycnogonida *(subphylum), L.Dev.-Rec.*
- Pantopoda *(order), Rec.*
- Palaeopantopoda, *L.Dev.*

FIG. 2. Diagram showing known stratigraphic distribution of main divisions of the Chelicerata and inferred phylogenetic relationships. The oldest discovered representatives of the Arachnida, consisting of Silurian scorpionids, are so strongly differentiated that divergence of the arachnid subclasses and orders is inferred to belong in the pre-Silurian part of early Palaeozoic time. Probably, also, the merostomes and arachnids comprised distinct stocks before the earliest Cambrian. The abrupt appearance of so many orders of Arachnida in the Late Carboniferous clearly reflects conditions affecting preservation of these arthropods, rather than origin in the latter part of the Paleozoic era.

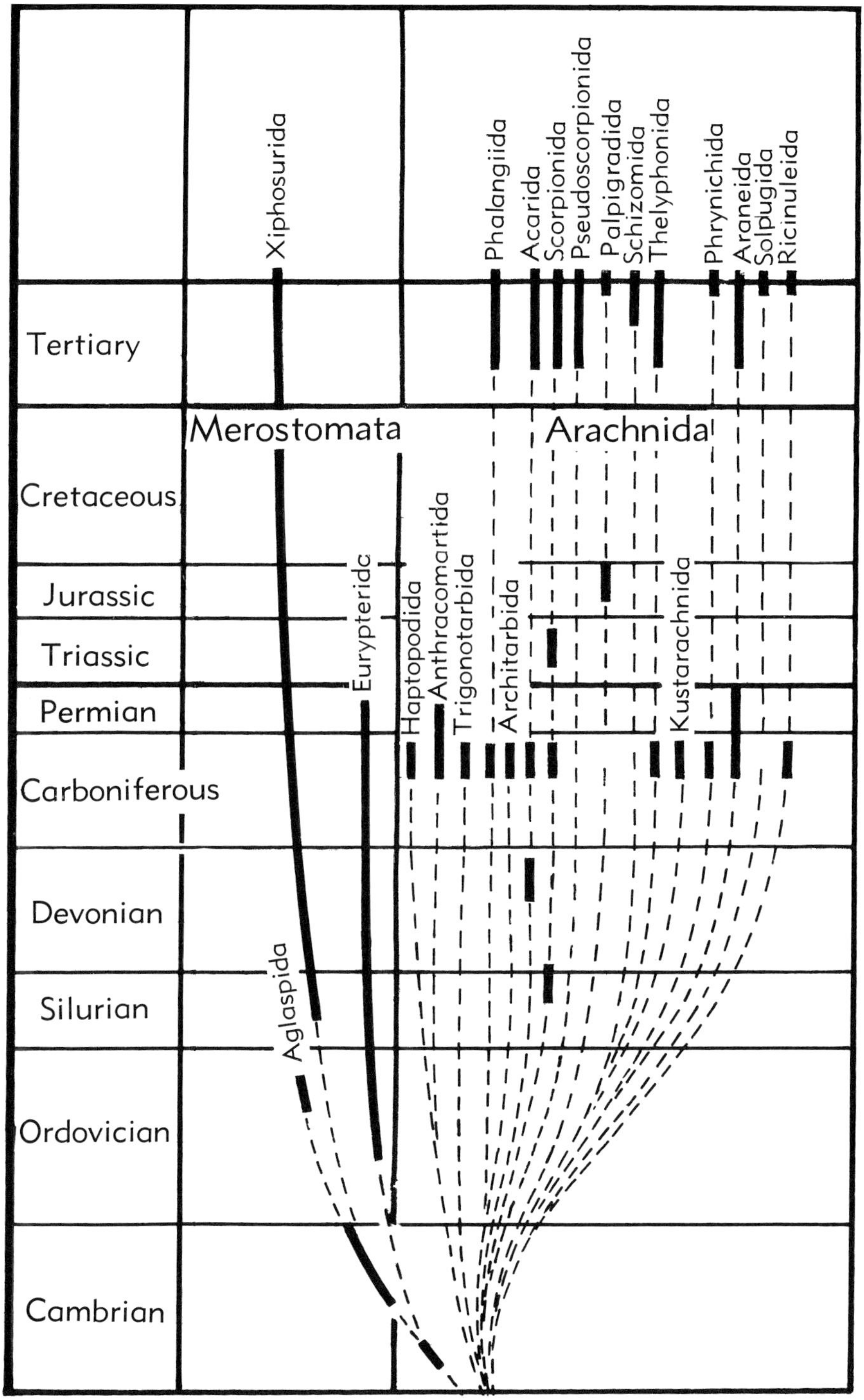

FIG. 2. *(See facing page.)*

MEROSTOMATA

By Leif Størmer

CONTENTS

Class MEROSTOMATA Dana, 1852

[=Poecilopoda M'Coy, 1849 (*non* Latreille, 1817); Palaeostraca Grobben, 1905]

Aquatic Chelicerata of widely varying size, with body divided into a prosoma or cephalothorax and opisthosoma or abdomen (thoracetron) with a powerful, mostly styliform telson. Prosoma with median ocelli and compound eyes (if not reduced), 6 pairs of prosomal appendages, the frontal one secondarily preoral, developed as 3- (or ?4-) jointed pincer-bearing limbs (chelicerae); other prosomal appendages serving mainly as walking legs, with proximal segments (coxae) acting partly as jaws. Opisthosomal segments varying in number, free or ankylosed into a continuous shield. Appendages of the 1st, partly reduced opisthosomal segment situated below the prosoma, form the chilaria or a single plate (metastoma); other opisthosomal appendages plate-shaped and biramous, each consisting of a considerably reduced median branch and a lateral broad, strongly flattened branch carrying the gills; instead of the 1st pair of appendages below the abdomen (belonging to the 8th somite), is an operculum which contains openings of the genital ducts. *Cam.-Rec.*

MORPHOLOGY OF LIVING MEROSTOMES

The largest of the Recent forms may attain a length of about 60 cm. The body is covered by a stout exoskeleton, of which the expanded dorsal portion forms an excellent protection for structures underneath. The body (Fig. 3) is divided into 3 parts, the **prosoma** or cephalothorax, the **opisthosoma** or abdomen (thoracetron), and the **telson.** The semicircular strongly vaulted prosoma has a median **cardiac lobe** with a median crest, provided with 3 spines in young specimens. The **axial furrows** bordering the cardiac lobe converge slightly forward and become obsolete. In front of the cardiac lobe lie the **median ocelli.** A longitudinal ridge extending forward and backward from each of the lateral compound eyes is called **ophthalmic ridge.** In young specimens each ridge bears 2 spines. The space between the ridges is termed the **cardiophthalmic region,** and that between the cardiac lobe and the ophthalmic ridges may be called the **interophthalmic region.** Here, and to some extent on the cardiac lobe, muscle markings suggest the segmentation.

The prosoma is separated from the opisthosoma by a transverse joint or hinge line. Embryological studies based on *Tachypleus gigas* (Müller) by Ivanov (11) show the joint line to cross the primary segmentation (Fig. 4*A*). The transverse joint line cuts through the 6th and 7th segments, the latter being reduced greatly. The broad steep rim surrounding the cardiophthalmic

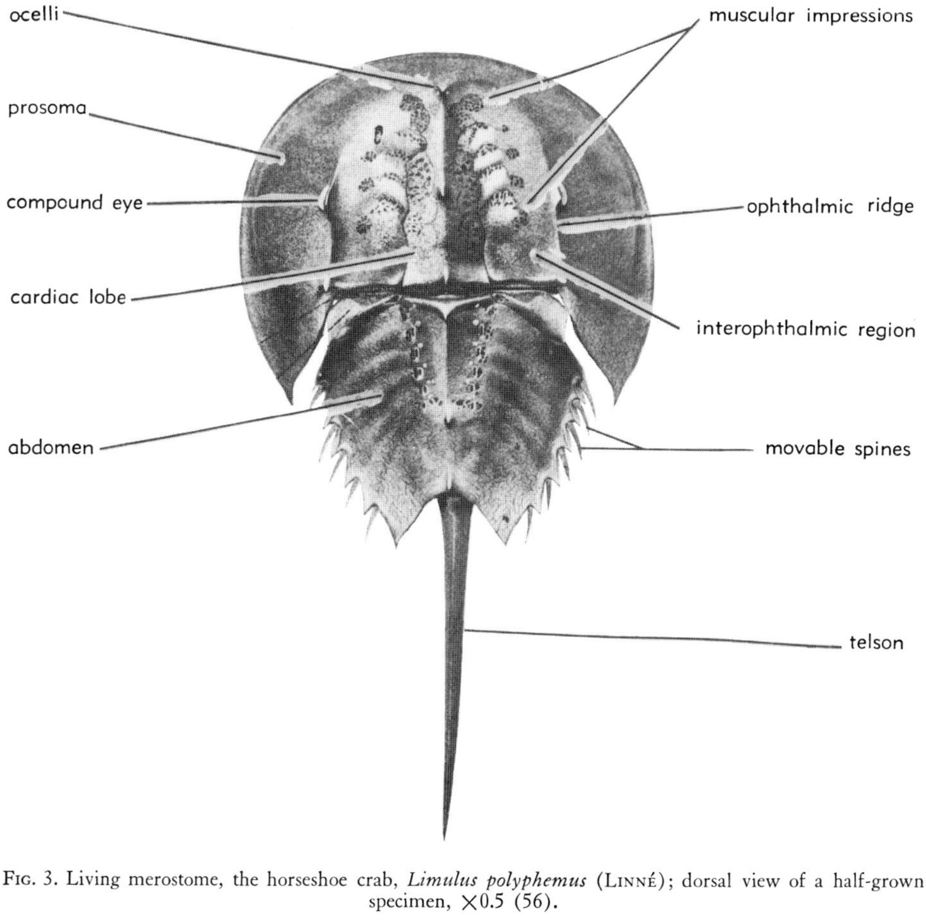

Fig. 3. Living merostome, the horseshoe crab, *Limulus polyphemus* (Linné); dorsal view of a half-grown specimen, ×0.5 (56).

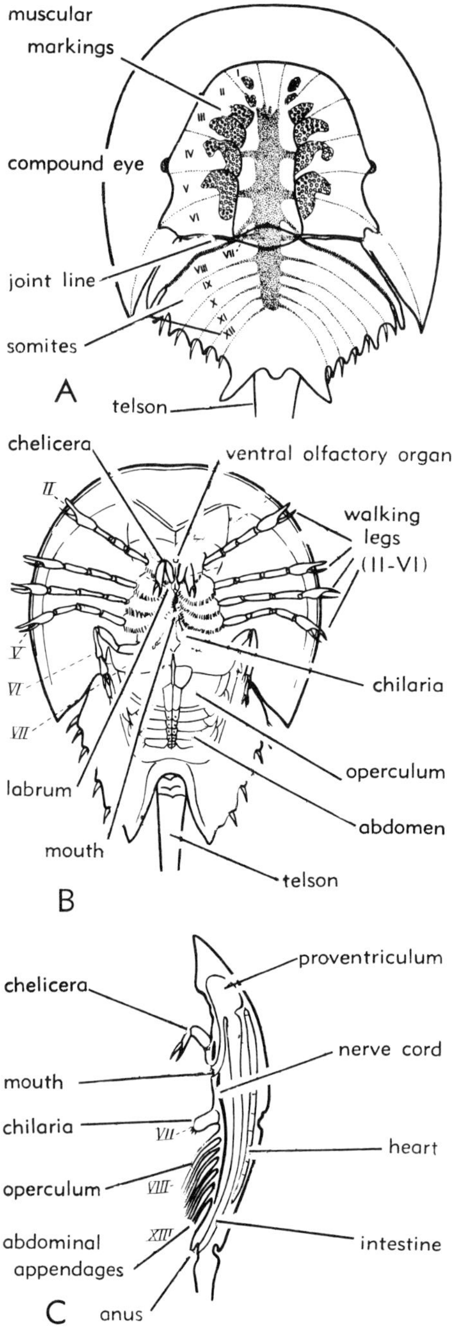

FIG. 4. Morphological features of living merostomes (64). *A, Tachypleus gigas* (MÜLLER), dorsal side of young specimen with segmentation indicated (enlarged). *B,C, Limulus polyphemus* (LINNÉ), ventral view and longitudinal section (much reduced) (64).

region and prolonged backward on each side may belong to an anterior segment of the prosoma.

The opisthosoma or abdomen forms a continuous, moderately vaulted shield. The serrate lateral margins, converging backward toward the telson, have intercalated movable spines. A trilobation of the abdomen is indicated by the axial furrows in which 6 pairs of invaginations, forming structures for muscle attachment **(apodemes)**, may be distinguished. Embryological studies indicate existence of several rudimentary segments behind the segments with apodemes. The long styliform telson, articulating by a special T-shaped process to the abdomen, evidently represents tergal outgrowth from the posterior rudimentary segments. The telson is able to move in any direction and serves apparently as a steering rod when the streamlined horseshoe crab moves forward on the bottom, and it also helps the animal to turn over when lying on its back.

The ventral side of the prosoma bears a broad **doublure,** which passes into the softer integument surrounding the appendages. The central mouth has a narrow upper lip **(labrum)** at the base of which occurs an olfactory organ (Fig. 4*B*).

The appendages surround the mouth. No multijointed preoral antennae occur. The secondarily preoral **chelicerae** or pincers are small and composed of 3 joints (Fig. 4*B,C*; 5). The 5 other pairs of appendages resemble one another generally. Their large oblong **coxae** are provided along the mid-line with spines that help to push prey into the gullet. A strong ridge or knob on the posterior coxa suggests some masticatory function. The **walking legs** are provided with **chelae,** except the posterior pair, which has several flat spines useful in digging. An unjointed lateral appendage **(flabellum)** is attached to the lateral base of the coxa. In the male, the first legs are modified into clasping organs. The mouth is bordered posteriorly by a small internal plate **(endostoma).** Two short vertical plates **(chilaria)** behind the mouth represent the rudimentary appendages of the pregenital (7th) segment (Figs. 4*B, C*; 5).

The opisthosomal shield has a broad doublure, but the major portion of the ven-

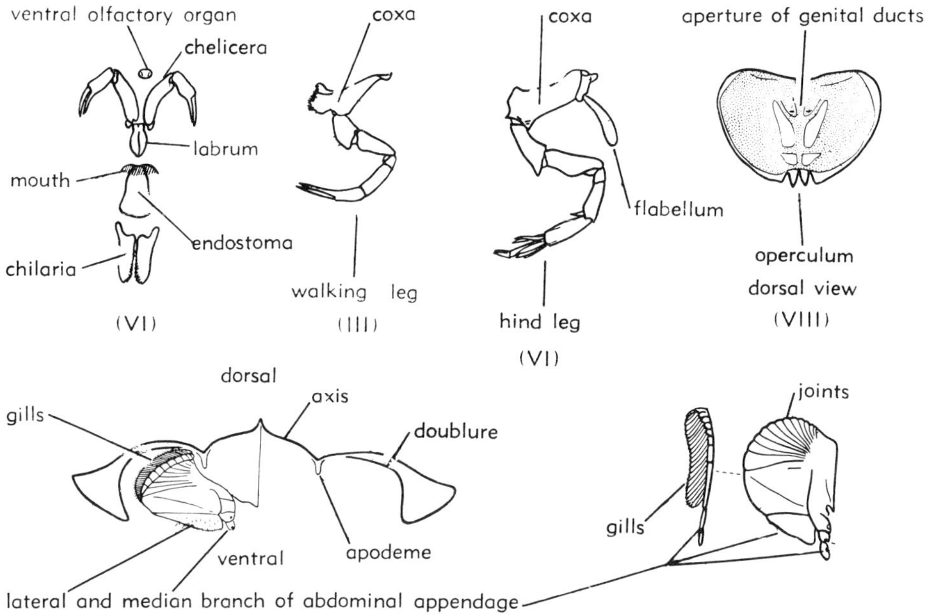

FIG. 5. Appendages of living xiphosurans; prosomal and abdominal appendages of *Limulus polyphemus* (LINNÉ) (much reduced) (64).

tral surface is occupied by the 6 overlapping plates representing the appendages. The first pair of appendages belong to the genital (8th) somite and form the **operculum** with openings of the genital ducts. The 5 succeeding pairs of plates **(gill appendages)** show more distinctly the biramous nature of the appendages. The jointed median branch, representing the "walking leg," is much reduced, whereas the lateral branch forms a powerful strongly flattened and multijointed appendage provided with the numerous book gills on the inner (dorsal) side (Fig. 5).

Internal organs include a strong "gizzard" and well-developed intestinal diverticula.

MORPHOLOGICAL TERMS APPLIED TO MEROSTOMES

The following alphabetically arranged glossary of morphological terms applied to merostomes is intended to give conveniently findable explanations of terms which may be unfamiliar to nonspecialists. Classification of the terms is indicated typographically so as to distinguish (1) those most commonly used (indicated by boldface capitalized words, as **CHELICERA**) and (2) those ranked as secondary in importance because less frequently used (indicated by boldface uncapitalized words, as **chilaria**).

Glossary of Morphological Terms

ABDOMEN. Posterior part of body behind cephalothorax (*see* opisthosoma) (Fig. 3).

apodeme. Inward deflection of sclerite serving for muscle attachment (Fig. 5).

axial furrow. Longitudinal groove separating median lobe (axis) from pleural area.

cardiac lobe. Median lobe of prosoma and opisthosoma (Fig. 3).

cardiophthalmic region. Space between ophthalmic ridges.

CEPHALOTHORAX. Fore part of body in front of opisthosoma (*see* prosoma).

chela. Pincer-like organ or claw of prosomal appendage.

CHELICERA. Preoral appendage modified for piercing or biting, composed of 3 or ?4 joints with distal ones forming a chela (pl., chelicerae) (Figs. 4*B,C*).

chilaria. Small plate forming rudimentary appendage of pregenital segment (Figs. 4*B,C*).

COMPOUND EYE. Lateral eye composed of many facets (Fig. 3).

coxa. Proximal (basal) joint of thoracic appendage, directly attached to body (Fig. 5).
doublure. Inwardly deflected marginal part of dorsal exoskeleton (Figs. 5, 18).
endostoma. Small plate on ventral side of body at posterior end of mouth (Fig. 5).
epicoxite. Small plate attached to median part of coxa (Fig. 18).
epimer. Lateral (pleural) expansion of tergite.
epistoma. Median plate on doublure in front of mouth (Fig. 7,*4b*).
epistomal suture. Longitudinal suture bordering epistoma.
exoskeleton. External more or less mineralized chitinous covering of body.
femur. Joint belonging to proximal part of prosomal appendage.
flabellum. Joint attached to lateral border of coxa (Fig. 5).
genal angle. Posterolateral corner of prosoma.
gill appendage. Opisthosomal appendage bearing respiratory organ (Figs. 5, 17).
glabellar area. Elevated part of prosoma between compound eyes, corresponding to interophthalmic region.
intercardiophthalmic region. Small rectangular area of prosoma embracing cardiac lobe and minor part of interophthalmic region.
interophthalmic region. Space between cardiac lobe and ophthalmic ridge (Fig. 3).
labrum. Upper lip, just in front of mouth (Figs. 4*B*, 5).
marginal suture. Ecdysial junction between exoskeletal elements at prosomal margin.
median suture. Longitudinal suture on doublure in front of mouth (Fig. 18).
MESOSOMA. Anterior part of opisthosoma (abdomen) carrying appendages.
METASOMA. Posterior part of opisthosoma (abdomen) lacking appendages.
METASTOMA. Plate at posterior edge of mouth (Figs. 17, 18).
ocellus. Median visual organ located on prosoma (Fig. 3).
occipital band. Rim along posterior border of prosoma.
occipital furrow. Groove in front of occipital band.
OPERCULUM. Plate adjoining appendages of genital segment (Figs. 4*B,C;* 17).
OPHTHALMIC RIDGE. Longitudinal ridge above compound eye and extending forward and backward from it (Fig. 3).
OPISTHOSOMA. Posterior part of body behind prosoma (same as abdomen).
PADDLE. Flat distal portion of last prosomal appendage (toward rear) (Fig. 17).
patella. Joint forming knee in prosomal appendage.
pleura. Lateral part of opisthosoma.
pleural furrow. Groove crossing pleura.
postabdomen. Narrow posterior part of opisthosoma (abdomen) (Fig. 17).
postventral plate. Plate below posterior part of abdomen.
preabdomen. Broad anterior part of opisthosoma (abdomen) (Fig. 17).
prefemur. Joint proximal to femur in prosomal appendage.
pretarsus. Joint distal to tarsus in prosomal appendage.
pretelson. Opisthosomal segment anterior to telson (Fig. 17*B*).
procephalic lobe. Anterior (preoral) part of embryo.
PROSOMA. Fore part of body in front of abdomen (same as cephalothorax).
proventriculum. Anterior part of intestine (Fig. 4*C*).
sclerite. Chitinized cover of body segment forming part of exoskeleton.
somite. Body segment.
STERNITE. Plate forming ventral cover of somite (Fig. 17*B*).
SWIMMING LEG. Hindmost prosomal appendage serving as swimming organ (Fig. 17*A*).
tarsus. Joint of distal part of prosomal appendage.
TELSON. Postanal spine or plate (Figs. 1, 17).
TERGITE. Plate forming dorsal cover of somite (Fig. 17*A*).
thoracetron. Opisothosoma or abdomen.
tibia. Joint of distal part of prosomal appendage.
trochanter. Joint of proximal part of prosomal appendage.
WALKING LEG. Prosomal appendage serving for walking.

CLASSIFICATION

Merostomes are divisible into 2 main groups which appropriately are ranked as subclasses; these are Xiphosura or horseshoe crabs, which are represented by modern *Limulus,* and the extinct Eurypterida. Classification of the Xiphosura is discussed in a recent paper (30) which explains considerations leading to recognition of 2 orders (Aglaspida, Xiphosurida), one without defined suborders and the other containing 2 suborders (Synziphosurina, Limulina). Families of the Limulina are grouped in 3 superfamilies. A tabulation of suprageneric units, showing the number of described genera and subgenera in each, follows.

Suprageneric Divisions of Merostomata

(First figures in parentheses indicate number of genera and second the number of subgenera.)

Xiphosura *(subclass)* (64; 3), *L.Cam.-Rec.*
  Aglaspida *(order)* (12), *L.Cam.-U.Ord.*
    Aglaspididae (8), *U.Cam.*
    Strabopidae (2), *U.Cam.-U.Ord.*
    Beckwithiidae (1), *M.Cam.*

Xiphosurida *(order)* (27), *Sil.-Rec.*
Synziphosurina *(suborder)* (9), *U.Sil.-L.Dev.*
Limuloididae (1), *U.Sil.*
Bunodidae (3), *U.Sil.*
Neolimulidae (1), *U.Sil.-L.Dev.*
Pseudoniscidae (3), *U.Sil.*
Weinberginidae (1), *L.Dev.*
Limulina *(suborder)* (18), *L.Dev.-Rec.*
Belinuracea *(superfamily)* (3), *Dev.-U.Carb.*
Belinuridae (3), *Dev.-U.Carb.*
Euproopacea *(superfamily)* (8), *L.Dev.-Perm.*
Euproopidae (4), *U.Carb.-Perm.*
Liomesaspididae (2), *Penn.-Perm.*
Elleriidae (1), *U.Dev.-Penn.*
Kiaeriidae (1), *?L.Dev.*
Limulacea *(superfamily)* (7), *Perm.-Rec.*
Paleolimulidae (1), *Perm.*
Mesolimulidae (3), *L.Trias.-Jur., ?Cret.*
Limulidae (3), *Rec.*
Limulinae *(subfamily)* (1), *Rec.*
Tachypleinae (2), *Rec.*
Uncertain superfamily and family (2), *U.Carb.-Perm.*
Eurypterida *(subclass)* (25; 3), *Ord.-Perm.*
Eurypteracea *(superfamily)* (14; 3), *Ord.-Perm.*
Hughmilleriidae (6), *Ord.-Perm.*
Pterygotidae (1; 3), *Ord.-Dev.*
Eurypteridae (3), *Ord.-U.Carb.*
Carcinosomatidae (1), *Ord.-Sil.*
Mixopteridae (1), *Sil.-L.Dev.*
Megalograptidae (2), *Ord.*

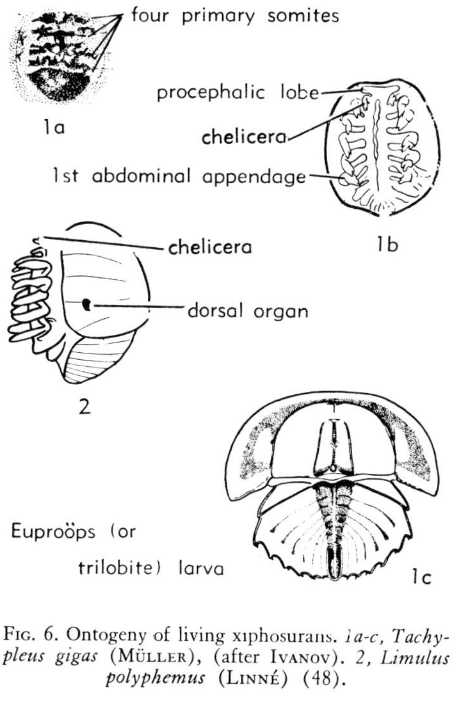

FIG. 6. Ontogeny of living xiphosurans. *1a-c, Tachypleus gigas* (MÜLLER), (after IVANOV). *2, Limulus polyphemus* (LINNÉ) (48).

Stylonuracea *(superfamily)* (11), *Ord.-L.Perm.*
Stylonuridae (7), *Ord.-L.Perm.*
Dolichopteridae (2), *Sil.-Dev.*
Rhenopteridae (1), *L.Dev.*
Mycteropidae (1), *U.Carb.*

ONTOGENY

The ontogeny of Recent limulids is of considerable interest in connection with study of the phylogeny of the Merostomata. The earliest stages (Fig. 6,*1a*) indicate only 4 primary segments (as inferred in the cephalon of trilobites), but in later ontogeny the secondary segments and appendages develop. A dorsal organ, probably sensory in function, is found in the last embryonic stages. The 1st free larva (Fig. 6,*1c*) has been called the trilobite stage because of its general resemblance to such forms. The prosoma has more segments than the trilobite cephalon, however, and all segments of the abdomen in *Limulus* are fused, being indicated only by rows of pigment. The 1st larva is more like the fossil Euproopacea.

During ecdysis (molting) the exoskeleton opens along a **marginal suture** on the prosoma. The animal creeps out and leaves the abandoned exoskeleton (exuvia) with a closed suture: this may be preserved in the sediment as a "complete" specimen.

OCCURRENCE

Although grown-up limulids are benthonic forms crawling and digging in the sand and mud, the young larvae (and to some extent the adults) are able to swim on their backs by rapid strokes of the abdominal appendages. The Recent representatives of the Merostomata are marine forms, but *Carcinoscorpius* is also able to endure brackish and fresh water of estuaries and rivers.

Inasmuch as most older-than-Recent merostomes are nonmarine forms, fossil remains of these arthropods are chiefly confined to the less common fresh-water and brackish-water sediments and are not common. Because of the rare and scattered finds, the number of species known is relatively small compared with the number of genera described.

The large majority of known species of merostomes occur in Paleozoic rocks and only a small group continues to Recent time.

The morphological changes that may be noticed during development of the Merostomata offer valuable information on the evolution of one arthropod group during the long time span from Cambrian to Recent. The living forms (horseshoe crabs or king crabs) are not much different from their Mesozoic relatives and are thus excellent examples of persistent forms. The Recent forms, comprising 5 species previously referred to the genus *Limulus,* are bottom dwellers, inhabiting shallow water along the east coasts of North America and Asia.

SYSTEMATIC DESCRIPTIONS

Subclass XIPHOSURA Latreille, 1802

[*nom. correct.* STØRMER, 1944 (*pro* Xiphosures LATREILLE, 1802)] [=Xyphosures LATREILLE, 1802 (*nom. neg.*); Gnathopoda STRAUS-DÜRKHEIM, 1829; Poecilopoda CLAUS, 1868 (*non* Poecilopes LATREILLE, 1817)]

Merostomata with trilobate dorsal shield. Prosoma large, abdomen with a variable number of segments and long styliform or lanceolate telson. Appendages of pregenital segment not ankylosed, abdominal appendages biramous with book gills. *L.Cam.-Rec.*

Order AGLASPIDA Walcott, 1911

Xiphosura with elongate, more or less vaulted body. Prosoma with prominent compound eyes. Opisthosoma with 11 or 12 segments, all free, or the posterior ones ankylosed into a shield; trilobation absent or indistinct. Telson with broad, powerful basal portion. Postventral plates below the posterior portion of the abdomen. ?Chelicera 4-jointed, other 5 prosomal appendages, as well as abdominal appendages, developed as short walking legs. Exoskeleton phosphatic, outer surface with variable granulation. *L.Cam.-U.Ord.*

DISCUSSION

Morphological features. The Aglaspida have an average length of 2 to 6 cm., but specimens longer than 21 cm. have been described. The subovate body has no true axial furrows, but a more or less distinct trilobation is suggested by different convexity of the axial and pleural regions. The comparatively large, mostly broad prosoma may have the genal angles prolonged into genal spines. The compound eyes generally have an antemedian position: ocelli may have been present but are not recognized surely. In some forms a glabellar area between the eyes and the posterior border, corresponding to the interophthalmic region in later forms, rises above the more flattened portion of the prosoma. The elevated glabellar area, which may be lobed (Fig. 7,*1,2*), has a subtriangular outline, the broadest part posterior.

The opisthosoma has 12 free articulating segments in the Paleomeridae and 11 in the Aglaspididae but 8 free anterior segments with ?3 ankylosed posterior segments in the Beckwithiidae. The pleurae, some with an anterior band or pleural furrow (or both) and mostly prolonged into pleural spines, overlap each other backwards. Their curved anterior border, not permitting a transverse straight joint line, suggests a somewhat imperfect articulation, the succeeding tergites sliding over each other. Inasmuch as only the 11th segment lacks pleurae in trilobed forms, no postabdomen is clearly demonstrated, except in the Beckwithiidae, where it forms the posterior shield.

The powerful telson has a broad basal portion, which is possibly a separate 12th segment united with the telson, as suggested by RAASCH (20). In *Beckwithia* (Fig. 9, *1*), the posterior abdominal shield seemingly has a powerful but short dorsal spine rising in the middle. A telson may have been present at the terminal margin.

On the ventral side, behind the doublure, a 4-sided plate possibly may be interpreted as an epistoma (Fig. 7, *4b*). Below the last abdominal segments and the base of the telson, the Aglaspididae have characteristic **postventral plates** (Fig. 7, *3*). The plates evidently give support to powerful muscles leading from the telson.

The appendages are known to a certain extent in *Aglaspis* (Fig. 15,*4ab*). The 1st pair seem to have 4-jointed chelicerae, whereas other prosomal appendages are developed as single walking legs, of which 4 joints have been noticed. The same kind of legs have been found on the abdomen. This suggests primitive features compared with later Xiphosura and Eurypterida, in which the abdominal walking legs are much reduced.

The exoskeleton of the Aglaspididae has proved to be phosphatic, but whether this is characteristic of all members of the order is not certain. The ornamentation is characteristic, consisting particularly of pustules, tubercles, and puncta.

Ecology. The Aglaspida were marine animals, for they occur together with trilobites and brachiopods. The sediments in which they are found suggest a quiet and fairly shallow marine environment.

Affinities. In general shape of the body, the Aglaspida recall the Trilobita and particularly some of the Merostomoidea (*Leanchoilia, Emeraldella*) of the Middle Cambrian. In the forms mentioned, however,

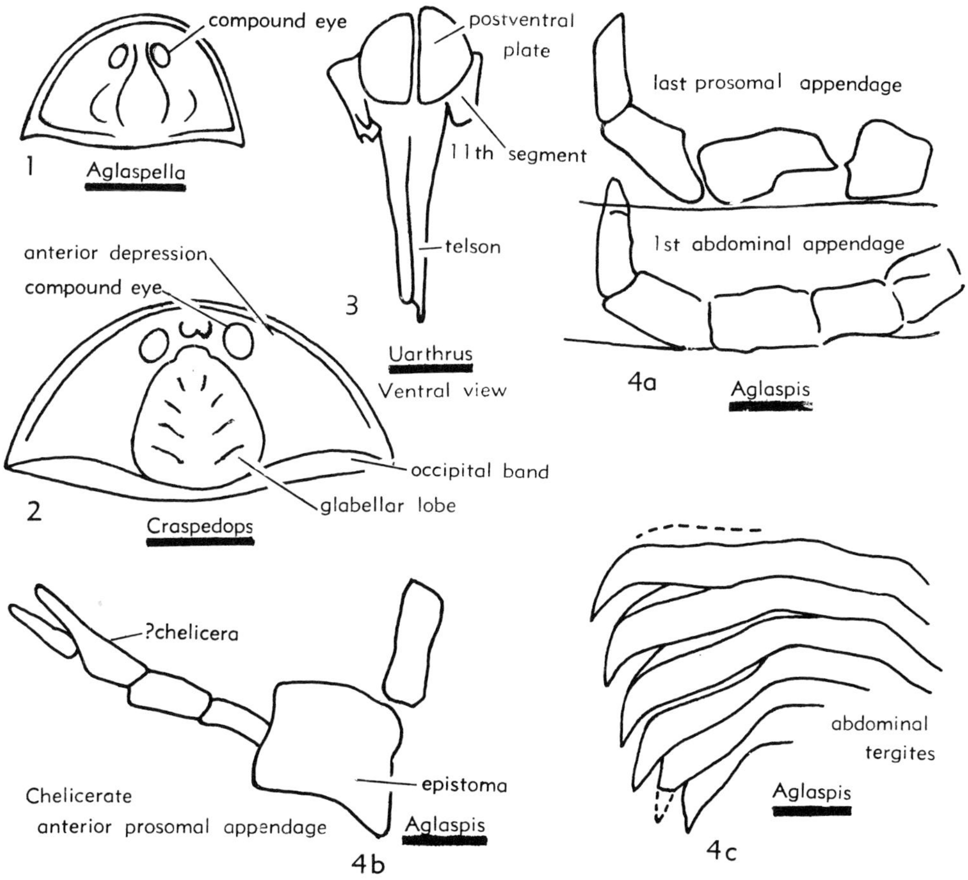

FIG. 7. Morphological features of aglaspids, U.Cam., Wis. (59) (p. P12-P14).

1, Aglaspella eatoni (WHITFIELD), prosoma (reconstr.), ×1.

2, Craspedops modesta RAASCH, prosoma, showing anteriorly placed eyes and elevated glabellar area (reconstr.), ×2.

3, Uarthrus instabilis RAASCH, posterior extremity from ventral side, ×2.

4a-c, Aglaspis spinifer (RAASCH); *4a,* last prosomal and 1st abdominal appendage; *4b,* chelicerate anterior prosomal appendage with epistoma; *4c,* freely articulating abdominal segments with laterally extended pleurae; all ×1.3.

the cephalon has only 4 appendages behind the mouth, as compared with ?6 in the Aglaspida; also antennae, such as possessed by Trilobita and Merostomoidea, are lacking in the Aglaspida. The Synziphosurina have much in common with the Aglaspida. Particularly, the Pseudoniscacea have a very similar body, differing mainly in possession of a large prosoma without compound eyes and an abdomen of 10 segments only. In the elongate nature of their body, the Aglaspida resemble the Eurypterida. One of the aglaspids from Upper Cambrian rocks (*Strabops,* Fig. 9,2) actually was interpreted as a primitive eurypterid by CLARKE & RUEDEMANN (5), partly on the basis of erroneous assumption that the abdomen contains 12 segments, as in the Eurypterida. On the other hand, the genus *Paleomerus* (Fig. 8), from the Lower Cambrian, seems to represent a transition between the Xiphosura of aglaspid form and the Eurypterida, for the body shape and presence of compound eyes correspond to both the Aglaspida and eurypterids such as *Hughmilleria* (Fig. 21,*1*); also, the seemingly loose articulation of the tergites in *Paleomerus* suggests the aglaspids, whereas the presence of 12 abdominal segments is a eurypterid character. The exact taxonomic placement of *Paleomerus* is uncertain because the nature of the appendages is unknown, but provisionally it is assigned to the Aglaspida.

Family PALEOMERIDAE Størmer, 1955

Exoskeleton of medium size; opisthosoma with 12 free segments, the 12th trapezoid; telson unknown. *L.Cam.*

Paleomerus STØRMER, 1955 [**P. hamiltoni*]. Body (exclusive of telson) ovate, strongly vaulted; prosoma short, with antemedian compound eyes; 12th segment long, trapezoid (31). *L.Cam.,* Sweden.——FIG. 8,*1*. **P. hamiltoni; 1a,b,* side and dorsal views (reconstr.), ×0.8 (64n).

Family AGLASPIDIDAE Miller, 1877

[*nom. correct.* STØRMER, herein (*pro* Aglaspidae MILLER, 1877)]

Medium-size forms with 11 free abdominal segments, abdomen moderately elongate. *U.Cam.*

Aglaspis HALL, 1862 [**A. barrandei*]. Dorsal shield pustulose; eyes submedian; prosoma with conspicuous rim, occipital band, and interophthalmic region set off by relative convexity, occipital band, and tergites bearing paired postaxial nodes (20). *U.Cam.,* N.Am.——FIG. 7,*4;* 9,*5a*. *A. spinifera* RAASCH, Trempeal., Wis.; 7,*4a,* prosomal appendages, ×2; 7,*4b,* last prosomal and 1st abdominal appendage, ×2; 7,*4c,* abdominal segments, ×0.5; 9,*5a,* prosoma, ×0.7 (59).——FIG. 9,*5b*. *A. simplex* RAASCH, Trempeal., Wis.; dorsal side, ×0.7 (59).

Aglaspella RAASCH, 1939 [**A. granulifera*]. Like *Aglaspis* but with smaller, more elevated eyes; no marginal rim on prosoma and no postaxial nodes (20). *U.Cam.,* N.Am.——FIG. 7,*1*. *A. eatoni* (WHITF.), Trempeal., Wis.; prosoma (reconstr.), ×1 (64).

Glypharthrus RAASCH, 1939 [**Eurypterus thomasi* WALTER, 1924]. Dorsal shield smooth or pitted; eyes large, subcentral; marginal rim present, occipital and pleural furrows distinct; telson slender (20). *U.Cam.,* N.Am.——FIG. 9,*3*. **G. thomasi* (WALTER), Trempeal., Wis.; dorsal side, ×2 (59).

Aglaspoides RAASCH, 1939 [**A. sculptilis*]. Dorsal shield striated, pitted, or smooth; prosoma flat, without rim and occipital furrows; large subcentral convex eyes; no pleural furrows (20). *U.Cam.,* N.Am.——FIG. 9,*6*. **A. sculptilis,* Trempeal., Wis.; telson, ×3 (59).

Uarthrus RAASCH, 1939 [as *U-arthrus*] [**U. instabilis*]. Dorsal shield obscurely pitted; prosoma convex, without rim or occipital furrow; large

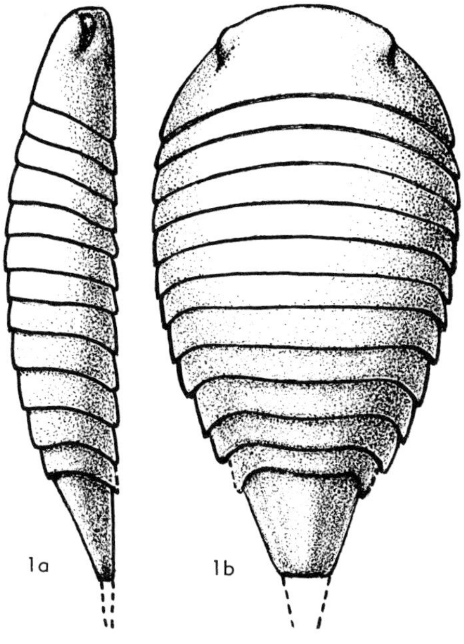

FIG. 8. *Paleomerus hamiltoni* STØRMER, from Lower Cambrian of Sweden; reconstr., side and dorsal views, ×0.8 (64 n) (p. P12).

close-set eyes near anterior margin; 11th segment U-shaped (20). *U.Cam.*, N.Am.——FIG. 7,*3;* 9,*4*. **U. instabilis*, Trempeal., Wis.; 7,*3*, ventral view of 11th segment with telson, ×2; 9,*4*, prosoma, ×3 (59).

Cyclopites RAASCH, 1952 [*pro Cyclopina* RAASCH, 1939 (*non* AGASSIZ, 1846; *nec* CLAUS, 1862; *nec* BRADY, 1872)] [**Cyclopina vulgaris* RAASCH, 1939]. Dorsal shield pitted; prosoma strongly convex; anterior very close-set eyes forming bilobed node; rim and occipital furrow absent; pleura with furrow near anterior margin (20). *U.Cam.*, N.Am. ——FIG. 9,*7*. **C. vulgaris* (RAASCH), Trempeal., Wis.; prosoma, ×1.3 (59).

Craspedops RAASCH, 1939 [**C. modesta*]. Dorsal shield nearly smooth; prosoma moderately convex;

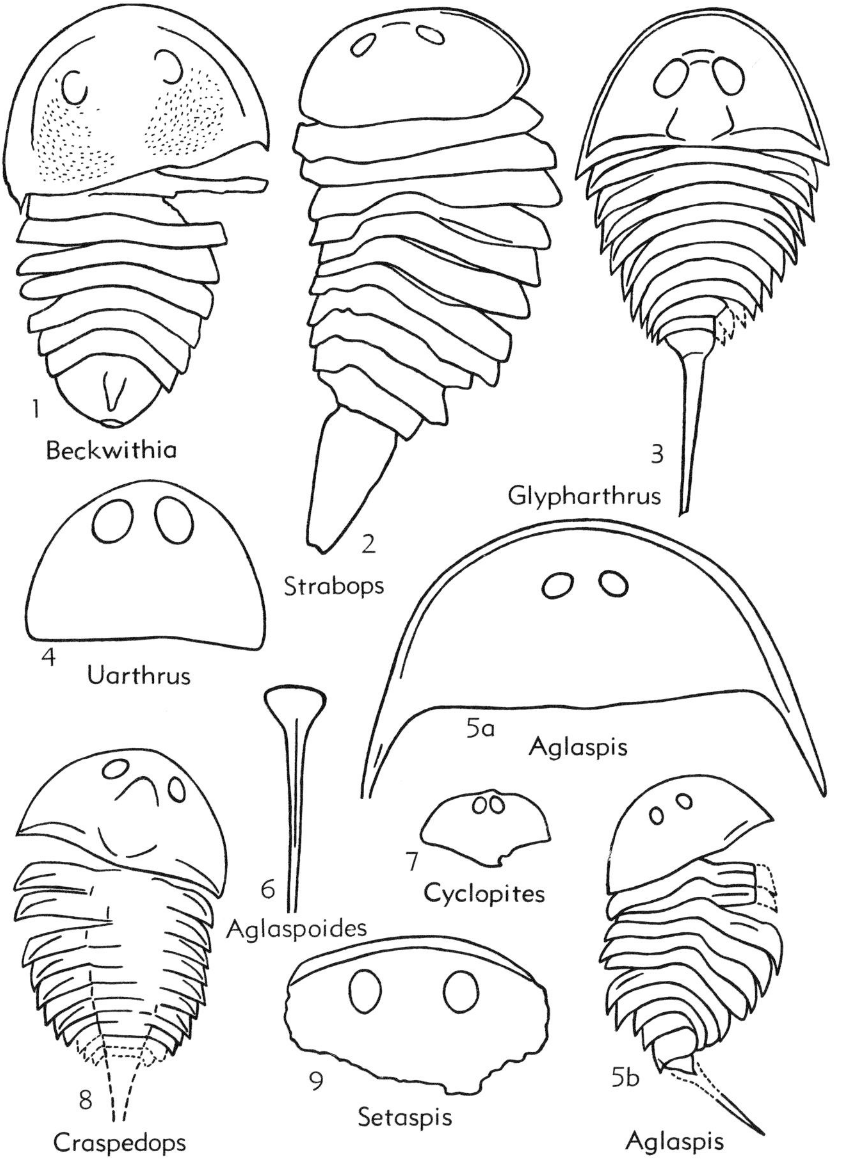

FIG. 9. Aglaspida (p. P12-P14).

eyes large, anterior, moderately far apart; occipital band present; interophthalmic region distinct, elevated lobate; pleural furrows on abdominal tergites (20). *U.Cam.*, N.Am.——FIG. *7,2; 9,8.* **C. modesta,* Trempeal., Wis.; *7,2,* prosoma (reconstr.) ×2; *9,8,* dorsal side, ×3 (59).

Setaspis RAASCH, 1939 [**S. spinulosa*]. Dorsal shield pustulose; eyes large, widely separated, near anterior marginal rim; no occipital furrow (20). *U.Cam.*, N.Am.——FIG. 9,*9*. **S. spinulosa,* Trempeal., Wis.; anterior part of prosoma, ×0.7 (59).

Family STRABOPIDAE Gerhardt, 1932

[*nom. correct.* STØRMER, herein (*pro* Strabopsidae GERHARDT, 1932)]

Medium-size forms with 11 free abdominal segments; prosoma short, elliptical; eyes antemedian; abdomen elongate or ovate; no pleural spines; telson broad. *U.Cam.-U.Ord.*

Strabops BEECHER, 1901 [*non* JORDAN, 1904] [**S. thacheri*]. Dorsal shield with indication of scales near posterior border of abdominal segments; body elongate. *U.Cam.*, N.Am.——FIG. 9,*2*. **S. thacheri,* Mo.; ×0.7 (34).

Neostrabops CASTER & MACKE, 1952 [**N. martini*]. Dorsal shield seemingly smooth; body ovate; prosoma with quadrate glabellar elevation and occipital furrow (4). *U.Ord.*, N.Am.——FIG. 10. **N. martini,* Ohio; ×1.5 (36).

Family BECKWITHIIDAE Raasch, 1939

Aglaspida with some posterior abdominal segments ankylosed. *M.Cam.*

Beckwithia RESSER, 1931 [**B. typa*]. Dorsal shield pustulose; prosoma large, with flattened rim; eyes situated about midway between center and margin; 8 free abdominal segments; posterior triangular shield with dorsal spines; telson probably styliform. *M.Cam.*, N.Am.—FIG. 9,*1*. **B. typa,* Utah; ×1 (61).

Order XIPHOSURIDA Latreille, 1802

[*nom. correct.* STØRMER, 1952 (*pro* Xiphosures LATREILLE, 1802)] [=Xyphosures LATREILLE, 1802; Xiphosura STØRMER, 1944]

Xiphosura with a subovate to subcircular, flat or vaulted body and styliform telson; prosoma generally with a distinct cardiac lobe and ophthalmic region bordered by ridges; abdomen with 10 or fewer segments, all free, or some or all ankylosed. Chelicerae 3-jointed, abdominal appendage seemingly plate-shaped. *Sil.-Rec.*

DISCUSSION

The Synziphosurina and Limulina, comprising the present order, have generally

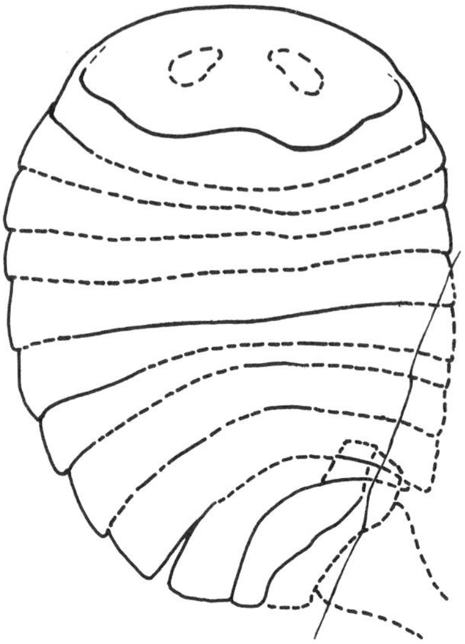

FIG. 10. *Neostrabops martini* CASTER & MACKE, U. Ord., Ohio, ×1 (36) (p. P14).

been regarded as separate orders (Synziphosura and Limulida, 22, 28), presumed to have taxonomic rank equal to the Aglaspida. Present knowledge of the Xiphosurida, however, indicates that the Synziphosurina and Xiphosurina are linked more intimately, whereas connection with the Aglaspida is somewhat remote. For these reasons, the Synziphosurina and Limulina are here regarded as suborders of the Xiphosurida and together ranked as coordinate with the Aglaspida.

Suborder SYNZIPHOSURINA Packard, 1886

[*nom. correct.* STØRMER, 1952 (*pro* Synziphosura, PACKARD, 1886)] [=Bunodomorpha ZITTEL-E., 1913]

Xiphosurida with a medium to large prosoma, lacking distinct traces of compound eyes, cardiac lobe present or absent; abdomen with 9 to 10 segments, of which all may be free, or the 6th and 7th ankylosed; appendages (as far as known) with flattened distal spines instead of chelae. *U. Sil.-L.Dev.*

DISCUSSION

The Synziphosurina form a small group of Xiphosura that are found associated with eurypterids and primitive vertebrates in the

Upper Silurian and Lower Devonian strata of several continents. They are small forms, measuring generally about 5 cm. in length (telson included).

Morphological features. The body (telson excluded) is subcircular to subovate. As shown in *Pseudoniscus,* the Synziphosurina were able to roll up. The large prosoma is semicircular to semielliptical in outline, the genal angles being slightly prolonged in certain forms. The prosoma is fairly large and smooth in *Pseudoniscus* (Fig. 11, *3*) and *Weinbergina* (Fig. 11,*9*), but smaller and more vaulted and sculptured in other genera. A cardiac lobe is distinct in *Bunodes, Bunaia,* and *Neolimulus* (Fig. 11, *2,4,10*), in particular. The lobe has an anterior swelling in the 2 former genera. In *Limuloides* and *Neolimulus* (Fig. 11, *1, 2*) ophthalmic ridges enclosing a cardiophthalmic region are indicated, features similar to those found in more recent forms. A segmentation of the area outside the cardiac lobe is suggested by the radial furrows in *Bunodes* and *Bunaia* and by the radial ridges in *Limuloides.* In *Bunodes,* the marginal portion of the prosoma is very steep. Differences in character of the prosoma of *Pseudoniscus* and *Weinbergina* on one side and remaining known genera on the other suggest recognition of 2 superfamilies, but such division probably should await more evidence.

Most genera lack lateral eyes. Possible traces of eyes have been described in *Neolimulus* and *Pseudoniscus.* Early reconstruc-

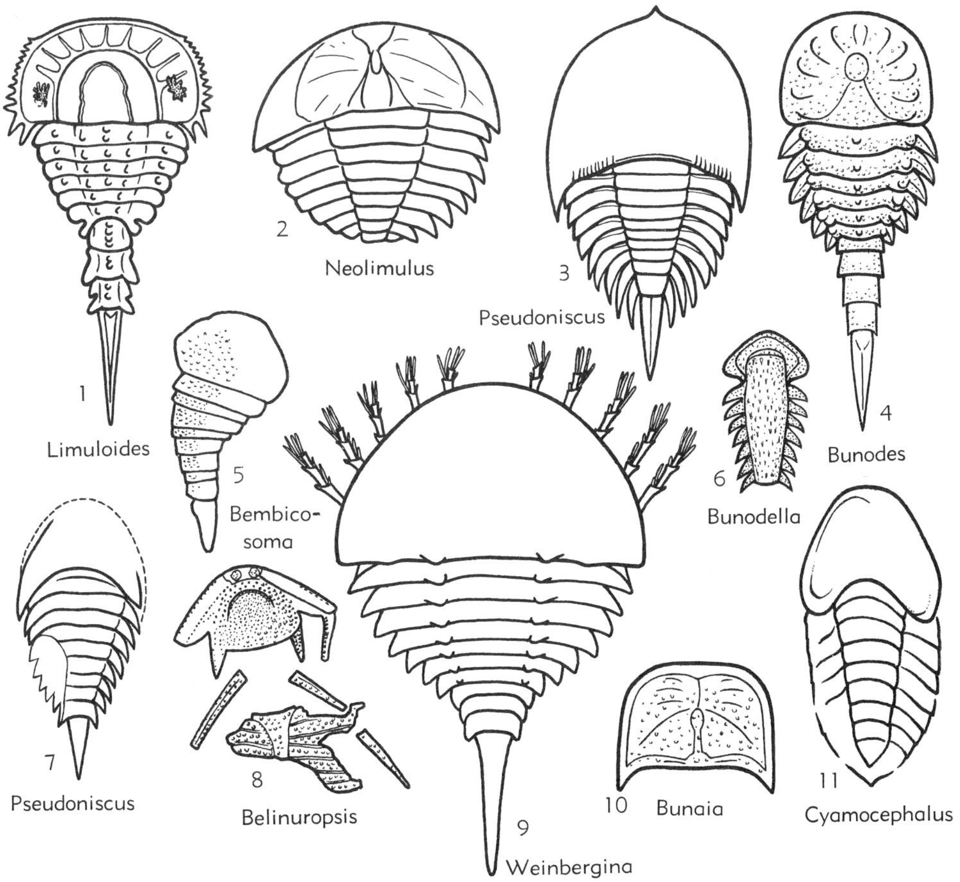

FIG. 11. Synziphosurina (p. P16-P17, P41).

tions of a Baltic species of *Pseudoniscus* indicated large eyes, but CLARKE found no trace of eyes in an American species, although RUEDEMANN has described certain structures that he interprets as small compound eyes and facial sutures. Because specimens commonly are somewhat crushed in the shale, it is difficult to be certain of the structures. A small stellate ornamentation outside the ophthalmic ridges in *Limuloides* (Fig. 11, *1*) may be explained as a sensory organ seemingly homologous with the dorsal organ in the embryo of Recent Xiphosura (Fig. 6,*2*).

The abdomen has 9 or 10 tergites, 10 segments being distinct in *Weinbergina* and assumed in *Pseudoniscus*. In *Limuloides* and *Bunodes,* there are 9 (fewer in an Australian species seemingly due to poor preservation), but here the 6th seems to be double, formed by fusion of the 6th and 7th. A trilobation is very distinct in *Pseudoniscus,* but may also be recognized in other genera having no axial (dorsal) furrows. The pleurae may be grooved and prolonged into short pleural spines. A postabdomen of 3 segments without pleurae is more or less conspicuous in some genera. In *Bunodes,* the 7th segment of the ankylosed 6th-7th segment may be included in the postabdomen. The telson is lanceolate.

Appendages are known only in *Weinbergina* (Figs. 11, 9, 15*A*). The 5 pairs of legs attached to the prosoma (chelicerae not known) closely resemble the posterior legs of Recent Limulina; the peculiar spines on the distal joints are very similar. The lack of traces of abdominal appendages in a well-preserved specimen might suggest the presence of plate-shaped appendages, such as in Limulina and Eurypterida.

The exoskeleton of the Synziphosurina is smooth (*Pseudoniscus*) or more or less strongly ornamented (*Bunodes, Bunaia*).

Ecology. The probably benthonic Synziphosurina are not constituents of typical marine faunas. All the Silurian-Downtonian species belong to the eurypterid faunas probably inhabiting brackish or fresh water. The only known Lower Devonian genus (*Weinbergina*) occurs in the Hunsrück fauna, which also includes marine forms such as trilobites.

Affinities. In general shape of the body, the Synziphosurina have much in common with the antecedent Aglaspida. Both have a prominent abdomen with free tergites. On the other hand, the presence of a cardiac lobe in *Bunodes, Bunaia,* and *Neolimulus,* and also of ophthalmic ridges in the last, points strongly toward the Limulida. In fact, it may be somewhat difficult to draw a distinct line between these groups. The appendages of *Weinbergina* also indicate relationship with the more recent Xiphosura.

Family LIMULOIDIDAE Størmer, 1952

[=Hemiaspididae ZITTEL, 1881 (*nom. correct.* STØRMER, herein, *pro* Hemiaspidae ZITTEL, 1881)]

Small forms with cardiac lobe, ophthalmic ridges, radial ridges and marginal spines on the prosoma. Abdomen with broad axis, 9 free segments, of which the 6th is possibly double. Postabdomen 3-segmented without pleurae. *U.Sil.*

Limuloides STØRMER, 1952 [*pro Hemiaspis* WOODWARD, 1864 (*non* FITZINGER, 1861)] [**Hemiaspis limuloides* WOODW.] (30). *U.Sil.,* Eu.——FIG. 11,*1*. **L. limuloides* (WOODW.), Eng.; ×1 (67).

Family BUNODIDAE Packard, 1886

Synziphosura with vaulted and radially lobed prosoma; abdomen with cardiac lobe forming broad axis, 9 free segments, of which the 6th seems to be double; narrow postabdomen of 3 to 4 circular segments. *U.Sil.*

Bunodes EICHW., 1859 [**B. lunula*] [=*Exapinurus* NIESZKOWSKI, 1859]. Dorsal shield granulated, strongly vaulted; prosoma distinctly lobed radially, without genal spines. *U.Sil.-?L.Dev.(Downton.),* Balt.-Norway.——FIG. 11,*4*. **B. lunula, U.Sil.,* Balt.; reconstr., ×1.3 (64).

Bunaia CLARKE, 1919 [**B. woodwardia*]. Dorsal shield smooth or granulated; prosoma with narrow cardiac lobe, marginal rim, and genal spines. *L.Dev.(Downton.),* N.Am.-Spitzb.——FIG. 11,*10*. *B. heintzi* STØRMER, Spitzb.; ×1 (64).

Bembicosoma LAURIE, 1899 [**B. pomphicus*]. Dorsal shield granulated; cardiac lobe not known; postabdomen indistinct. May be eurypterid larva or related to *Bunodes* (16). *U.Sil.,* Eu.——FIG. 11,*5*. **B. pomphicus,* Scot.; ×0.7 (53).

Family NEOLIMULIDAE Packard, 1886

Small forms with broad prosoma having cardiac lobe, ophthalmic ridges uniting in front, with acute genal spines; abdomen broad, with axial furrows, no postabdomen recognized. *U.Sil.-L.Dev.*

Neolimulus Woodward, 1868 [**N. falcatus*]. *U.Sil.-L.Dev.*, Eu.——Fig. 11,*2*. **N. falcatus*, Scot.; ×3 (64).

Family PSEUDONISCIDAE Packard, 1886

Small forms with large, somewhat flat smooth prosoma; abdomen with distinct axial furrows, 9 to 10 free segments prolonged laterally into pleural spines; no postabdomen recognized. *U.Sil.*

Pseudoniscus Nieszkowski, 1859 [*non* Costa, 1882] [**P. aculeatus*]. Sixth and 7th abdominal segments possibly ankylosed (5). *U.Sil.*, Eu.-N.Am. ——Fig. 11,*7*. **P. aculeatus*, Balt.; ×1.3 (63).——Fig. 11,*3*. *P. roosevelti* Clarke, N.Y.; ×2 (39).

?Cyamocephalus Currie, 1927 [**C. loganensis*]. Prosoma with rounded genal angles; abdomen with axial furrows, probably 10 segments. *U.Sil.*, Eu.——Fig. 11,*11*. **C. loganensis*, Scot.; ×0.7 (41).

Bunodella Matthew, 1889 [*non* Pfeffer, 1889; *nec* Verrill, 1889] [**B. horrida*]. Abdomen elongate as in Pseudoniscidae but prosomal features and ornamentation different. *U.Sil.*, N.Am.——Fig. 11,*6*. **B. horrida*, Can.; ×0.7 (54).

Family WEINBERGINIDAE Richter & Richter, 1929

Medium-size forms with large semicircular smooth prosoma and short 10-segmented trilobate abdomen, with last 3 segments forming postabdomen; ?chelicerae and 5 pairs of walking legs with spines. *L.Dev.*

Weinbergina Richter-R., 1929 [**W. opitzi*] (22). *L.Dev.*, Eu.——Fig. 11,*9*; 15*A*. **W. opitzi*, Ger.; 11,*9*, dorsal side (reconstr.), ×0.7; 15*A*, prosomal appendages, ×0.7 (62).

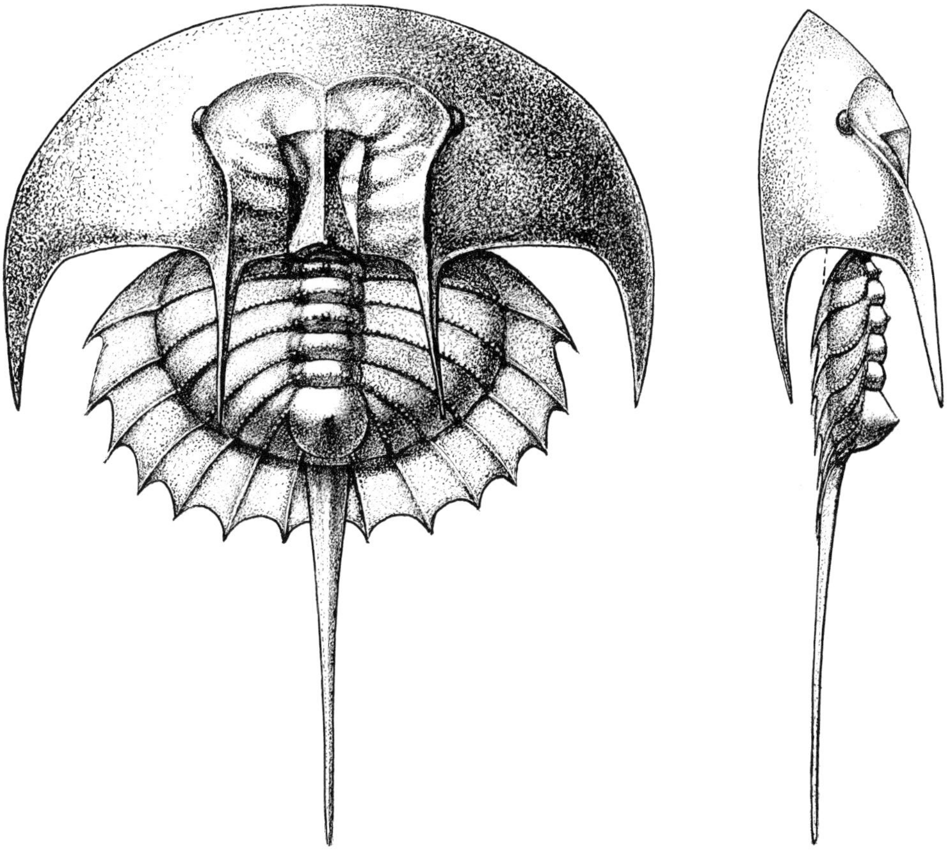

Fig. 12. *Euproops rotundatus* (Prestwich), U.Carb., Eng.; side and dorsal views (reconstr.), ×2.5 (64 n).

Suborder LIMULINA Richter & Richter, 1929

[*nom. correct. et transl.* STØRMER, 1952 (*ex* Limulida RICHTER & RICHTER, 1929)] [=Limulada RAYMOND, 1944]

Prosoma large, bearing cardiac lobe, ophthalmic ridges, ocelli, and lateral eyes; abdomen with as many as 9 well-developed segments, posterior ones or all of them ankylosed, or all free; telson styliform, longer than rest of abdomen (22). *?L.Dev. (Downton.)-Rec.*

DISCUSSION

Morphological features. The basic, characteristic morphological features of this suborder were established already in the oldest known representatives, the Recent ones representing true persistent forms.

Whereas most Paleozoic Limulina were small forms, measuring about 3 to 5 cm. in length, Mesozoic ones are medium-sized, and Cenozoic species are large, measuring 30 to 60 cm. In general, the size of adult individuals belonging to the Limulina increases during development of the suborder.

The Limulina (Figs. 3, 12, 13, 16) have a large semicircular prosoma, generally prolonged into genal spines. The wedge-shaped, commonly keeled cardiac lobe, which may bear an anterior boss or node, extends forward to a re-entrant of the confluent ophthalmic ridges. These ridges run laterally from the re-entrant, turn posteriorly above lateral eyes forming strong keels which generally terminate in ophthalmic spines at the posterior margin of the prosoma. The cardiophthalmic region may have interophthalmic ridges bordering a rectangular intercardiophthalmic area which embraces the cardiac lobe and a narrow tract on either side of it (Fig. 12). Where interophthalmic ridges are present, the cardiophthalmic region is divided into 5 parts instead of 3. The 5-fold division is significant in forms which have been referred to *Prestwichianella* (now provisionally considered to be a synonym of *Euproops,* because the same structure seemingly occurs in the type species of the latter) (Figs. 12; 13, *3*), as suggested by STUBBLEFIELD in 1947. The previous reconstruction of *Prestwichianella* (Fig. 13, *3,* designated as *Euproops*) was based on a badly preserved specimen; a new reconstruction (Fig. 12) based on English specimens, shows typical characters of *Euproops.*

The opisthosoma has a semicircular, subtriangular, or irregular hexagonal outline. The larvae of certain species at least were able to roll up. The abdomen is commonly more or less ankylosed into one continuous shield. *Neobelinuropsis* (Dev.) (Fig. 14) seems to have 8 free segments in front of the telson, the last segment probably double. In most species of *Belinurus* (Fig. 13,*1a,c,d*) the 2 to 3 posterior segments probably are ankylosed. In almost all Euproopacea and Limulacea, all abdominal tergites are ankylosed into one continuous shield. The segmentation is suggested by annulation of the axis and in Paleozoic forms by pleural ridges prolonged into spines along the margin. Intercalated movable spines are present in members of the Limulacea. In certain genera, such as *Kiaeria, Elleria,* and *Paleolimulus,* one or more free segments evidently were present between the telson and the abdominal shield in front (21). In genera of the Euproopacea the annulated axis of the abdomen terminates in a broad segment and is provided with a conical spine at the top (Fig. 12).

Prosomal appendages are known in *Euproops* (Fig. 15*B*), *Paleolimulus* (Fig. 15*C*) and certain Mesozoic Limulacea. The appendages seem to be strikingly like those of Recent *Limulus.* In *Paleolimulus* the chelicerae are small and the walking legs slender and chelate, except the last pair, which have a whorl of spines; even a flabellum and plate-shaped abdominal appendages are demonstrated. Upper Devonian trails (named *Paramphibius*) have been referred to *Belinurus*-like forms, possibly *Protolimulus* (1).

The exoskeleton is very thin and but slightly impregnated with calcium carbonate in the Euproopidae (21). The surface of the shell is smooth (tuberculated in the doubtful genus *Bunodella*).

Little is known of the larvae of fossil Limulina. Possibly some of the smaller species described represent larvae of larger forms. The larvae seem to have been more spinous than the adults.

Ecology. The habitat of the Limulina changed between Devonian and Recent time. The Devonian species were almost confined to marine waters probably outside deltaic deposits, but *Belinurus kiltorkensis* occurs in the Old Red Sandstone, suggesting that some species were able to invade brackish or fresh waters. With few exceptions the many species of Carboniferous-Permian time evidently were fresh-water forms. In the Triassic, a few specimens from the Zechstein suggest marine conditions, whereas the rest belong to brackish- or fresh-water faunas. The common *Mesolimulus walchi* from the Jurassic of Germany inhabited marine coastal waters, and a Liassic specimen from Scania suggests a brackish- or fresh-water environment. *Tachypleus decheni* of the Miocene lignite beds of Germany may have lived in brackish waters.

The Limulina were bottom-dwellers. Like Recent larvae of Limulus, some small species of *Belinurus* possibly were swimmers. The dorsal shield of the Limulina is well adapted (streamlined) for crawling rapidly through the water on the bottom. The telson chiefly served as a steering-rod.

The taxonomy of the Limulina is based mainly on development of the ophthalmic ridges and general structure of the abdomen.

Phylogeny. The fossil record permits us to follow evolution of the Limulina from Paleozoic to Recent time. The line of development may be traced from the Silurian *Neolimulus* (placed in the Synziphosurina) through *Belinurus* to the Euproopacea, and further to the Limulacea of Mesozoic and Cenozoic time. Characteristic of the phylogenetic development is shortening of the abdomen and gradual fusion of the abdominal segments. Gradual increase in size also

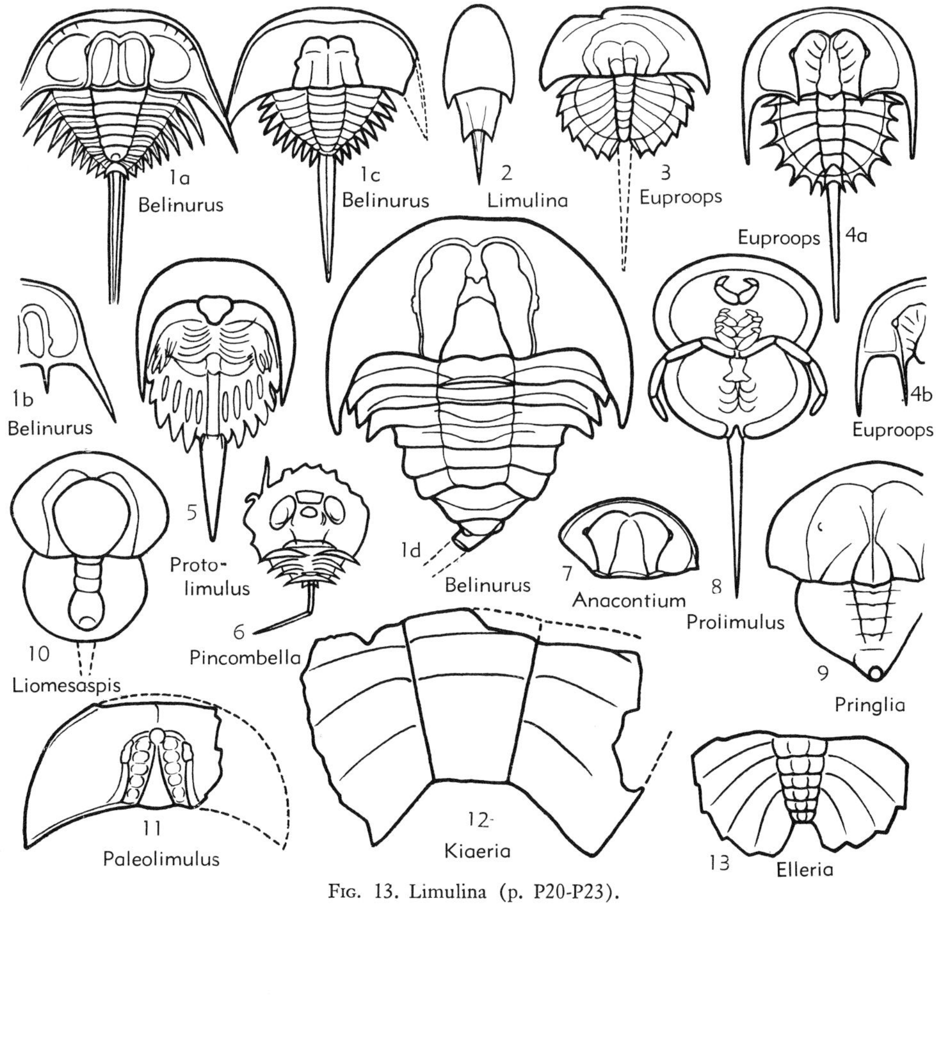

FIG. 13. Limulina (p. P20-P23).

is suggested. *Kiaeria* and *Elleria* seem to form a special line of development. The former probably had free segments in the posterior portion of the abdomen, a tendency also demonstrated in *Paleolimulus.*

Superfamily BELINURACEA Zittel & Eastman, 1913

[*nom. transl.* RAYMOND, 1944 (*ex* Belinuridae ZITTEL-E., 1913)]

Anterior abdominal segments (or possibly all) movable but 2 or more posterior ones may be ankylosed. *M.Dev.-U.Carb.*

Family BELINURIDAE, Zittel & Eastman, 1913

[=Bellinuridae PACKARD, 1886]

Characters of superfamily. *M.Dev.-U.Carb.*

Belinurus KÖNIG, 1820 [**B. bellulus*] [=*Bellinurus* MEEK & WORTHEN, 1865; *Koenigiella* RAYMOND, 1944]. Small to medium in size, prosoma with genal spines and cardiac lobe of variable width (21). *U.Dev.-U.Carb.,* Eu.-N.Am.——FIG. 13,*1a*. *B. regina* BAILY, U.Carb., Ire.; dorsal side, ×2 (67).——FIG. 13,*1b*. *B. arcuatus* BAILEY, U.Carb., Ire.; part of prosoma, ×3 (67).——FIG. 13,*1c*. *B. baldwini* WOODWARD, U.Carb., Eng.; dorsal side, ×2 (67).——FIG. 13,*1d*. *B. alleganyensis* ELLER, U.Dev., N.Y.; dorsal side, ×1 (44).

Neobelinuropsis ELLER, 1938 [**Bellinuroopsis rossicus* TSCHERNYSCHEW, 1933] [=*Bellinuroopsis* TSCHERN., 1933]. Medium in size, prosoma with distinct wedge-shaped cardiac lobe; rounded abdomen with 8 movable segments, the last one seemingly double, pleural spines present. *U.Dev.,* Russ. ——FIG. 14. **B. rossicus* (TSCHERN.); dorsal side, ×0.5 (65).

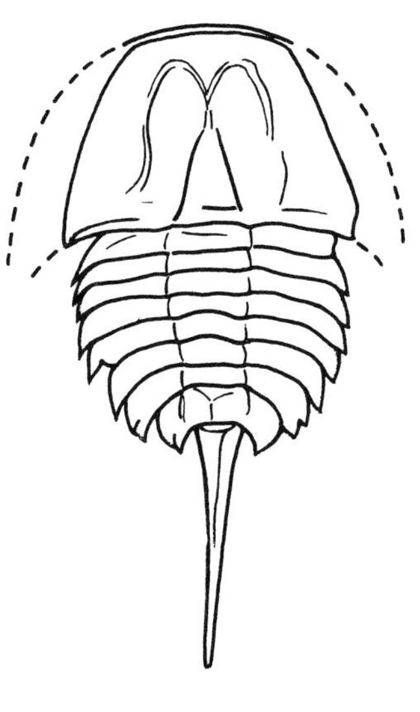

FIG. 14. *Neobelinuropsis rossicus* (TSCHERNYSCHEW), U.Dev., Russ., ×0.5 (65).

Protolimulus PACKARD, 1886 [**Prestwichia eriensis* WILLIAMS, 1885]. Small, only ventral surface known; abdominal segments terminating in blunt spines; telson broad and short (17). *M.Dev.,* N.Am.——FIG. 13,*5*. **P. eriensis* (WILLIAMS), N.Y.; ventral side, ×0.3 (66).

Superfamily EUPROOPACEA Eller, 1938

[*nom. transl.* RAYMOND, 1944, *ex* Euproopidae ELLER, 1938)]

Cardiophthalmic area well marked, ophthalmic ridges converging backward from eyes; abdomen broad and rounded, composed of 6 or 7 ankylosed segments, with or without lateral spines. *?L.Dev.(Downton.)-Perm.*

Family EUPROOPIDAE Eller, 1938

Small forms with wedge-shaped cardiac lobe bordered by distinct axial furrows, abdominal shield with annulated axis bearing a high boss on last segment. *U.Carb.-Perm.*

Euproops MEEK, 1867 [**Bellinurus danae* MEEK & WORTHEN, 1865] [=*Anthracopeltis* BOULAY, 1880; *Prestwichianella* WOODWARD, 1918 (*pro Prestwichia* WOODW., 1867; *non* LUBBOCK, 1863)]. Prosoma with flat genal spines and carinate ophthalmic spines; cardiophthalmic region with or without intercardiophthalmic area; abdomen with raised pleural ridges that cross flattened rim and are prolonged as marginal spines; annulated axis with knob on 1st and 3rd segments and elevated boss or short spine on hindmost segment; telson long. *U.Carb.-Perm.*——FIG. 13,*4*. *E. thompsoni* RAYMOND, Penn., Ill.; *4a,* dorsal side, ×1; *4b,* part of prosoma showing genal and ophthalmic spines, ×1 (60).——FIG. 15*B*. **E. danae* (MEEK-W.), Penn., Ill.; ventral view showing remnants of prosomal legs, ×1.6 (55).——FIGS. 12; 13,*3*. *E. rotundatus* (PRESTWICH), U.Carb., Eng.; 12, dorsal and side views, new reconstr. based on better-preserved specimen than available to WOODWARD, ×2.5 (64n); 13,*3*, old reconstr. referred to *Prestwichianella* by WOODWARD, ×0.5 (67).

Pringlia RAYMOND, 1944 [**Prestwichia birtwelli* WOODWARD, 1872]. Genal spines vestigial or lacking, cardiophthalmic region tripartite, abdomen triangular, with faint axial furrows, high posterior boss or spine on axis, seemingly no raised pleural ridges and probably no marginal spines. *U.Carb.,* N.Am.-Eu.——FIG. 13,*9*. **P. birtwelli* (WOODW.), Eng.; dorsal side, ×2 (67).

Anacontium Raymond, 1944 [**A. carpenteri*]. Genal spines vestigial, no ophthalmic spines, cardiophthalmic region tripartite. *Perm.*, N.Am.——Fig. 13,*7*. **A. carpenteri*, Okla.; dorsal side of prosoma, ×3 (60).

Family LIOMESASPIDIDAE Raymond, 1944

[*nom. correct.* Størmer, herein (*pro* Liomesaspidae Raymond, 1944)]

Prosomal cardiac lobe bulbous; adult lacking genal spines; abdomen without marginal spines and raised pleural ridges; posterior axial segment with high boss (21). *Penn.-Perm.*

Liomesaspis Raymond, 1944 [**L. laevis*] (21). *Penn.*, N.Am.——Fig. 13,*10*. **L. laevis*, Ill.; dorsal side, ×1.7 (60).

Prolimulus Fritsch, 1899 [**P. woodwardi*]. Prosoma and abdomen broad, well rounded, both without marginal spines. *Perm.*, Eu.——Fig. 13,*8*. **P. woodwardi*, Czech.; ventral side, ×2 (45).

Family ELLERIIDAE Raymond, 1944

[*nom. correct.* Størmer, herein (*pro* Elleridae Raymond, 1944)]

Abdominal axis and pleural area distinctly segmented, no boss on posterior axis; posterior part of abdominal shield deeply indented; prosoma unknown. *U.Dev.-Penn.*

Elleria Raymond, 1944 [**Euproöps morani* Eller, 1938]. *U.Dev.-Penn.*, N.Am.——Fig. 13,*13*. **E. morani*, U.S.A.; dorsal side of abdominal shield, ×1 (44).

Family KIAERIIDAE Størmer, 1952

[*nom. correct.* Størmer, herein (*pro* Kiaeridae Størmer, 1952)]

Large abdominal shield with axial furrows; segmentation and broad posterior indentation suggest posterior free segments; prosoma unknown. *?L.Dev.*

Kiaeria Størmer, 1934 [**K. limuloides*]. *?L.Dev.*, Eu.——Fig. 13,*12*. **K. limuloides*, ?Downton., Norway; dorsal side of abdominal shield, ×0.5 (64).

Superfamily LIMULACEA Zittel, 1885

[*nom. transl.* Raymond, 1944 (*ex* Limulidae Zittel, 1885)]

Small to large forms with cardiophthalmic region not well defined, posterior branches of ophthalmic ridges parallel; abdomen mostly subtrapezoidal with movable marginal spines. (Because of mode of preservation, ventral structures are commonly traced on dorsal surface, which has caused confusion.) *Perm.-Rec.*

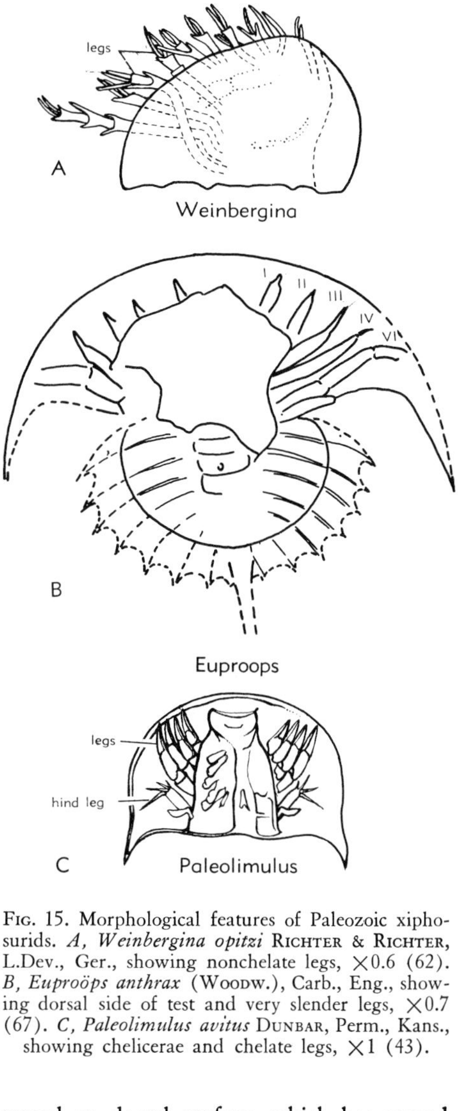

Fig. 15. Morphological features of Paleozoic xiphosurids. *A*, *Weinbergina opitzi* Richter & Richter, L.Dev., Ger., showing nonchelate legs, ×0.6 (62). *B*, *Euproöps anthrax* (Woodw.), Carb., Eng., showing dorsal side of test and very slender legs, ×0.7 (67). *C*, *Paleolimulus avitus* Dunbar, Perm., Kans., showing chelicerae and chelate legs, ×1 (43).

Family PALEOLIMULIDAE Raymond, 1944

Small forms with ophthalmic ridges meeting in front of cardiac lobe; abdominal axis distinctly annulated with movable segment in front of telson (21). *Perm.*

Paleolimulus Dunbar, 1923 [**P. avitus*] (7). Interophthalmic area conspicuously lobed. *Carb.-Perm.*, Eu.-N.Am.——Fig. 15*C*; 16,*1*. **P. avitus*, Perm., Kan.; 15*C*, ventral side of prosoma with appendages, ×1; 16,*1*, dorsal side (reconstr.) ×2 (43).——FIG. 13,*11*. *P.? randalli* (BEECHER), U. Dev., Pa.; ×1 (34).

Family MESOLIMULIDAE Størmer, 1952

Small to medium in size; prosoma with ophthalmic ridges not meeting in front of cardiac lobe, genal angles moderately prolonged backwards; axial furrows distinct; no distinct annulation of abdominal axis, first pair of marginal spines within anterior third of abdomen. *L.Trias.-Jur., ?Cret.*

Psammonlimulus LANGE, 1923 [**P. gottingensis*]. Genal angles prolonged into blunt genal spines: abdomen narrow trapezoidal with postlateral prolongations. *L.Trias.*, Eu.——FIG. 16,*2*. **P. gottingensis;* dorsal side, ×0.7 (52).

Limulitella STØRMER, 1952 [*pro Limulites* SCHIMPER, 1850 (*non* KRÜGER, 1823)] [**L. bronni* SCHIMPER, 1850]. Inner margin of genal angle forming a distinct angle with anterolateral margin of narrow subtriangular abdomen; axis may bear a median carina (30). *L.Trias.-U.Trias.*, Eu.-N.Am.——FIG. 16,*3*. **L. bronni* (SCHIMPER), L.Trias, Fr.; dorsal side, ×0.4 (64).

Mesolimulus STØRMER, 1952 [**Limulus walchi* Demarest, 1822]. Inner margin of genal angle more or less parallel to anterolateral margins of semicircular to broad trapezoid abdomen, axis with median carina. *Jur., ?Cret.*, Eu.-?AsiaM.——FIG. 16,*4*. **M. walchi* (DEMAREST), Jur., Ger., ×0.4 (49).——FIG. 16,*5*. *M. syriacus* (H.WOODWARD), Cret., Asia M.; ×0.2 (67).

Family LIMULIDAE Zittel, 1885

[=Xiphosuridae POCOCK, 1902]

Medium to large forms: prosoma with ophthalmic ridges not meeting in front of cardiac lobe, genal angles considerably prolonged backwards; abdomen subhexagonal, axial furrows indistinct, first pair of marginal spines about midway between anterior and posterior margin of abdomen. *Tert.-Rec.*

Subfamily LIMULINAE Zittel, 1885

[*nom. transl.* POCOCK, 1902 (*ex* Limulidae ZITTEL, 1885)]

Prosoma vaulted; abdomen hexagonal with prominent posterior prolongations, movable lateral spines decreasing in length backwards. *Rec.*

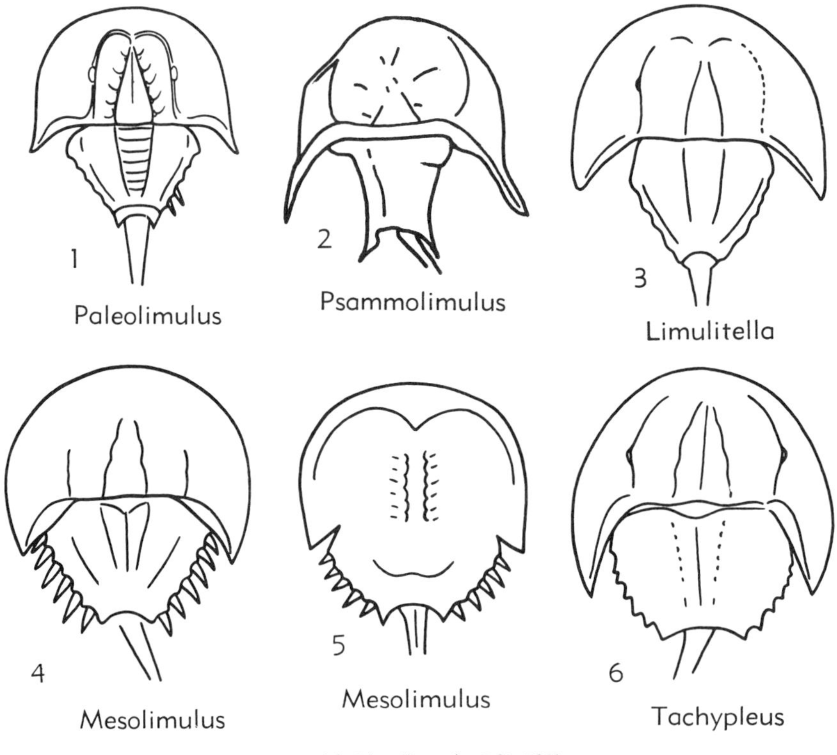

FIG. 16. Limulicae (p. P21-P23).

Limulus MÜLLER, 1785 [*pro Xiphosura* BRÜNNICH, 1771; ICZN 1952] [**Monolucus polyphemus* LINNÉ, 1758]. *Rec.*, E.N.Am.——FIGS. 3; 4*B,C;* 5; 6,*2*. **L. polyphemus* (LINNÉ); 3, half-grown specimen, about ×1 (56); 4*B,C*, ventral side and long. sec., ×0.3 (64); 5, prosomal appendages (64); 6,*2*, embryo, enlarged (48).

Subfamily TACHYPLEINAE Pocock, 1902

Prosoma less vaulted than in Limulinae; hexagonal abdomen with less prominent posterior prolongations, movable lateral spines not decreasing in length backwards. *?Tert., Rec.*

Tachypleus LEACH, 1819 [**Limulus gigas* MÜLLER, 1785]. Movable lateral spines of abdomen different in sexes, long and equal in male and young female, posterior 3 short and apically acute in adult female. *?Tert., Rec.*, E.Asia-?Eu.——FIG. 4*A;* 6,*1*. **T. gigas* (MÜLLER), Rec.; 4*A*, young specimen with segments indicated; 6,*1a,b*, embryological stages; 6,*1c*, first free larva (IVANOV).——FIG. 16,*6*. *T. decheni* (ZINKEN), Mio., Ger., ×0.13 (35).

Carcinoscorpius POCOCK, 1902 [**Limulus rotundicaudatus* LATREILLE, 1802]. Movable spines of abdomen short, 2nd and 3rd longest, others progressively decreasing backwards in male, becoming abruptly shorter in female. *Rec.*, E.Asia.

Superfamily and Family UNCERTAIN

Pincombella CHAPMAN, 1932 [**P. belmontensis*]. Anterior margin of prosoma spinose, no true genal spines; abdomen with 5 transverse segments, all but last with marginal spines. *Perm.*, Austral.——FIG. 13,*6*. **P. belmontensis*, N.S.W., ×1 (38).

Genus undet. Body narrow, abdominal shield without visible segmentation. *U.Carb.*, Eng.——FIG. 13,*2*. Gen. and sp. undet., ×2 (42).

Subclass EURYPTERIDA Burmeister, 1843

[=Gigantostraca HAECKEL, 1896]

Small to very large merostomes with elongate lanceolate, rarely trilobed body; prosoma of moderate size; opisthosoma with 12 movable segments and styliform to spatulate telson, with division commonly into a 7-segmented preabdomen and 5-segmented postabdomen; prosomal appendages 6, comprising 3-jointed chelicerae, walking legs, the last pair commonly transformed into swimming legs. Mouth central, bordered posteriorly by endostoma and metastoma. Operculum with median genital appendage, abdominal appendages plate-shaped with nonlaminate gills. *Ord.-Perm.*

DISCUSSION

Like the Xiphosura, the Eurypterida are not typical marine forms. Their remains generally are confined to sediments deposited in brackish or fresh water and are not abundant.

Fragments of the giant *Pterygotus* in the Scottish Old Red Sandstone early attracted attention of workmen in the quarries and by reason of their characteristic sculpture were interpreted as petrified seraphim. Descriptions of the eurypterids were published first in America by DE KAY in 1825 and HARLAN in 1825, but more extensive knowledge of the group is obtained from publications by HALL in 1859, HUXLEY & SALTER in 1859 (10), and WOODWARD in 1866-78 (33). Important morphological details were described by SCHMIDT in 1883 (24), and LAURIE in 1892 (16), 1893, and 1898; by means of a special method, HOLM in 1898 (9) was able to isolate and study in minute details the excellently preserved exoskeletons of *Eurypterus fischeri* from Silurian rocks of the Baltic. Knowledge of American eurypterids was considerably extended by CLARKE & RUEDEMANN in 1912 (5).

In spite of considerable variation in development of the exoskeleton, particularly the prosomal appendages, the Eurypterida exhibit a very definite plan of construction, thus forming a well-defined group.

Morphological features. The length of the eurypterid body generally ranges from 10 to 20 cm., but very large forms occur also in each geological system from Ordovician to Carboniferous. *Megalograptus walchi* MILLER and *Echinognathus clevelandi* (WALC.), from the Ordovician, *Eurypterus pustulosus* HALL and *Slimonia acuminata* (SALTER), from the Silurian, and *Tarsopterella scoticus* (WOODW.), from the Devonian, attained a length of at least 100 cm.; *Ctenopterus? lacoana* (CLAYPOLE), Devonian, probably 140 cm., and large species of *Pterygotus* at least 180 cm., this being the largest arthropod known.

The body is covered by a chitinous exoskeleton of variable thickness. Most "complete specimens" probably represent cast exuviae from the ecdysis. Besides the more solid exoskeleton of the body and appendages, parts of the softer integument, such as intersegmental membranes, may be preserved (*Eurypterus, Rhenopterus*). The soft integument is provided with numerous hairs (Fig. 17*C*), and the presence of tactile hairs also on the more solid test is indicated by numerous pores, particularly in the doublure of the abdominal plate-shaped appendage (Fig. 22, *3f*).

A distinct ornamentation is characteristic of the exoskeleton of the Eurypterida. The surface may be smooth or provided with scales, tubercles, knobs, or spines of different kinds. In *Hughmilleria* (Fig. 21, *1*) numerous parallel integumental folds ("terraced lines") are found on the anterior portions of prosoma and tergites. This is apparently the more primitive ornamentation (not unlike structures in trilobites), which develops into rows of scales or tubercles. The scales may be crescentic, as in *Pterygotus* (Fig. 20, *2a,c;* 22, *3f*) and to some extent in *Eurypterus* (Fig. 17*A,B*); or tongue-shaped, as in *Echinognathus* (Fig. 29, *5d*); or acute, as in *Lepidoderma* (Fig. 21,*3a*) and *Campylocephalus* (Fig. 29, *8*). The scales tend to develop into round or oblong raised tubercles, as in *Carcinosoma* (Fig. 23; 29, *3c*), *Ctenopterus* (Fig. 27, *4b*), and *Tarsopterella* (Fig. 27, *6b*), or undulating as in *Tylopterella* (Fig. 29, *4a*). The scales and tubercles may be closely set or scattered, and commonly are confined to distinct parts of the body. Prominent knots tend to occur along the posterior margins of the prosoma and tergites, as in *Tarsopterella* (Fig. 27, *6a*), or *Rhenopterus* (Fig. 27, *5a*). A peculiar, reticulate ornamentation is found in *Mycterops* (Fig. 30). A color pattern is noticed in *Megalograptus.*

The elongate, mostly lanceolate body may have a scorpion-like appearance, a feature particularly characteristic of the Carcinosomatidae (Fig. 23) and Mixopteridae (Fig. 25). The body is divided into a prosoma or cephalothorax, and an opisthosoma comprising 12 movable segments and a telson. The abdomen of many eurypterids is divided into a broader **preabdomen** of 7 segments and a narrower nearly cylindrical **postabdomen** of 5 segments. When the term opisthosoma is applied, one may distinguish the limb-bearing **mesosoma** of 6 segments from the 6-segmented **mesosoma** without ventral appendages.

The prosoma is moderately convex, with a subquadratic, subtriangular to semicircular outline. A pair of median ocelli, which may be mounted on a node (Fig. 19,*3d;* 21,*1a;* 29,*3a*), is generally situated on the highest part of the prosoma.

The lateral compound eyes have a marginal to subcentral or anterior position. The eyes vary considerably in size and shape. The large eyes of the Pterygotidae (Fig. 19, *2*; 22, *1a,2a,3a*) have an elliptical outline. In most eurypterids, the eyes are kidney-shaped, and in some strongly curved (Stylonuracea, Fig. 27,*1,2b,3a,4a,6b,7b*), with the more or less steep visual surface facing outward and forward. Small eyes are found in *Mixopterus* (Fig. 25*A*). The visual surface is composed of numerous closely set pits representing the individual facets, visible in *Pterygotus* (Fig. 19, *2*) and some species of *Hughmilleria.*

The margin of the prosoma is inflected, forming a broad doublure that passes into the more flexible softer integument surrounding the appendages and the mouth. A **marginal suture** (or line of weakness), which opens during ecdysis, is characteristic of the Eurypterida. The doublure is traversed by one or more sutures. One **median suture,** which may be more or less obsolete, is characteristic of *Eurypterus* (Fig. 17) and *Rhenopterus* (Fig. 27, *5b*). In most eurypterids, however, a median **epistoma** is bordered by a pair of **epistomal sutures** (Fig. 22, *3c*). Possibly one extra pair of sutures crosses the doublure in *Hughmilleria* (Fig. 19, *1a*; 21, *1g*). Inasmuch as a small narrow epistoma is found in *Lepidoderma* and the plate seems to be reduced in *Rhenopterus,* a tendency toward reduction of the epistoma in evolution of the Eurypterida is suggested.

No separate labrum is developed among eurypterids. Impressions near the posterior border of the epistoma possibly may represent traces of an olfactory organ, such as in Recent Xiphosura (Fig. 19, *1a*).

The abdomen is attached to the prosoma by a special articulation formed by the doublure of the prosoma and 1st tergite. This is characteristic of *Eurypterus,* but may be found in all eurypterids. The preabdomen has a moderately vaulted median (axial) portion and a flattened or slightly concave pleural portion. A distinct trilobation is found in *Mixopterus* (Fig. 25*A*) and *Megalograptus* (Fig. 24). The first tergite is generally shorter than the others. Probably, it represents the 7th (pregenital) segment, which is reduced in most arachnids. As an exception, the large 1st (and possibly 2nd) tergite in *Mycterops* (Fig. 29,*1*; 30) should be mentioned. The 1st to 6th tergites have a more or less rectangular outline, and the 7th or last preabdominal

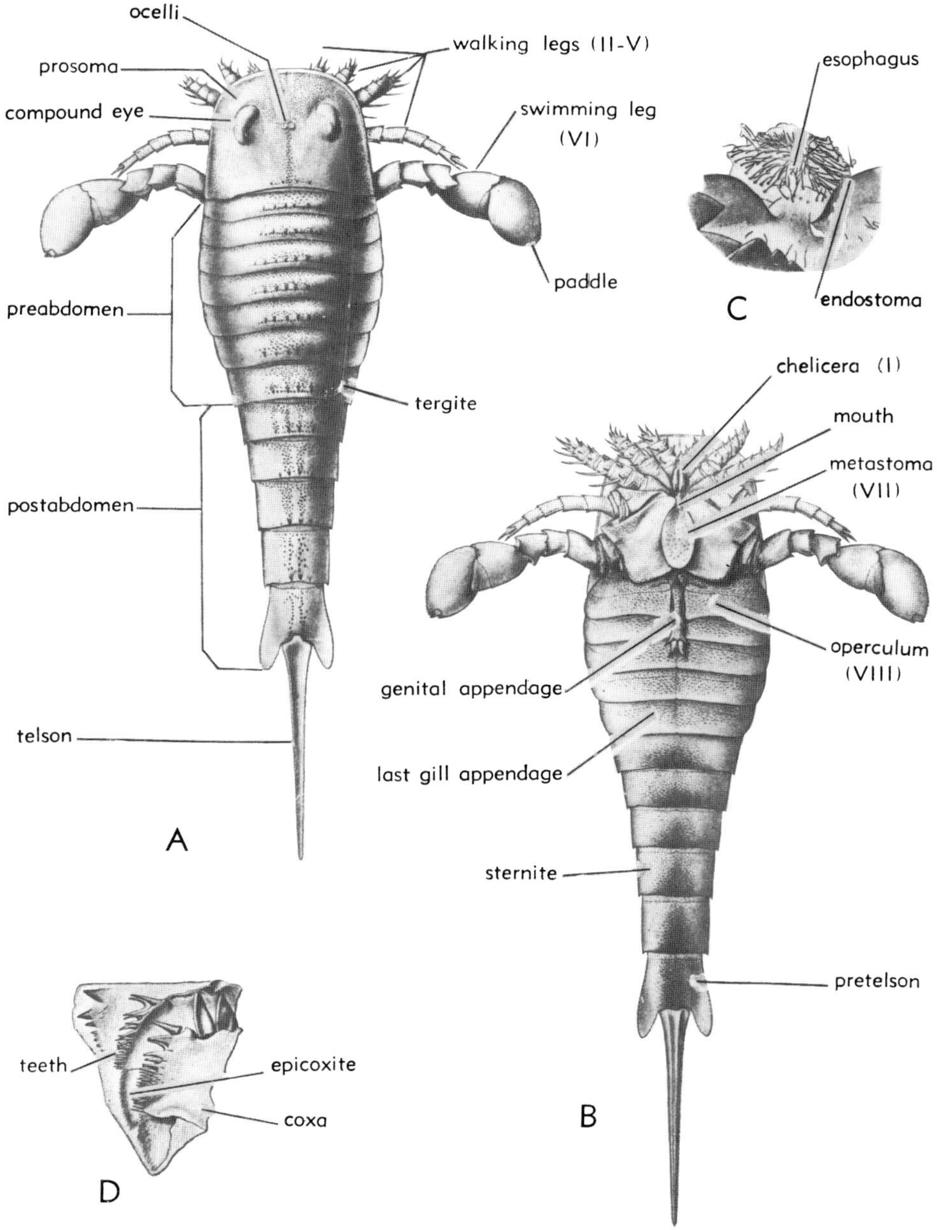

FIG. 17. Morphological features of eurypterids as shown by *Eurypterus fischeri* EICHW., M.Sil., Balt. (47). *A,B,* Dorsal and ventral views of entire specimen (reconstr.), ×0.5. *C,* Gullet, ×13. *D,* Coxae around mouth, showing denticulate margins, ×6 (47).

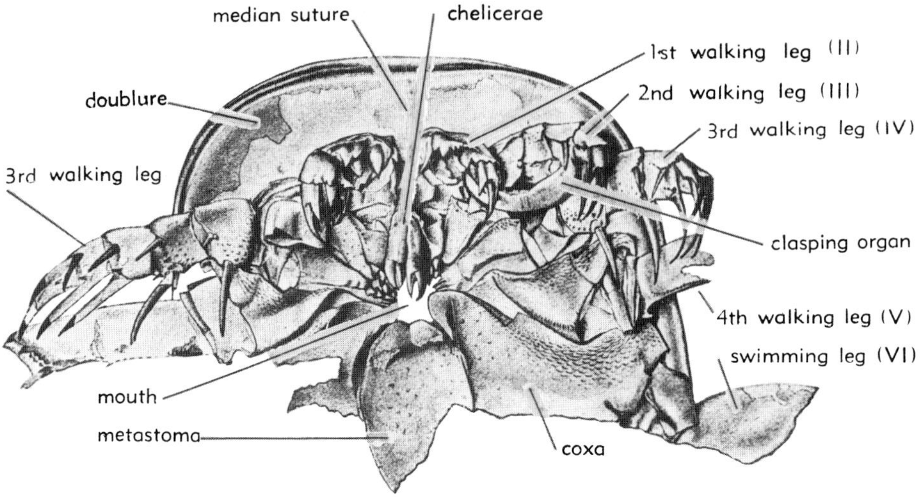

FIG. 18. Ventral surface of prosoma of *Eurypterus fischeri* EICHW., Sil., Balt., showing well-preserved appendages of etched-out specimen, ×1.3 (47).

tergite may have postlateral fins or spurs. Each tergite is attached to the one in front by means of an intersegmental membrane attached to the anterior and posterior doublure.

Because of the powerful plate-shaped appendages on the first 6 segments of the abdomen, corresponding powerful sternites are not developed. Such sternites, on the other hand, are developed on the posterior 6 segments and are fused with the tergites on the dorsal surface, forming cylindrical rings telescopically inserted into each other. Lateral spurs or fins may be present on the postabdominal segments (*Lepidoderma,* Fig. 21, *3a*; *Tarsopterella,* Fig. 27, *6a*). The 12th segment, or **pretelson,** may differ somewhat from the other abdominal segments (*Eurypterus,* Figs. 17, 18; *Pterygotus,* Fig. 22, *1a,b, 2a, 3a; Megalograptus,* Fig. 24). In *Megalograptus* a peculiar pair of flat ?cerci are developed.

A powerful telson articulates with the last abdominal segment. The most primitive type seems to be a lanceolate telson with a flat ventral surface and a median dorsal keel (*Hughmilleria,* Fig. 21, *1a,b*). From this type the longer styliform telson of *Lepidoderma* was possibly developed (Fig. 21, *3a*), and similarly that of certain stylonurids (Fig. 27, *4a*), and the curved, possibly poisonous tail-spines of *Carcinosoma* (Fig. 23) and *Mixopterus* (Fig. 25). Another line of development possibly led to the telsons of *Slimonia* (Fig. 21, *2a*) and *Salteropterus* (Fig. 21, *5a,b*), in which the lanceolate telson bears lateral fins. In *Pterygotus* (Fig. 22, *1a,b, 2a, 3a*), nothing is left of the lanceolate telson. The telson forms a broad tail-fin or horizontal rudder, which in *Pterygotus (Pterygotus)* (Fig. 22, *2a*) may have a raised median keel, probably serving also as a competent vertical rudder.

Appendages are attached to the ventral surface of the prosoma and preabdomen. Below the prosoma, the 6 pairs of appendages are radially arranged around the central mouth. The first pair of appendages is invariably developed as 3-jointed chelicerae (Fig. 19, *3a*), in which the 3rd joint forms the opposing movable finger. In most forms the chelicerae are very small, but in the Hughmilleriidae (Fig. 21, *1b*) and the Stylonuracea (Fig. 27), they are somewhat larger, and in the Pterygotidae (Fig. 22, *1a-c, 2a,b, 3a,b*), they are enormously developed, the basal joint being greatly prolonged and the inside of the chelae being provided with small and large striated teeth, the large ones measuring several cm. in length. Studies of the brain in recent Chelicerata indicate the chelicerae to be innerved from the tritocerebrum, thus representing the first postoral segment, the actual preoral position of the chelicera being a secondary phenomenon due to a posterior migration of the mouth in the early ontogenetic stages.

The 2nd to 5th pairs of appendages are generally developed as walking legs, more or less alike and increasing in size backward (Fig. 18; 19, *3b,c*). Each walking leg has an elongate, subtriangular flat coxa with a small **epicoxite** (Fig. 17; 19, *3b,c*) at the base of the serrate oral margin. The 5th coxa of *Eurypterus* (Fig. 19, *3c*) has a small circular foramen, possibly covered by a thin membrane and functioning as an auditory organ. The walking leg attached to the distal portion of the coxa has 7 to 8 joints, 8 when a double trochanter is present in the 5th leg. In *Megalograptus* (Fig. 24) 8 joints seem to be present in the 2nd walking leg, owing to addition of an extra segment between the 2nd and 3rd. The joints of the walking legs have been interpreted as trochanter (1-2), prefemur, femur, patella, tibia, tarsus, and pretarsus. The more or less cylindrical joints generally are inserted telescopically into each other. Each joint is provided with fixed (or rarely movable) primarily ventral spines. In less primitive legs, the spines may be absent, as in the 5th walking leg ("balancing leg") of *Eurypterus* (Fig. 19, *3c*) and *Megalograptus* (Fig. 24) and in other legs of *Pterygotus* (Fig. 22, *1a, 3a, 3g*) and stylonurids (Fig. 27, *1, 2c, 4a, 5a,b, 7a*); or the spines may be strongly developed and specialized, as in *Carcinosoma* (Fig. 23), *Mixopterus* (Fig. 25), *Megalograptus* (Fig. 24), and the 1st and 2nd pairs of legs in *Ctenopterus* (Fig. 27, *4a*).

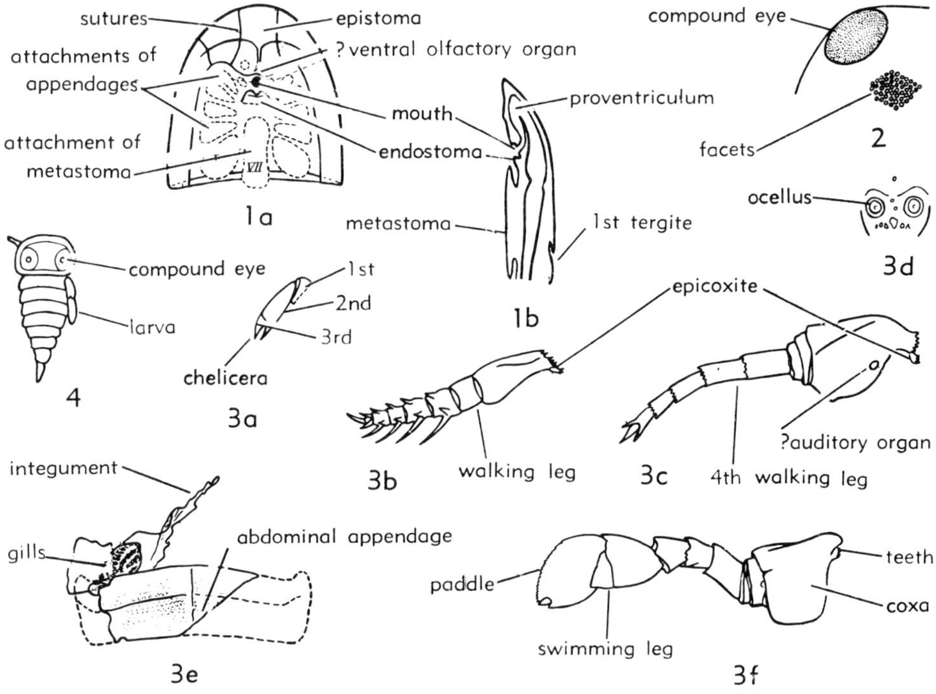

Fig. 19. Morphological features of eurypterids.

1a,b, Hughmilleria, L.Dev., Norway; *1a,* ventral side of prosoma with appendages removed; *1b,* longitudinal section; both ×1 (64).

2, Pterygotus sp., Sil., Balt., showing compound eye and enlarged view of eye facets (47).

3a-f, Eurypterus fischeri Eichw., Sil., Balt.; *3a,* chelicera; *3b,* walking leg; *3c,* last walking leg ("balance leg"); *3d,* median ocelli, ×4; *3e,* abdominal plate-shaped appendage with gills; *3f,* swimming legs; all except *3d,* ×1 (47).

4, Stylonurus myops Clarke, Sil., N.Y., young larva, ×11 (40).

Special lobes on the 1st and 2nd walking legs in *Eurypterus* (Fig. 18), *Mixopterus* (Fig. 25), and *Brachyopterus* (Fig. 26) are interpreted as sexual clasping organs.

The 5th pair of legs (6th pair of prosomal appendages) is either developed as walking legs (Stylonuracea) or converted into broad and flat swimming and balancing legs (Eurypteracea) (Fig. 19, *3f*). The swimming legs of the Eurypteracea are formed by the strong expansion of the 2 penultimate joints into an oar-blade or paddle. Also, among the Stylonuracea certain forms show a tendency toward converting the last legs into swimming legs. In *Dolichopterus* (Fig. 27, *3a,b*; 24,*1b*), the ultimate joint takes part in the formation of the paddle. The coxae of the last pairs of legs are strongly developed as large plates, which cover half the ventral surface of the prosoma. The oral margin is provided with powerful teeth and knobs.

The mouth is bordered dorsoposteriorly by a short plate, cleft in front. This plate, called the **endostoma** (Figs. 17*C*; 19,*1a,b*), is probably formed by prosomal sternites.

The **metastoma** is a larger plate covering (in ventral view) the endostoma and median portions of the posterior coxae (Figs. 14, 17*B*). The metastoma, attached at its dorsal surface, is probably homologous with the chilaria of Xiphosura and hence represents the fused and strongly modified appendages of the 7th (pregenital) somite. The metastoma thus actually belongs to the first segment of the abdomen. The plate is elliptical to trapezoid in outline, with a cleft or transverse anterior margin and a transverse or cleft *(Megalograptus)* posterior margin (Fig. 22,*1e, 3d;* 25,*2b*). Broader and narrower forms of one species may denote different sexes.

The operculum and 4 succeeding plate-shaped appendages (Fig. 19, *3e;* 20, *1a,c*) are built on the same plan. The appendages (not sternites) are attached along the anterior border and overlap each other backwards. The posterior plates are fused along the median line, transverse rows of pigmented scales suggesting a rudimentary segmentation. A broad posterior doublure passes into a more flexible integument with elliptical branchial areas containing the gills (Fig. 19, *3e*; 21, *1b,d*). The nature of the gills is uncertain, but unpublished plates by Holm, based on specimens of *Eurypterus fischeri,* suggest minute tufts instead of book gills. In large forms, such as *Pterygotus* and *Slimonia,* more prominent leaflike appendages have been interpreted as gills.

The operculum (Fig. 17*B*; 20,*1a,b*) represents appendages of the 8th (genital) segment corresponding to the 2nd abdominal tergite. Sexual dismorphism is expressed by 2 types of genital appendages (Fig. 20, *1a,b, 2a-d*; 21, *1e,f, 2b,c*). The appendage, which probably is formed by fusion of the inner branches of a primarily biramous appendage (as in *Limulus*), is either long and narrow (male) or short and broad (female). In *Pterygotus* (Fig. 20, *2a,b*; 22, *2d*), the pear-shaped median appendage shows traces of dorsal apertures (possibly female) of oviducts or a small combined aperture (possibly male) (Fig. 20, *2d*). Both the female and male appendage has a hastate basal portion, generally with 3 separate joints. In *Eurypterus,* the median appendages of the 9th segment also take part in formation of the genital appendage (Fig. 20, *1c*).

Remains of the intestine, reaching from the mouth backward to the base of the telson, were described in *Carcinosoma* by Ruedemann in 1921. Possible coprolites have been found in connection with *Megalograptus* by Caster & Kjellesvig-Waering (1955).

Ontogeny. The smallest larva known (*Stylonurus,* Fig. 19, *4*) measures 2 to 3 mm. in length. The specimen is not well preserved; but it deviates from the adult in having a smaller number of abdominal segments, and the compound eyes evidently are mounted on ovate nodes. In a specimen of *Eurypterus* 7 mm. long, the basal portion of the telson is much broader than in the adult.

Ecology. Eurypterid remains are scarce, and particularly so in marine faunas. Most described species belong to brackish- and fresh-water faunas. In general, one may say that the eurypterids inhabited fresh and brackish waters near the coast and occasionally visited the sea. With their well-protected gills, it is possible that the eurypterids were able to spend short intervals of time on land.

Most eurypterids were benthonic forms, crawling around on the bottom and digging in mud and sand. On the other hand, streamlined forms (as *Hughmilleria* and *Pterygotus*) probably belong to the nekton. The swimming was performed by means of the swimming legs and movements of the body, but it is also possible that a rapid shooting forward through the water may have been accomplished by powerful strokes of the abdominal appendages, as in early larvae of *Limulus*. In *Pterygotus*, the walking legs are developed only slightly, and swimming abilities are indicated by modification of the telson into a tail-fin and rudder.

Because of powerful teeth on the coxae and chelicerae of Pterygotidae, the eurypterids were probably able to feed on larger animals which possessed a solid exoskeleton, perhaps including primitive vertebrates, common in the same fossil faunas.

Phylogeny. The geological appearance of the Eurypterida gives little information on phylogeny of the group. The earliest known forms occur in Lower Ordovician (Deepkill) rocks, and the Ordovician species comprise representatives of most of the families. During evolution of different branches of the eurypterid stock, a tendency toward increase in size and in elaboration of ornamentation may be noticed.

The genus *Hughmilleria*, with rather uniform walking legs and seemingly primitive ornamentation, probably occupies a central position in the eurypterid stock. In this and all other genera belonging to the superfamily Eurypteracea, the swimming legs are similarly developed. In the Stylonuracea, on the other hand, the last prosomal legs are still walking legs, although in some forms (Dolichopteridae) a tendency toward formation of swimming legs is demonstrated.

The ancestors of the Eurypterida are probably to be found in forms such as *Paleomerus* (Fig. 8) from Lower Cambrian deposits. As already mentioned this genus combines xiphosuran and eurypterid characters. The earliest eurypterids, possibly from Lower Cambrian (or perhaps Precambrian), probably had 5 pairs of simple walking legs. The main eurypterid branch developed characteristic swimming legs, and the further branching of this stock manifests itself in adaptations to different habitats. In another branch of the Eurypterida, the last walking legs were kept more or less unchanged in most forms and only a few developed some kind of swimming legs.

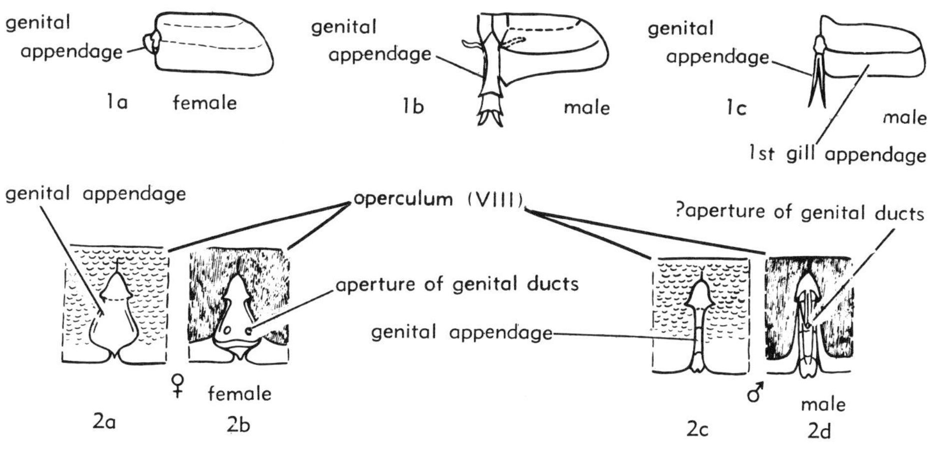

FIG. 20. Genital appendages of eurypterids.

1a-c, Eurypterus fischeri EICHW., Sil., Balt., from ventral side; *1a*, female; *1b,c*, male; ×1 (47).

2a-d, Pterygotus (Pterygotus) rhenaniae JAEKEL, Dev., Ger.; *2a,b*, female, ventral and dorsal views; *2c,d*, male, ventral and dorsal views (64).

Superfamily EURYPTERACEA Burmeister, 1845

[*nom. transl.* STØRMER, 1951 (*ex* Eurypteridae BURMEISTER, 1845)]

Eurypterida with last prosomal appendages developed as swimming legs that carry paddles formed by expansion of the 2 penultimate joints. *Ord.-Perm.*

Family HUGHMILLERIIDAE Kjellesvig-Waering, 1951

[*nom. correct.* STØRMER, herein (*pro* Hughmilleridae KJELLESVIG-WAERING, 1951)]

Small to large forms with outer surface smooth or partly provided with transverse integumental folds, or semilunar to pointed scales; compound eyes small, antelateral, marginal or subcentral; telson lanceolate or somewhat expanded; epistoma present; chelicerae small to medium, without teeth; walking legs stout, with spines or denticles; swimming paddles narrow; metastoma ovate to cordate; genital appendage of male narrow, without lateral points. *Ord.-Perm.*

Hughmilleria SARLE, 1902 [**H. socialis*]. Small to medium forms, with smooth surface or with transverse folds or semilunar scales; body lanceolate, with postlateral prolongations of first postabdominal segment, and lanceolate telson; chelicerae medium-sized; metastoma ovate; genital appendage short and broad in female, long and narrow in male. *Ord.-Perm.*, Eu.-N.Am.——FIG. 19,*1*; 21,*1a-g*. *H. norvegica* (KIAER), ?L.Dev. (Downton.), Norway; 19,*1a*, ventral side of prosoma with appendages removed, attachments of appendages outlined; 19,*1b*, long. sec. of prosoma and 1st segment of abdomen, both about ×1.3; 21,*1a*, dorsal side; 21,*1b*, ventral side with branchial area outlined; 21,*1c,d*, sections of body; all ×0.5; 21,*1e,f*, median genital appendages of male and female, ×1; 21,*1g*, ventral doublure of prosoma, ×0.8 (all 64).——FIG. 21,*1h*. *H. bellistriata* KJELLESVIG-WAERING, Sil., Wis.; prosoma with sculpture, ×0.7 (50).

Slimonia PAGE, 1856 [**Himantopterus acuminatus* SALTER, 1856]. Large forms with smooth outer surface; prosoma quadratic, with small compound eyes at antelateral corners; postabdomen narrow, anterior half of telson strongly expanded; chelicerae small, walking legs with denticles but no spines; metastoma narrow cordate, genital appendages fairly long and narrow in both sexes. *U.Sil.*, Eu.——FIG. 21,*2*. **S. acuminata* (SALTER), Sil., Scot.; *2a*, dorsal side, ×0.1; *2b*, ?female genital appendage; *2c*, ?male genital appendage, about ×0.13 (67).

Salteropterus KJELLESVIG-WAERING, 1951 [**Eurypterus abbreviatus* SALTER, 1859]. Small forms; outer surface with pointed triangular scales; anterior part of carinate telson broadly trigonal with serrated posterior edges and followed by a long, flat median stem which expands into a flattened posterior end (13). *L.Dev.(Downton.)*, Eng.——FIG. 21,*5*. **S. abbreviatus* (SALTER); anterior and posterior part of telson, ×1 (50).

Grossopterus STØRMER, 1934 [**Eurypterus? overathi* GROSS, 1933]. Medium forms; outer surface with minute semilunar scales; prosoma subquadratic with expanded, rounded antelateral corners, compound eyes small, intermarginal; walking legs powerful, with spines; genital appendage of male long lanceolate, with serrated margins (26). *L.Dev.-M.Dev.*, Eu.-N.Am.——FIG. 21,*4*. **G. overathi* (GROSS), L.Dev., Ger.; *4a*, dorsal side, about ×0.25; *4b*, genital appendage of ?male, ×0.3 (64).

Lepidoderma REUSS, 1855 [*non* WATERHOUSE, 1875; *nec* ZELINKA, 1889] [**L. imhofi*] [=*Adelophthalmus* JORDAN & MEYER, 1856; *Anthraconectes* MEEK-W., 1868]. Small to medium forms; outer surfaces with pointed scales and striae; body elongate, with spurs, contracted postabdomen, and long and styliform telson; compound eyes intramarginal; walking legs mostly devoid of spines, genital appendage of male long, of female short, with spatulate lateral lobes (12). *?U.Dev., Carb.-M.Perm.*, Eu.-N.Am.-Asia.——FIG. 21,*3a*. *L. mansfieldi* (C. E. HALL), Carb., Penn., U.S.A.; about ×1.3 (46).——FIG. 21,*3b,c*. *L. mazonense* (MEEK-W.), Penn., Ill.; about ×1; *3b*, metastoma and swimming leg; *3c*, female genital appendage (50).

?**Hastimima** WHITE, 1908 [**H. whitei*]. Large forms; test with scattered oblong scales or closely set mucrones; telson acute, hastate, with lateral expanded portion well separated from narrow median plate. *Carb.*, S.Am.-N.Am.——FIG. 21,*6*. **H. whitei*, Brazil; *6a*, coxa with proximal joints; *6b*, telson, ×0.3 (after CLARKE & RUEDEMANN).

Family PTERYGOTIDAE Clarke & Ruedemann, 1912

Small to very large exoskeleton with distinct semilunar scales; large marginal compound eyes; telson strongly expanded; epistoma present; chelicerae very large and long, provided with strong teeth in chelae; walking legs generally small, slender, without spines; metastoma subovate to cordate; genital appendage short and pear-shaped in female, somewhat short, narrow and blunt in male. *Ord.-Dev.*

Pterygotus AGASSIZ, 1839 [**P. problematicus*]. *Ord.-Dev.*, Eu.-Spitz.-N.Am.-Austral.

P. (Pterygotus) [=*Curviramus* RUEDEMANN, 1935]. Prosoma subtrapezoid; compound eyes adjoining antelateral margin; chelicerae with curved distal margin, coxae of swimming legs with 11 to 13 teeth; telson with pronounced median dorsal carina passing into short terminal spine. *?Ord., Dev.* Eu.-N.Am.——FIGS. 20,*2;* 22,*2a-d. P. (P.) rhenaniae* JAEKEL, L.Dev., Ger.; 20,*2a,b,* genital appendage of female, dorsal, ventral, ×0.14; 20, *2c,d,* genital appendage of male, dorsal, ventral, ×0.14; 22,*2a,* dorsal, ×0.45; 22,*2b,* chelicera, ×0.7; 22,*2c,d,* genital appendages, male, female, ×0.2 (64).——FIG. 22,*2e. P. (P.) anglicus* AGASSIZ, L.Dev.(Old Red Sandstone), Scot.; metastoma, ×0.3 (67).

P. (Acutiramus) RUEDEMANN, 1935 [**Pterygotus buffaloensis* POHLMAN, 1881]. Prosoma subquadrate; compound eyes at antelateral margin; chelicerae with acute distal margin, the large tooth of chelae distally inclined, coxae of swimming legs with 13 to 15 teeth; telson with low median carina or row of knobs. *?Ord., Sil.-Dev.,* Eu.-N.Am.-?Austral.——FIG. 22,*1a-e. *P. (A.) buffaloensis* POHLMAN, Sil., N.Y.; *1a,b,* dorsal, ventral, ×0.045; *1c,* chelicera, ×2; *1d,* swimming leg, ×0.3; *1e,* metastoma, ×0.3: (40).——FIG. 22,*1f. P. (A.) bohemicus* (BARRANDE), Dev., Czech., ?male genital appendage, ×0.17 (57).

P. (Erettopterus) SALTER, 1859 [*pro Himantopterus* SALTER, 1856 (*non* WESTMAIL, 1836)] [**Erettopterus bilobus*]. Prosoma semielliptical, compound eyes at antelateral margin; chelicerae with angular distal margin, coxae of swimming legs with 13 to 15 teeth, telson bilobed. *Sil.-Dev.,* Eu.-N.Am.——FIG. 22,*3a-f. P. (E.) osiliensis* Schmidt, Sil., Balt.; *3a,* dorsal, ×0.1; *3b,* chelicera, fixed joint, ×0.3: *3c,* prosomal doublure with epistoma, ×0.3: *3d,* metastoma, ×0.2: *3e,* ?male genital appendage, ×0.3: *3f,* ventral surface 1st postabdominal seg., ×0.7 (63).——FIG. 22,*3g, *P. (E.) bilobus* (SALTER), Sil., Scot.; walking leg, ×1 (67).

Family EURYPTERIDAE Burmeister, 1845

Small to large exoskeleton with inconspicuous, semilunar, tongue-shaped to

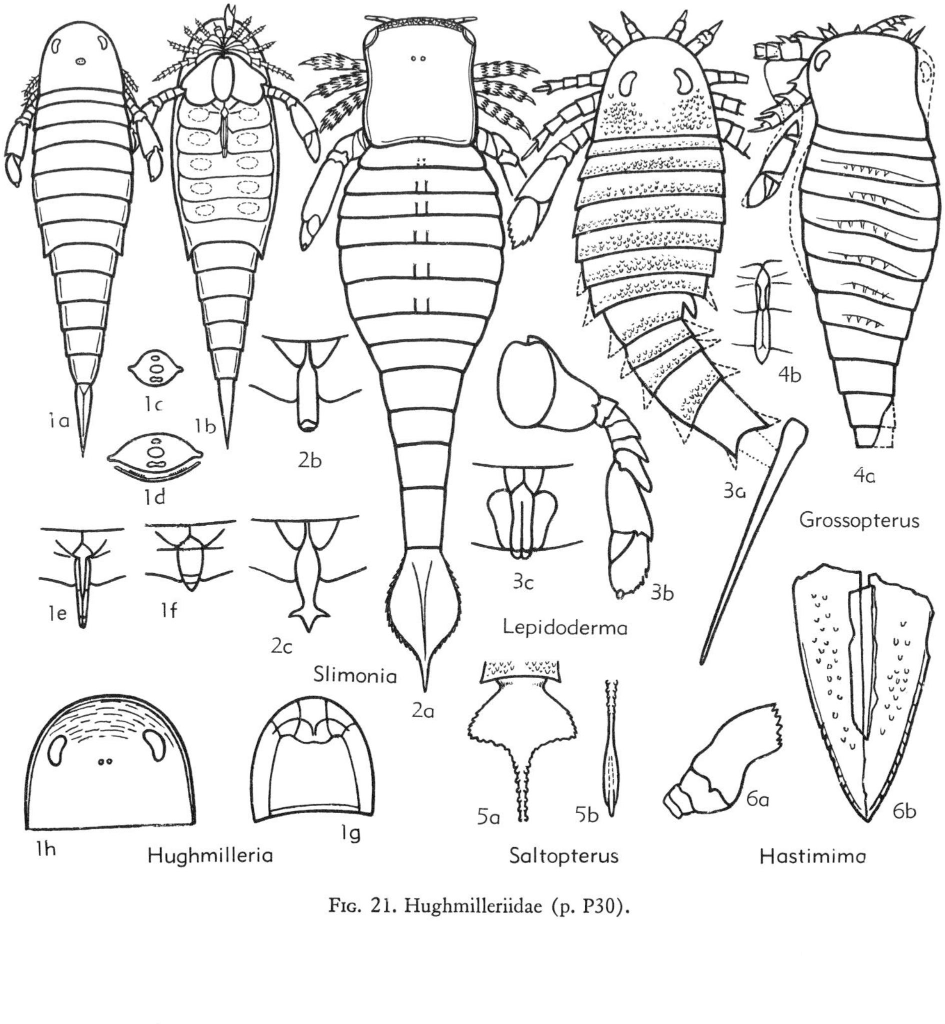

FIG. 21. Hughmilleriidae (p. P30).

pointed scales or curved ridges; prosoma subrectangular; telson styliform; compound eyes subcentral, slightly curved; no epistoma; chelicerae small, walking legs mostly spinose; metastoma ovate, genital appendage very short with covered lateral lobes in

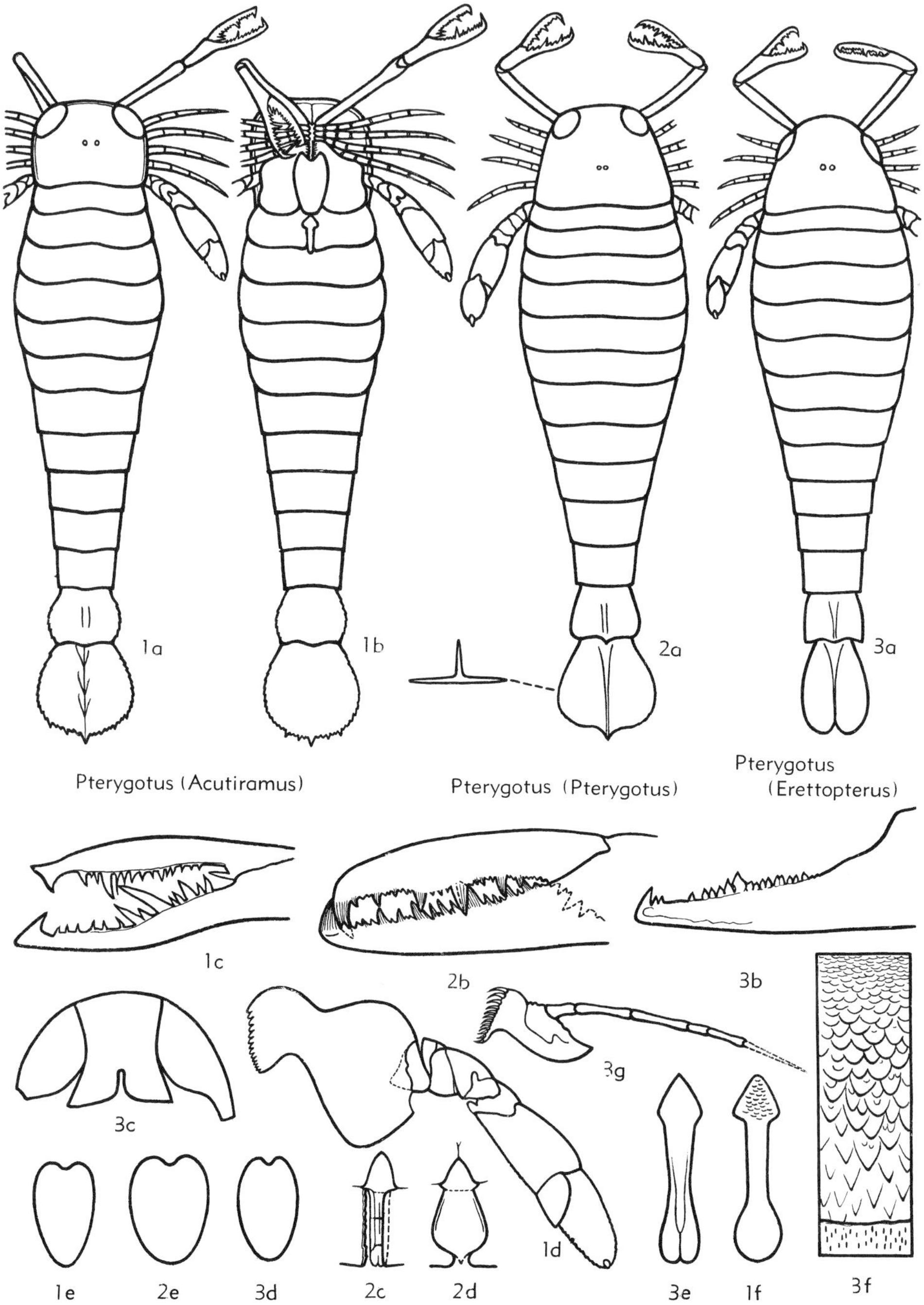

FIG. 22. Pterygotidae (p. P30-P31).

female, but long, with diverging distal spines in male. *Ord.-U.Carb.*

Eurypterus De Kay, 1825 [*non* Mabille, 1877; *nec* Sharpe, 1896] [**E. remipes*]. Exoskeleton

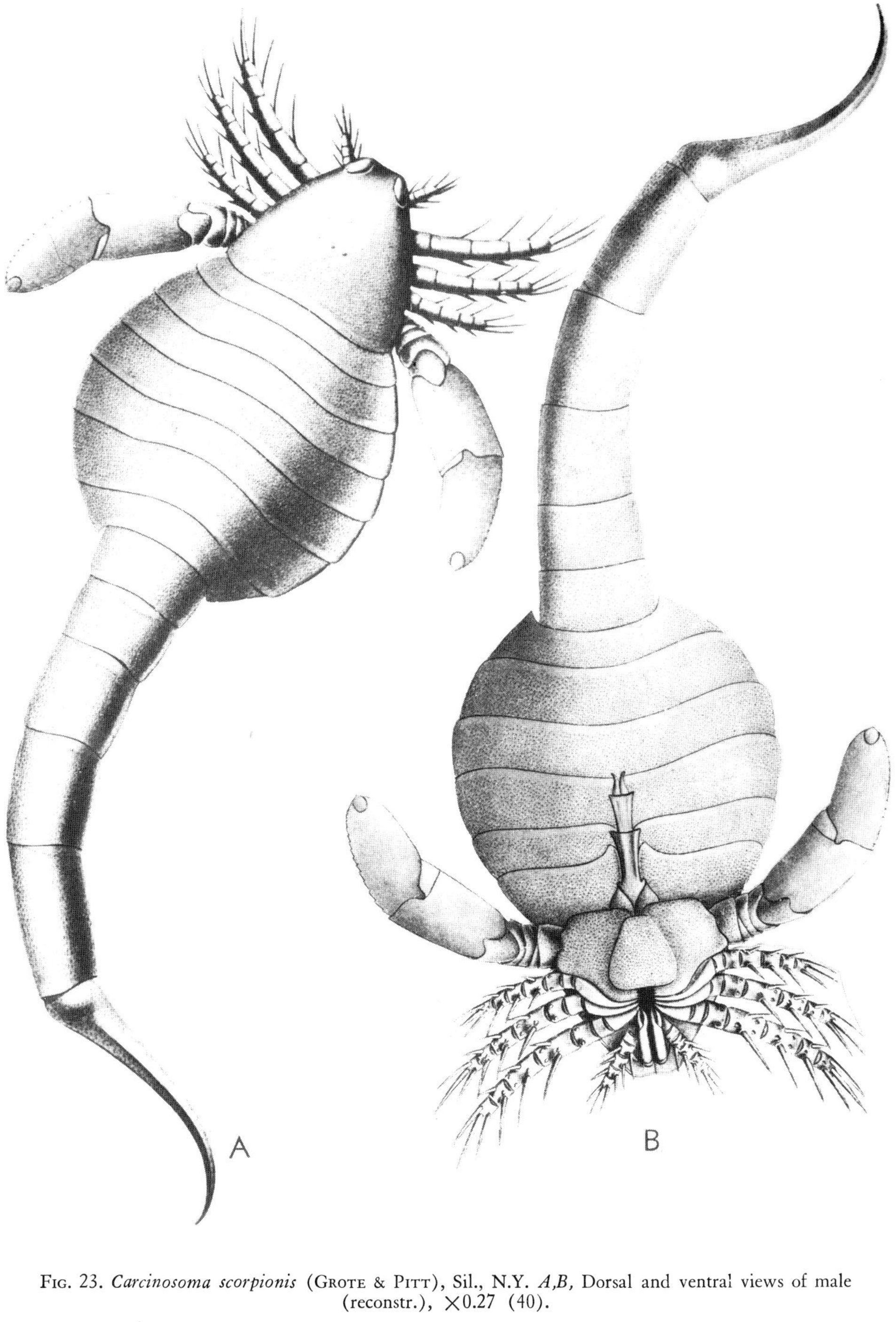

Fig. 23. *Carcinosoma scorpionis* (Grote & Pitt), Sil., N.Y. *A,B,* Dorsal and ventral views of male (reconstr.), ×0.27 (40).

locally provided with scales, metastoma not contracted, pretelson with postlateral prolongations: paddle of swimming leg strongly expanded. *Ord.-Carb.*, Eu.-N.Am.-Asia.——FIG. 17; 18; 19,*3a-c,e, f*; 20,*1*. *E. fischeri* EICHW., Sil., Balt.; 17*A,B*, dorsal and ventral side, about ×0.5 (reconstr.); 17*C*, gullet, with endostoma, ×13; 17*D*, denticular margin of coxae around mouth, ×6; 18, ventral side of prosoma of etched specimen, 2nd walking leg with clasping organ, probably female, ×1.3; 19,*3a*, chelicera; 19,*3b*, walking leg; 19,*3c*, 4th walking leg; 19,*3f*, swimming leg; 19,*3e*, abdominal appendages (plate), with gills; 20,*1a*, left half of operculum with median genital appendage of female; 20,*1b*, of male, 20,*1c*, median appendages of next abdominal appendages (plate), supporting male genital appendage of the operculum (all 47).

Onychopterella STØRMER, 1951 [*pro Onychopterus* CLARKE & RUEDEMANN, 1912 (*non* REICHENBACH, 1850)] [**Eurypterus kokomoensis* MILLER & GURLEY, 1896]. Medium size, outer surface with minute pointed scales; metastoma not contracted; last walking leg long, swimming leg with inconspicuous paddle ending in a spur (29). *Sil.*, N.Am. ——FIG. 29,*2*. **O. kokomoensis* (MILLER-G.), Ind.; *2a*, ventral side, ×0.5: *2b*, metastoma, ×1 (40).

Tylopterella STØRMER, 1951 [*pro Tylopterus* CLARKE & RUEDEMANN, 1912 (*non* CAPIOMONT, 1868; *nec* LECONTE, 1876)] [**Eurypterus boyli* WHITEAVES, 1884]. Small thick (partly calcareous) exoskeleton with curved ridges of confluent tubercles and large knobs near median line; prosoma with raised margin; metastoma contracted (29). *Sil.*, N.Am.——FIG. 29,*4*. **T. boyli* (WHITEAVES), Ont.; *4a*, dorsal side, about ×0.17; *4b*, ornamentation, enlarged (40).

Family CARCINOSOMATIDAE Størmer, 1934

[*nom. correct.* STØRMER, herein (*pro* Carcinosomidae STØRMER, 1934)]

Small to large exoskeleton with scattered minute tubercles or raised tongue-shaped scales; body scorpion-like, prosoma subtriangular, compound eyes small, intramarginal and antemedian; preabdomen broad, ovate, postabdomen narrow, cylindrical, with curved telsonic spine; epistoma short and narrow; chelicerae small, walking legs with long spines, 2nd pair of legs being the largest, swimming legs with long, slightly expanded 7th and 8th joints; metastoma subtriangular; genital appendage probably short and broad in female, long with distal spines in male. *Ord.-Sil.*

Carcinosoma CLAYPOLE, 1890 [*pro Eurysoma* CLAYPOLE, 1890 (*non* KOCH, 1839; *nec* DUJARDIN, 1845; *nec* GISTL, 1850)] [**Eurysoma newlini* CLAYPOLE, 1890] [=*Eusarcus* GROTE & PITT, 1875 (*non* PERTY, 1833)]. *Ord.-Sil.*, Eu.-N.Am.——FIG. 23; 29,*3b-d*. *C. scorpionis* (GROTE & PITT), Sil., N.Y.; 23*A,B*, dorsal and ventral side of male (reconstr.) ×0.27; 29,*3b*, metastoma, ×3; 29,*3c*, ornamentation of abdominal appendage, ×2: 29,*3d*, swimming leg, ×0.4 (FIG. 23, 40*; **FIG. 29,*3b-d*, 40**).——FIG. 29,*3a*, *C. vaningeni* CLARK & RUEDEMANN, Sil., N.Y.; prosoma and opisthosoma, about ×3 (40).

Family MIXOPTERIDAE Caster & Kjellesvig-Waering, 1955

Large exoskeleton with scattered tubercles or semicircular scales, body scorpion-like, prosoma subquadrate protruding antemedially; preabdomen fairly narrow, with axial furrows; postabdomen narrow, with curved

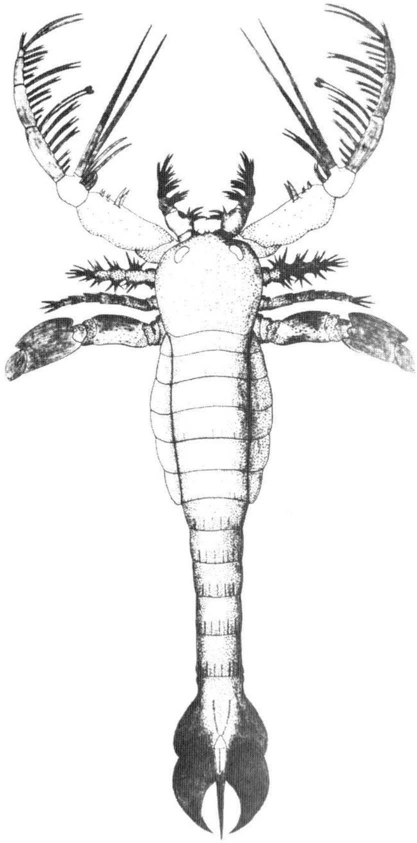

FIG. 24. *Megalograptus ohioensis* CASTER & KJELLESVIG-WAERING, U.Ord., Ohio, dorsal view (reconstr.), ×0.12 (37).

telsonic spine; 1st and 2nd walking legs strongly developed, with paired spines, 3rd and 4th moderate in size, with short spines, swimming legs with long 7th and short 8th joint; metastoma narrow cordate; genital appendage of male long. *Sil.-L.Dev.*

Mixopterus RUEDEMANN, 1921 [**Stylonurus (Ctenopterus) multispinosus* CLARKE-R., 1912] (33).——FIG. 25; 29,6. *M. kiaeri* STØRMER, *?L.Dev.* (Downton.), Norway; 25*A,B,* dorsal and ventral views of male (reconstr.), showing clasping organs on 1st walking legs, ×0.17 (64*); 29,*6a,b,* metastoma and part of swimming leg, ×0.3 (64).

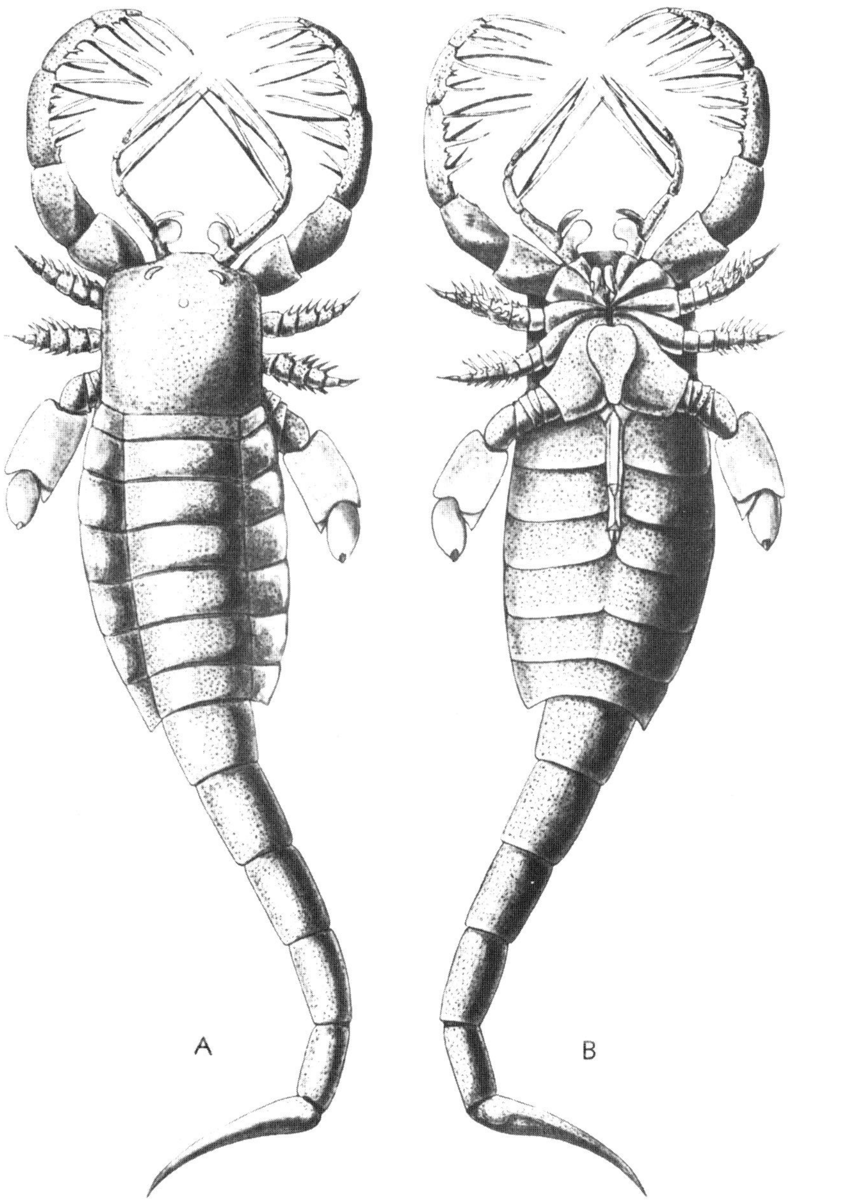

FIG. 25. *Mixopterus kiaeri* STØRMER, ?L.Dev.(Downton.), Norway. *A,* Dorsal view (reconstr.), ×0.17. *B,* Ventral view (reconstr.), showing clasping organs on 1st walking legs, ×0.17 (64).

Family MEGALOGRAPTIDAE Caster & Kjellesvig-Waering, 1955

Large exoskeleton with ovate to triangular scales; prosoma subquadrate, with tonguelike anterior process bearing marginal spines; compound eyes antemedian; preabdomen narrow, with axial furrows; postabdomen moderately narrow, with broad, curved and flat appendages; telson narrow lanceolate; epistoma short and broad; chelicerae small and short, 1st and 3rd walking legs short with diverging, or closely set spines, 2nd leg enormously developed, with long paired spines, 4th leg almost spineless, swimming leg with long 4th and 7th joints; metastoma subovate to cordate, with anterior cleft and with or without posterior indentation; genital appendage of female club-shaped, of male short and narrow. *Ord.*

Megalograptus Miller, 1874 [**M. welchi*]. Third walking leg with rather short diverging spines; metastoma subovate, with posterior indentation.

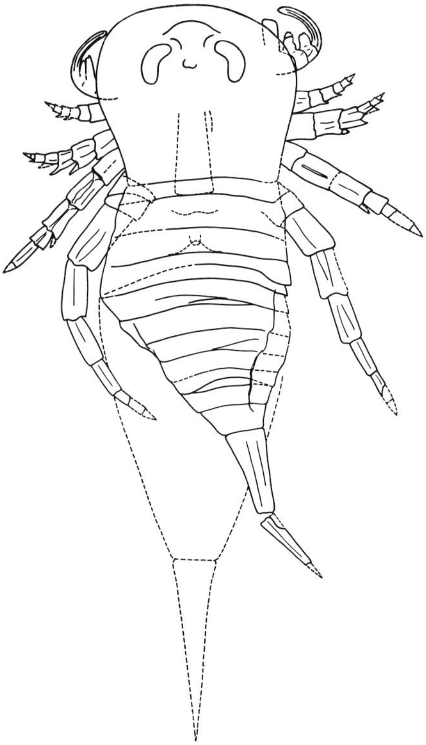

Fig. 26. *Brachyopterus stubblefieldi* Størmer, Ord., Wales, dorsal side, ×1.5 (64).

Ord., N.Am.——Fig. 24. *M. ohioensis* Caster & Kjellesvig-Waering, Ohio; reconstr., about ×0.12 (37*).——Fig. 29,7. **M. welchi*, Ohio; walking leg, about ×0.5 (40).

Echinognathus Walc., 1882 [**Eurypterus clevelandi* Walc., 1882]. Third walking leg with long closely set spines; metastoma broadly cordate, without posterior indentation. *Ord.*, N.Am.——Fig. 29,5. **E. clevelandi* (Walc.), N.Y.; *5a*, walking leg, ×0.5; *5b*, probably spine of walking leg, ×0.3; *5c*, metastoma, ×0.3; *5d*, ornamentation, ×0.75 (40).

Superfamily STYLONURACEA Diener, 1924

[*nom. transl.* Størmer, 1951 (*ex* Stylonuridae Diener, 1924)]

Eurypterida with last prosomal legs developed as walking legs, or less commonly modified into swimming legs in which the paddle is formed by the 2 to 3 ultimate joints. *Ord.-L.Perm.*

Family STYLONURIDAE Diener, 1924

Small to very large exoskeleton with scales developing into tubercles and knobs; prosoma of variable outline with subcentral to more anterior arcuate compound eyes; abdomen slender, without or with lateral epimers; telson styliform to clavate; prosomal doublure narrow, with epistoma; walking legs long and powerful, with or without spines; metastoma narrow; genital appendage of female short and broad, of male probably long. *Ord.-L.Perm.*

Stylonurus Page, 1856 [**S. powriei*]. Small to medium size; outer surface mostly smooth; prosoma semiovate to subrectangular, compound eyes with axes parallel or slightly converging forward; abdomen narrow, with long powerful styliform telson; first 3 pairs of walking legs with spines, last 2 pairs of walking legs long, keeled, tapering slightly in width, last pair of legs reaching as far as telson; metastoma with deep anterior notch. *?Ord., Sil.-Dev.*, Eu.-N.Am.——Fig. 27,*2b*. *S. dolichopteroides* Størmer, ?L.Dev.(Downton.), Norway; prosoma, ×0.6 (64).——Fig. 27,*2c*. *S. logani* Woodw., Sil., Scot.; about ×0.3 (67).——Fig. 27,*2a*. *S. macrophthalmus* Laurie, Sil., Scot.; metastoma, ×0.5 (53).

Drepanopterus Laurie, 1892 [**D. pentlandicus*]. Fairly large; outer surface with acute scales or tubercles; prosoma subrectangular to subovate, compound eyes with parallel axes; abdomen narrow, with styliform to clavate telson; first 3 pairs of walking legs powerful, with spines, last 2 pairs moderate in length, smooth, tapering slightly in width toward strong curved terminal claw, last pair of legs reaching to penultimate abdominal segment; metastoma unknown. *Sil.-U.Dev.*, Eu.-N.Am.——FIG. 27,*1. D. longicaudatus* CLARKE & RUEDEMANN, Sil., Ind.; dorsal side (reconstr., ×0.25 (40).

Brachyopterus STØRMER, 1951 [**B. stubblefieldi*]. Small; outer surface probably smooth; prosoma large, subtrapezoid to subpentagonal; compound eyes slightly anterior in position, with axes converging anteriorly; abdomen narrow, with short styliform telson; first 3 walking legs short, with spines except when modified into clasping organ; last 2 pairs moderate in length, keeled, tapering in width, last pair not reaching penultimate abdominal segment; metastoma narrow (29). *M.Ord., ?L.Dev.*, Eu.——FIG. 26.**B. stubblefieldi*, M.Ord., Wales; dorsal side showing appendages, 1st walking legs modified into clasping organs, ×1.5 (64).——FIG. 27,7. *B.? pentagonalis* (STØRMER), ?L.Dev.(Downton.), Norway; *7a*, ventral side of prosoma (reconstr.), ×0.4; *7b*, dorsal side of prosoma, ×0.4 (64).

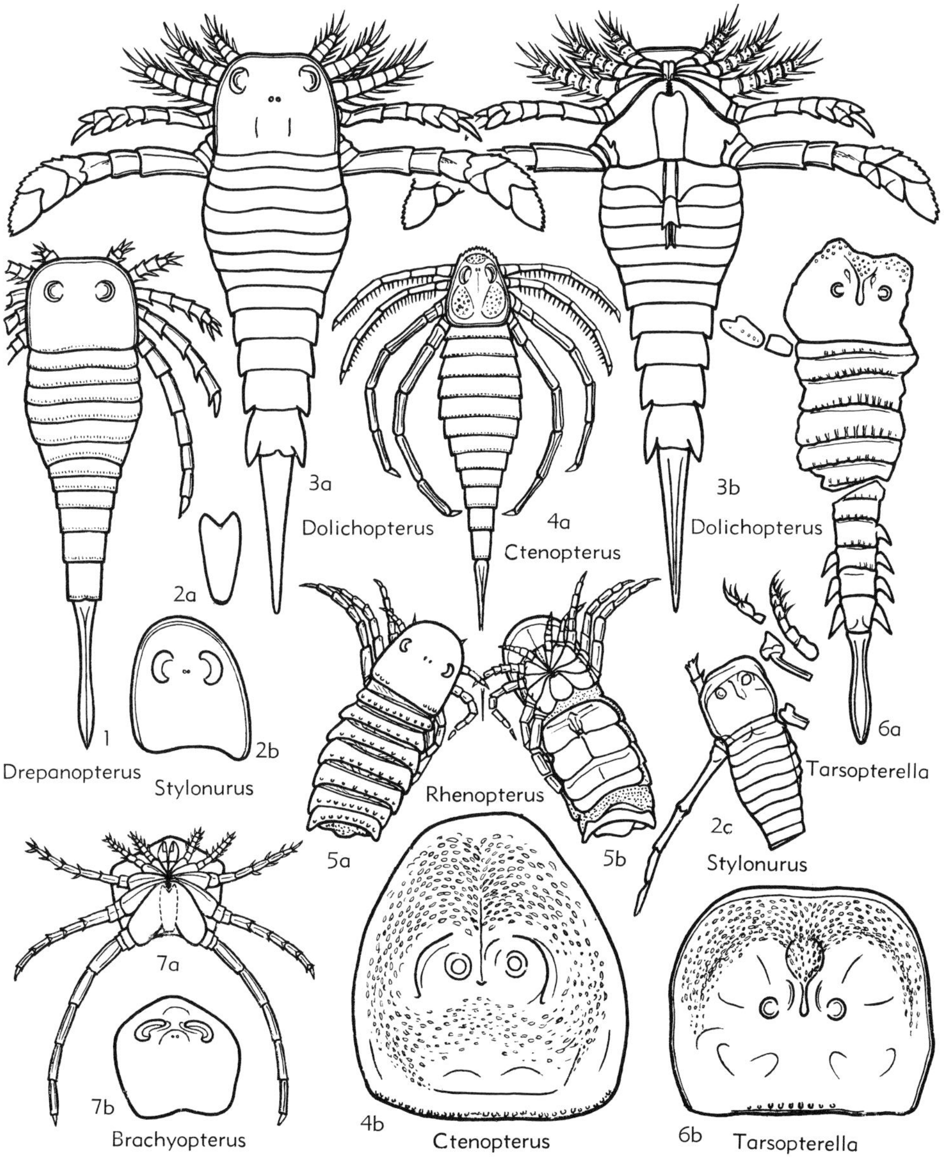

FIG. 27. Stylonuricae (p. P36-P38).

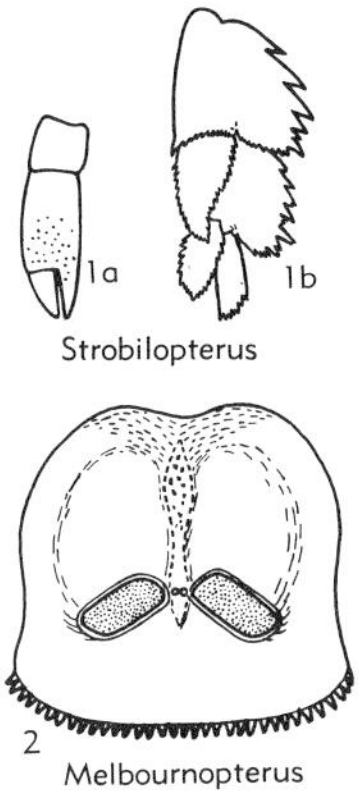

FIG. 28. Stylonuricae (p. P39).

Ctenopterus CLARKE & RUEDEMANN, 1912 [**Eurypterus cestrotus* CLARKE, 1907] [=?*Dolichocephala* CLAYPOLE, 1883 (*non* MACQUART, 1823)]. Small to very large; outer surface with numerous tubercles and scales; prosoma narrow in front, 2nd and 3rd pairs of walking legs strongly developed with double row of numerous flat spines, last pair of legs long, without spines (5). *Sil.-Dev.*, N.Am.-Eu.——FIG. 27,*4a*. **C. cestrotus* (CLARKE), Sil., N.Y.; dorsal side, ×0.5 (40).——FIG. 27,*4b*. *C.? lacoana* (CLAYPOLE), U.Dev., N.Y.; dorsal side of prosoma, ×0.17 (40).

Tarsopterella STØRMER, 1951 [*pro Tarsopterus* CLARKE & RUEDEMANN, 1912 (*non* REINHARDT & LUETKEN, 1862)] [**Stylonurus scoticus* WOODWARD, 1864]. Medium to large; outer surface with strongly developed knobs and scales; prosoma broadly subrectangular, slightly concave in front; compound eyes small; abdomen with pronounced lateral epimers, telson clavate; prosomal legs prob-

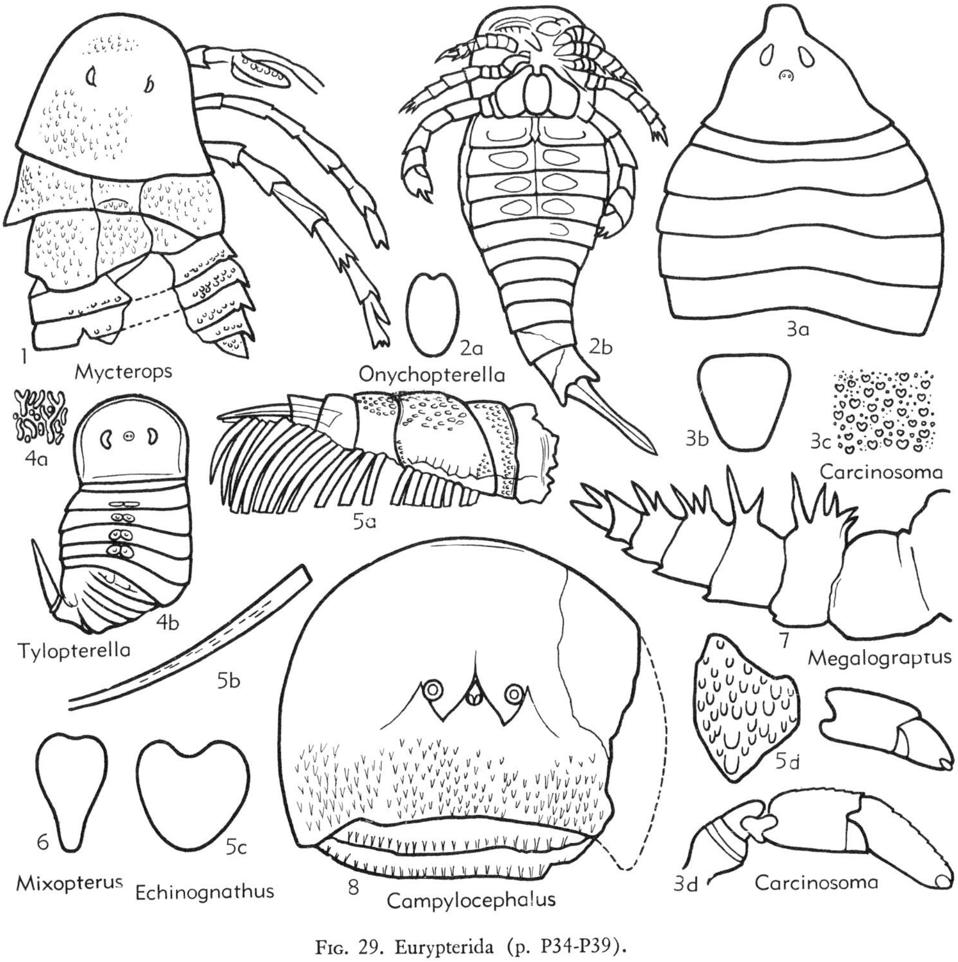

FIG. 29. Eurypterida (p. P34-P39).

ably stylonurid (29). *L.Dev.*, Eu.——FIG. 27,*6*. **T. scoticus* (WOODW.), L. Old Red Sandstone, Scot.; *6a*, dorsal side, ×0.07; *6b*, prosoma, ×0.2 (67).

Melbournopterus CASTER & KJELLESVIG-WAERING, 1953 [**M. crossotus*]. Small; prosoma bell-shaped, emarginate in front; subrectangular compound eyes in posterior half of prosoma, with axes strongly converging anteriorly; abdomen and appendages unknown (2). *U.Sil.*, Austral.——FIG. 28,*2*. **M. crossotus*, Vict.; prosoma, ×1 (37).

Campylocephalus EICHW., 1860 [**Limulus oculatus* KUTORGA, 1852] [=*Eidothea* SCOULER, 1831 (*non* RISSO, 1826; *nec* CHAMBERS, 1873)]. Large; outer surface with arcuate scales; prosoma subsemicircular, strongly convex; compound eyes subcentral, with inflated posteriorly acute lobes between them; abdominal tergites convex, with articular processes; appendages almost unknown. *L.Carb.-L.Perm.*, Eu.——FIG. 29,*8*. *C. scouleri* (HIBBERT), L.Carb., Scot.; prosoma and 2 abdominal tergites, ×0.25 (67).

Family DOLICHOPTERIDAE Kjellesvig-Waering & Størmer, 1952

Outer surface smooth or with pustules and semilunar scales; compound eyes arcuate, located anteriorly, axes nearly parallel; abdomen commonly with epimers, telson lanceolate; chelicerae small, 1st 3 walking legs stout, with powerful spines, last pairs with supplementary lobes, last legs with ultimate joint enlarged to form part of paddle; metastoma narrow, male genital appendage long. *Sil.-Dev.*

Dolichopterus HALL, 1859 [**Eurypterus (Dolichopterus) macrocheirus*]. Medium size; outer surface almost smooth; prosoma subquadrate; last legs with slightly serrate margin on distal joints and lobes. *Sil.*, N.Am.——FIG. 27,*3*. **D. machocheirus*, N.Y.; *3a,b*, dorsal and ventral side of male, reconstr., ×0.3 (40).

Strobilopterus RUEDEMANN, 1935 [**S. princetoni*]. Medium size; outer surface with pustules and semilunar scales; prosoma semiovate; compound eyes submarginal; last legs with strongly serrate margin on distal joints and lobes. *L.Dev.*, N.Am. ——FIG. 28,*1*. **S. princetoni*, Va.; *1a*, chelicera, ×1; *1b*, distal end of swimming leg, ×1 (51).

Family RHENOPTERIDAE Størmer, 1951

Small; outer surface with scattered tubercles and knobs; prosoma subtrapezoid, prosomal doublure with median suture, no epistoma; first 2 (or ?3) pairs of walking legs with spines, last 2 pairs long and powerful, without spines; metastoma pear-

FIG. 30. *Mycterops mathieui* (PRUVOST), U.Carb., Belg., prosoma with large 1st abdominal tergite, ×0.5 (58).

shaped, with slightly concave anterior margin; genital appendages of male short, with 2 distal spines. *L.Dev.*

Rhenopterus STØRMER, 1936 [**R. diensti*]. *L.Dev.*, Eu.-?N.Am.——FIG. 27,*5*. *R. diensti*, Ger.; *5a,b*, dorsal and ventral side of prosoma and preabdomen of ?male, ×0.75 (64).

Family MYCTEROPIDAE Størmer, 1951

[*nom. correct.* STØRMER, herein (*pro* Mycteropteridae STØRMER, 1951)]

Medium to fairly large forms; outer surface with numerous scales and reticulate ornamentation; prosoma subtrapezoid, 1st and ?2nd tergite of abdomen strongly developed; prosomal legs of stylonurid type. *U.Carb.*

Mycterops COPE, 1886 [**M. ornatus*] [=*Glaucodes* PRUVOST, 1923]. *Carb.*, N.Am.-Eu.——FIG. 29,*1*. *M.? scabrosus* (H.WOODWARD), L.Carb.,Scot.; ×0.5 (67).——FIG. 30. *M. mathieui* (PRUVOST), Belg.; prosoma with large 1st tergite of abdomen, about ×0.5 (30*).

Class UNCERTAIN

Glyptoscorpius PEACH, 1882 [**G. perornatus*]. Very large arachnids or merostomes possibly belonging with eurypterids, exoskeleton provided with tongue- or V-shaped scales; distal part of legs with 2 claws, scorpion-like; comb organs with 2 rows of numerous flattened filaments. *L.Carb.*, Scot.

REFERENCES

Publications cited in the following list are judged to be most helpful in furnishing additional information concerning the Merostomata and in offering a guide to more extensive literature. The index numbers enclosed in parentheses in the column at left are employed in the text for identification of the publications.

Caster, K. E.

(1) 1938, *A restudy of the tracks of Paramphibius:* Jour. Paleont., v. 12, p. 3-60, pl. 1-13, fig. 1-9.

——— & Kjellesvig-Waering, E. N.

(2) 1953, *Melbournopterus, a new Silurian eurypterid from Australia:* Jour. Paleont., v. 27, p. 153-156, pl. 20, fig. 1-2.

(3) 1955, *Upper Ordovician eurypterids of Ohio:* Paleont. Americana (Ithaca), *in press.*

——— & Macke, W. B.

(4) 1952, *An aglaspid merostome from the Upper Ordovician of Ohio:* Jour. Paleont., v. 26, p. 753-757, pl. 109, fig. 1.

Clarke, J. M., & Ruedemann, Rudolf

(5) 1912, *The Eurypterida of New York:* New York State Mus., Mem. 14, 439 p., 88 pl., 121 fig.

Diener, Carl

(6) 1924, *Eurypterida:* Fossilium Catalogus, 1: Animalia, pt. 25, p. 1-25 (Berlin).

Dunbar, C. O.

(7) 1923, *Paleolimulus, a new genus of Paleozoic Xiphosura, with notes on other genera:* Am. Jour. Sci., ser. 5, v. 5, p. 443-454, fig. 1-8.

Gerhardt, U.

(8) 1932, *Merostomata: in* Kükenthal, Handbuch der Zoologie (Berlin), ed. T. Krumbach, Bd. 3, Abt. 2, p. 1-96, fig. 1-74.

Holm, G.

(9) 1898, *Über die Organisation von Eurypterus fischeri Eichw.:* Acad. Sci. St. Petersbourg, Mém., ser. 8, v. 8, no. 2, p. 1-57, pl. 1-10.

Huxley, T. H., & Salter, J. W.

(10) 1859, *On the anatomy and affinities of the genus Pterygotus (Huxley) and description of the species of Pterygotus (Salter):* Geol. Survey United Kingdom, Mem., Mon. 1, p. 1-105, pl. 1-16.

Ivanov, P. P.

(11) 1933, *Die embryonale Entwicklung von Limulus moluccanus:* Zool. Jahrb., Abt. Anatomie und Ontogenie der Tiere, v. 56, pt. 2, p. 163-348, pl. 1-3, fig. 1-78.

Kjellesvig-Waering, E. N.

(12) 1948, *The Mazon Creek eurypterid; a revision of the genus Lepidoderma:* Illinois State Mus., Sci. Papers, v. 3, no. 4, p. 1-48, pl. 1-8, fig. 1.

(13) 1951, *Downtonian (Silurian) Eurypterida from Perton, near Stoke Edith, Herefordshire:* Geol. Mag., v. 88, p. 1-24, pl. 1-3, fig. 1-2.

——— & Størmer, Leif

(14) 1952, *The Dolichopterus-Strobilopterus group in the Eurypterida:* Jour. Paleont., v. 26, p. 659-661, fig. 1-2.

Lankester, E. R.

(15) 1881, *Limulus, an arachnid:* Quart. Jour. Micros., Sci., v. 23, p. 506.

Laurie, S.

(16) 1892, *On some eurypterid remains from the Upper Silurian rocks of Pentland Hills:* Roy. Soc. Edinburgh, Trans., v. 37, 6. 151-162.

Packard, A. S.

(17) 1886, *On the Carboniferous xiphosurous fauna of North America:* Natl. Acad. Sci. (Washington), Mem. 3, pt. 2, p. 143-157, pl. 5-7, fig. 9-19.

Pocock, R. F.

(18) 1902, *The taxonomy of Recent species of Limulus:* Ann. Mag. Nat. History, ser. 7, v. 9, p. 256-266.

Prantl, F., & Pribyl, A.

(19) 1948, *Revision of the Bohemian Silurian Eurypterida:* Rospravy Státníko Geologickéko Ustavei Republ. Československe, v. 10, p. 1-116, pl. 1-8, fig. 1-30.

Raasch, G. O.

(20) 1939, *Cambrian Merostomata:* Geol. Soc. America, Spec. Paper 19, p. 1-146, pl. 1-21, fig. 1-14.

Raymond, P. E.

(21) 1944, *Late Paleozoic xiphosurans:* Mus. Compar. Zool. Harvard Coll., Bull., v. 94, p. 475-508, pl. 2, fig. 1-11.

Richter, Rudolf, & Richter, Emma

(22) 1929, *Weinbergina opitzi, n. g., n. sp., ein Schwertträger (Merost., Xiphos.) aus dem*

Devon (Rheinland): Senckenbergiana, v. 11, p. 193-209, fig. 1-4.

Ruedemann, Rudolf

(23) 1926, *The Utica and Lorraine formations of New York, Part 2:* New York State Mus., Bull. 272, p. 1-227, pl. 1-28, fig. 1-26.

Schmidt, F.

(24) 1883, *Die Crustacéenfauna der Eurypteren-schichten von Rootziküll auf Oesel:* Acad. Imp. St. Petersbourg, Mém., ser. 7, v. 31, p. 28-88, pl. 1-7, fig. 1-5.

Simpson, S.

(25) 1951, *A new eurypterid from the upper Old Red Sandstone of Portishead:* Ann. Mag. Nat. History, ser. 12, v. 4, p. 849-861, pl. 15-16.

Størmer, Leif

(26) 1934, *Merostomata from the Downtonian sandstone of Ringerike, Norway:* Skr. Vid. Akad. Oslo, I, Math.-Natur. Kl., 1933, no. 10, p. 1-125, pl. 1-12, fig. 1-39.

(27) 1935, *Über den neuen, von W. Gross beschriebenen Eurypteriden aus dem rheinischen Unterdevon von Overath im Rheinland:* Preuss. Geol. Landesanst., Jahrb. für 1934, Bd. 55 [N. F., Heft 175], p. 284-291, fig. 1-3.

(28) 1944, *On the relationships and phylogeny of fossil and Recent Arachnomorpha:* Skr. Vid. Akad. Oslo, I. M.-N. Kl. 1944, no. 5, p. 1-158, fig. 1-30.

(29) 1951, *A new eurypterid from the Ordovician of Montgomeryshire, Wales:* Geol. Mag., v. 88, p. 409-422, fig. 1-2.

(30) 1952, *Phylogeny and taxonomy of fossil horse-shoe crabs:* Jour. Paleont., v. 26, p. 630-639, fig. 1-3.

(31) 1955, *A new merostome from the Lower Cambrian of Kinnekulle, Sweden:* Arkiv f. Zoologie (Stockholm), in press.

Tschernyschew, B. F.

(32) 1933, *Arthropoda from the Urals and other regions of the U.S.S.R.:* Centr. Sci. Geol. and Prospecting Inst., Paleontology and Stratigraphy, Mag. 1, p. 15-24, pl. 1.

Woodward, H.

(33) 1866-78, *Monograph of British fossil Crustacea belonging to the Order Merostomata:* Palaeontograph. Soc. (London), p. 1-263, pl. 1-36, fig. 1-81.

SOURCES OF ILLUSTRATIONS

(34) **Beecher, C. E.**
(35) **Böhm, H.**
(36) **Caster, K. E., & Macke, W. B.**
(37) **——— & Kjellesvig-Waering, E. N.**
(38) **Chapman, F.**
(39) **Clarke, J. M.**
(40) **——— & Ruedemann, Rudolf**
(41) **Currie, E. C.**
(42) **Dix, E., & Pringle, J.**
(43) **Dunbar, C. O.**
(44) **Eller, E. R.**
(45) **Fritsch, A. J.**
(46) **Hall, James**
(47) **Holm, G.**
(48) **Kingsley, J. S.**
(49) **Kirchner, H.**
(50) **Kjellesvig-Waering, E. N.**
(51) **——— & Størmer, Leif**
(52) **Lange, W.**
(53) **Laurie, S.**
(54) **Matthew, G. F.**
(55) **Packard, A. S.**
(56) **Patten, W.**
(57) **Prantl, F., & Pribyl, A.**
(58) **Pruvost, Pierre**
(59) **Raasch, G. O.**
(60) **Raymond, P. E.**
(61) **Resser, C. E.**
(62) **Richter, Rudolf, & Richter, Emma**
(63) **Schmidt, F.**
(64) **Størmer, Leif**
(65) **Tschernyschew, B. F.**
(66) **Williams, H. S.**
(67) **Woodward, H. B.**

Incertae Sedis

Belinuropsis MATTHEW, 1909 [**B. wigodensis*]. Little-known form. *U.Sil.*, Can.—FIG. 11,*8*. **B. wigodensis*, ×1 (after MATTHEW).

ARACHNIDA

By Alexander Petrunkevitch

CONTENTS

INTRODUCTION

Many arachnids were known and described before LINNÉ's time, but the history of arachnology begins with LAMARCK, who was the first to propose for them in 1801 a separate class which he named Arachnidae. Before him they were always treated as insects. In the 10th edition of LINNÉ's *Systema Naturae* the genera *Acarus, Aranea, Phalangium* and *Scorpio* are placed with the wingless insects, crustaceans, centipedes and millipedes in the 7th order, Aptera, a name which was still used by WALCKENAER in 1837-41. LINNÉ did not recognize families. In his classification genera were placed directly under orders. DE-GEER seems to be one of the first zoologists who used the "family" in his classification, but he divided the genera into families and in 1778 listed 7 families under the genus *Aranea*. His classification also persisted for a number of years and was adopted by WALCKENAER in his celebrated *Histoire Naturelle des Insectes Aptères*. However, at the same time with DEGEER, BATSCH (1778) introduced the concept of the family as a group below the order and above the genus. This concept also found followers and we find LATREILLE using the term "family" in this sense in 1806 and LEACH in 1815. Thus two different concepts of the term "family" existed side by side for more than half a century until BATSCH's concept prevailed and was incorporated in the report of the Committee of the British Association for the Advancement of Science in 1842. Since the 14th International Congress of Zoology gave a precise definition of the Family-Group and Ordinal-Group categories, family names proposed by DEGEER and WALCKENAER cannot be used even when they have been latinized by the authors themselves, because according to WALCKENAER's own statement his families have been placed "under more general divisions called *Genera*." This complicates the problem of family priority in the case of all families proposed before 1842. I intend to discuss this subject in a separate paper which will be published elsewhere. Here I mention it only because certain changes in the names of higher categories introduced in the present paper may give the impression that I have changed the principles of my classification as laid down in my various studies of fossil Arachnida. The principles of classification remain the same. The changes involve only some names in conformity with the rules which I tried to follow to my best ability by tracing the priority of familial and ordinal names by careful examination of the original publications all the way back to LINNÉ's 10th edition. In doing so I had, of course, to depend upon the author's own statement as to which category he applied a certain name and to accept it as a family name when he called it "Family" and placed it under an order and above the genus, or as an order or class if he called it so. Such classes, orders and families rarely, if ever, correspond to the classes, orders and families which we now recognize. Thus, for example, Class Arachnida LAMARCK, 1801, does not mean at all that this class is used here in LAMARCK's sense, who included in it crustaceans, wingless insects, pycnogonids and myriapods, as well as spiders, scorpions, etc., which constitute only a small portion of it. It means only that LAMARCK was the first to propose the name Arachnida. In reality, his class was whittled down and added to and again trimmed many times in the course of the 150 years since its inception, until it assumed the content given it independently by STØRMER in Norway and myself in America.

Although the orders recognized by LAMARCK and his contemporaries differed greatly by their content from the orders which are recognized now, representatives of all but 3 of the latter were already known. The first representative of the order Ricinuleida was described by GUÉRIN in 1838 under the name of *Cryptostemma* (now *Ricinoides*) *westermanni* and placed in the order Phalangiida (harvestmen), in which it remained until the order Ricinulei was proposed by THORELL in 1892. The first representative of the order Schizomida was described by CAMBRIDGE in 1872 under the name of *Nyctalops* (now *Schizomus*) *crassicaudatus* and placed into a new family Tartarides in the order Pedipalpi. This family was raised to the rank of an order by me in 1945. Finally, the first representative of the order Palpigradida was described by GRASSI & CALANDRUCCIO in 1885 and named *Koenenia* (now *Eukoenenia*) *mirabilis*.

Fossil Arachnida found in Baltic amber must have been seen in ancient times, for TACITUS mentions insects enclosed in transparent amber. Since the middle of the past century many spiders were described, first in the Baltic amber (Fig. 31) and later in such Tertiary deposits as those of Aix-en-Provence and Florissant. However, the first scientific mention of fossil Arachnida was published in 1819 when SCHWEIGGER described and figured a spider, later named *Entomocephalus formicoides* by HOLL (1829), and a scorpion which HOLL named *Scorpio schweiggeri*.

The first Paleozoic arachnid, a scorpion, was described by CORDA in 1835 from the Carboniferous of Czechoslovakia. He named it *Cyclophthalmus senior*. No other fossil scorpions were found until 1868 when MEEK & WORTHEN described *Mazonia woodiana* and *Eoscorpius carbonarius*, both specimens from the Pennsylvanian of Mazon Creek, Illinois.

In the year 1836, following upon the description of *Cyclophthalmus,* BUCKLAND figured in the first edition of his Bridgewater Treatise 2 fossils from the ironstone of Coalbrook Dale, naming them *Curculioides ansticii* and *C. prestvicii.* The former was recognized as an arachnid related to *Poliochera* by SCUDDER in 1884, who placed his family Poliocheridae in the order Anthracomarti (here Anthracomartida), from which it was removed by POCOCK in 1902 into the order Ricinulei (here Ricinuleida), established by THORELL in 1892 for Recent arachnids of the family Cryptostemmidae (now Ricinoididae), formerly placed in the same order with harvestmen. *C. prestvicii* BUCKLAND was removed to the genus *Eophrynus* by WOODWARD in 1877 and became the first representative of that order, which in 1949 I proposed to call Trigonotarbi (here Trigonotarbida) (Fig. 32). The first spider, i.e., the first representative of the order Araneida, was described by RÖMER in 1866 under the name *Protolycosa anthracophila,* from the Carboniferous of Upper Silesia. Eight years later, in 1874, the first American fossil spider *Arthrolycosa antiqua* was described by HARGER from the Pennsylvanian of Mazon Creek (Fig. 32).

The first anthracomartid, *Brachypyge carbonis,* was discovered by WOODWARD in 1878, but he thought it to be a crab, and its true nature was recognized only in 1885 by SCUDDER, after *Anthracomartus völkelianus* had been described and figured by KARSCH in 1882.

The first fossil whip scorpion was described by KUŠTA in 1884, who named it *Prothelyphonus bohemicus.* It was found in Rakonitz, Czechoslovakia. In the same year an American whip scorpion, *Geralinura carbonaria,* was described by SCUDDER.

The first representative of the fossil order Architarbida, *Architarbus rotundatus,* was discovered in America and described by SCUDDER in 1868. Four years later, in 1872, the first European species was discovered and named by WOODWARD *Architarbus* (now *Phalangiotarbus*) *subovalis.* In 1878 SCUDDER described and figured a fossil which he thought to be an insect and named *Libellula carbonaria.* In 1890 he realized his mistake, proposed for the fossil the new generic name *Graeophonus* and placed it in the arachnid order Pedipalpi. At his time, as indeed even much later, LATREILLE's order Pedipalpi was still generally accepted, although already reduced to include only whip scorpions, schizonotids (schizomids) and phrynichids. POCOCK was the first to elevate the phrynichids to the rank of an order in 1900, for which he retained THORELL's suborder name Amblypygi. Thus, *Graeophonus carbonarius* became the first fossil representative of the order called here Phrynichida. A fossil representative of the former small suborder of Pedipalpi Tartarides, later elevated by me to the rank of an order which I named Schizomida (a *nom. transl.* from the family name of the type genus) was discovered by me in 1945 in the onyx marble of Arizona and named *Calcitro fisheri.*

A fossil arachnid described by SCUDDER in 1890 under the name of *Kustarachne tenuipes* was placed by me in 1913 in an order of its own, Kustarachnae (=Kustarachnida). The first fossil phalangiids (harvestmen), *Eotrogulus fayoli* and *Nemastomoides elaveris,* were described by THEVENIN in 1901 from the Coal Measures of Commentry, France. In 1911 POCOCK described the only fossil representative of his order Haptopoda (=Haptopodida) from the Coal Measures of England and named it *Plesiosiro madeleyi.* The only fossil representative of the order Solpugida was described by me in 1913 from the Pennsylvanian of Mazon Creek, under the name of *Protosolpuga carbonaria.*

Although in the middle of the past century several fossils from the Carboniferous of Europe were described as pseudoscorpions, their identification proved to be incorrect. We now know that fossil pseudoscorpions have been found so far only in Cenozoic deposits. They are well represented in the Baltic amber. The only fossil representative of the order Palpigradida was found by OPPENHEIM in the Jurassic lithographic shale of Germany in 1887. He mistook the fossil for an aquatic bug and called it *Halometra minor.* Its true nature was recognized by HAASE in 1890 and as the name *Halometra* was preoccupied, HAASE gave it its present name, *Sternarthron zitteli.*

Fossil Acarida were first discovered in Baltic amber. BERENDT & KOCH described several species of them in 1854. The only known Paleozoic species, *Protacarus crani,* was described by HIRST in 1923 from the Old Red Sandstone of Devonian age.

The number of fossil arachnid species discovered since the first Carboniferous scorpion was found has increased steadily at a fairly rapid rate, which slowed down

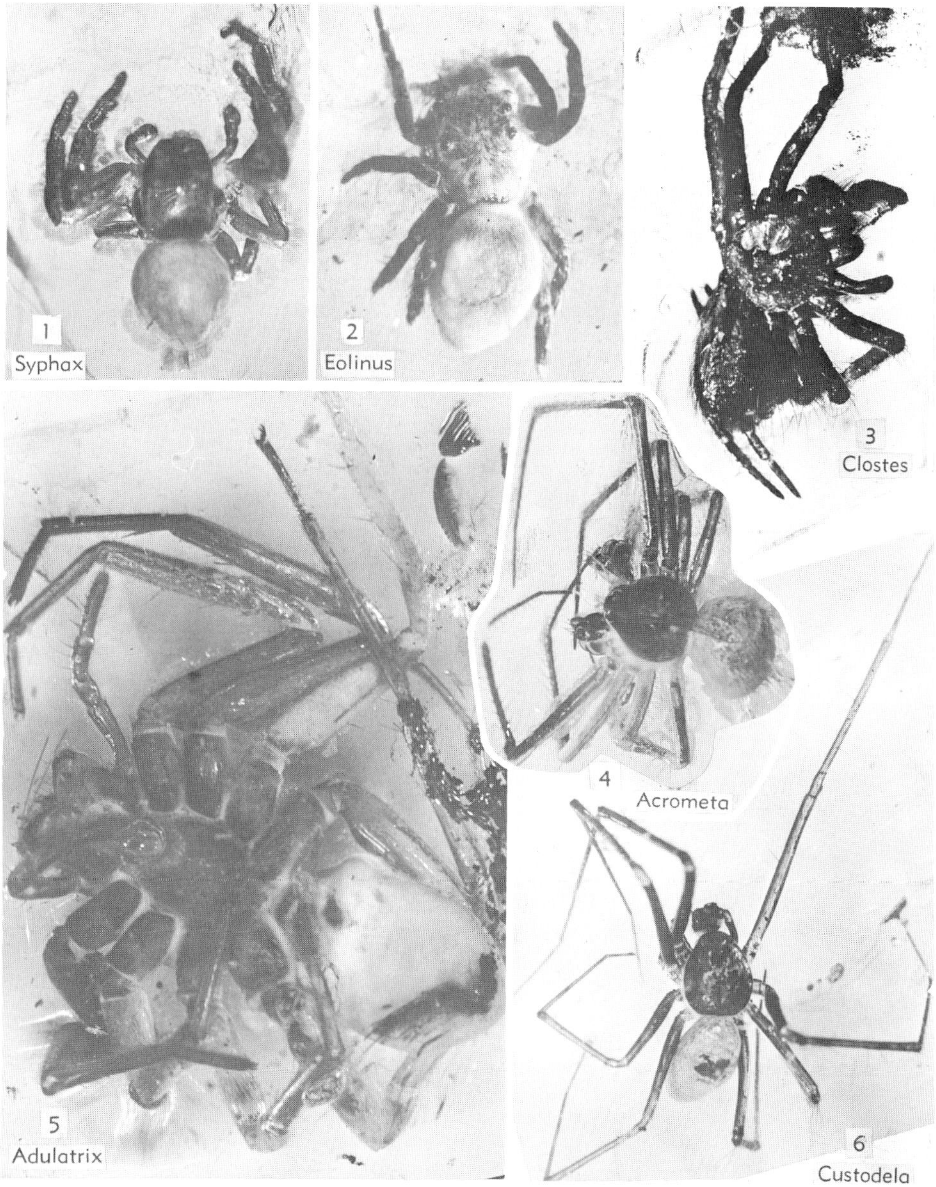

FIG. 31. Spiders (order Araneida) preserved in Oligocene amber of the Baltic region of Europe, photographs. *1, Syphax crassipes,* ×6.3. *2, Eolinus theryi,* ×11. *3, Clostes priscus,* ×8. *4, Acrometa cristata,* ×12. *5, Adulatrix fusca,* ×8.5. *6, Custodela cheiracantha,* ×6 (*76).

only in the second quarter of the present century. By now, every Recent order of Arachnida has one or more fossil representatives and in addition we know 5 Paleozoic orders which became extinct. The total number of species which can be recognized from descriptions and figures has reached about 500 or a few more. One-third of them are Paleozoic. The accompanying table shows their distribution in systematic orders and main geological eras. The table does not include unrecognizable species which were described at a time when indispensable specific characters were not yet understood or by investigators who were not professional arachnologists and who were unfamiliar with the distinctive characters even of orders. The figures given in the table for the number of Recent genera and species are only approximate. It is probable that they will continue to increase for some time to come and at a much more rapid pace than the number of new fossil species.

Numbers of Known Arachnid Genera and Species

Order	Paleozoic		Mesozoic		Tertiary		Quaternary	
	Gen.	Sp.	Gen.	Sp.	Gen.	Sp.	Gen.	Sp.
Scorpionida	26	42	2	7	2	2	70	600
Pseudoscorpionida	..	..	..	..	13	18	234	1,000
Phalangiida	2	4	..	..	9	16	640	2,350
Architarbida	14	25	..	..	..	..	..	..
Acarida	1	1	..	..	60	107	1,389	6,000
Haptopodida	1	1	..	..	..	..	..	..
Anthracomartida	11	20	..	..	..	..	..	..
Trigonotarbida	24	38	..	..	..	..	..	..
Palpigradida	..	..	1	1	..	..	4	21
Thelyphonida	2	5	..	..	1	1	10	70
Schizomida	..	..	..	..	2	2	3	28
Kustarachnida	1	3	..	..	..	..	..	..
Phrynichida	3	4	..	..	..	..	18	60
Araneida	16	18	..	..	94	180	2,735	25,000
Solpugida	1	1	..	..	..	..	134	600
Ricinuleida	2	9	..	..	..	..	2	16
Total	102	169	3	8	181	326	5,239	35,745

NOTE. The figures for Tertiary genera include extinct and surviving genera. Extinct Tertiary genera are restricted to the following 4 orders: Pseudoscorpionida, 5; Phalangiida, 1; Acarida, 10; Araneida, 67. Total extinct Tertiary genera, 83; surviving genera included in the Quaternary figures, as well as in the Tertiary ones, 98.

CLASSIFICATION

As explained in the introduction, the classification used in this article is essentially the same as that proposed by me in 1933 for families of spiders and in 1949 for orders of Arachnida. The latter differs only slightly from the generally accepted classification, inasmuch as it omits from the list of orders the name of Pedipalpi because that name stands for an incongruous group which I have divided into 3 natural orders, Thelyphonida, Schizomida, and Phrynichida. It also differs by the substitution of 2 older names, Phalangiida and Solpugida, for Opiliones and Solifugae, names which became more generally used in the 20th century. The definition of all orders takes into account structural differences shown by their fossil representatives, but in orders which survived, is primarily based on characters found in Recent species, the anatomy and embryology of which is by now known in such great detail. In extinct orders the definitions are of necessity limited to their known characters, but when these exhibit trends similar to those observed in living Arachnida, the evidence derived from the latter is taken into account.

The classification of the order Araneida, proposed by me in 1933 and supplemented in 1939 and 1942, is based on a correlation of external characters with comparative anatomy of internal organs. It differs from

FIG. 32. Order Trigonotarbida. *Eophrynus prestvici*, U.Carb., Eng.; *A,B*, photographs of dorsal and ventral sides of lectotype, ×3 (*76) (p. P112).

other classifications mainly in the grouping of families, in the splitting up of some older families into several independent families for each original one, and in their reassignment to other divisions of the same suborder or even to another suborder. Some features of my classification have been objected to by other arachnologists, but I consider it superior to their own substitute classifications which disregard internal anatomy. The introduction of superfamilies was first proposed by CAPORIACCO in 1938 and I have incorporated them in my present article, but their definitions and groupings had to be somewhat changed to conform with my views.

The classification of scorpions is the same as in my monograph published in 1953. I had however to introduce superfamilies for the purpose of bringing my classification of fossil scorpions in line with the generally accepted classification of Recent scorpions. This presented a difficulty only in the case of the Carboniferous family Eoscorpiidae (=Eoscorpionidae) from which, as I have shown, all Recent families have evolved, until finally I succeeded in establishing the priority of the family Scorpionidae, proposed as a family Scorpionides by LEACH in 1815 and including LINNÉ's genus *Scorpio*. Of course, the definition of the family Scorpionidae as recognized by arachnologists at present is that of POCOCK (1893) and not of LEACH, who recognized but a single family, yet I am given to understand that it is the policy of the *Treatise* to use the *oldest* family name for a superfamily (Fig. 33).

In the classification of pseudoscorpions I followed BEIER. For the order Acarida I used the classification of VITZTHUM, which I prefer to all others, except that I had to change the names of several categories and use older ones in their place. A similar change applies to some other orders, but since the more usual names are mentioned in every case, the reader will have no difficulty in recognizing familiar categories, notwithstanding their old, disused names.

A tabular outline of suprageneric categories of Arachnida is given here with the omission of most of those which contain no known fossil representatives. Taxonomic units with unmentioned genera are distinguished by an asterisk (*)

Suprageneric Divisions of Arachnida

Figures in parentheses indicate number of recognized genera; for the taxa of the class/order group the number includes Recent genera; for the taxa of the family group it refers only to fossil genera. Families represented only by Recent genera are omitted.

Arachnida *(class)* (5,432) *Sil.-Rec.*
Latigastra *(subclass)* (2,394) *Sil.-Rec.*
Scorpionida *(order)* (98) *Sil.-Rec.*
Protoscorpionina *(suborder)* (4) *Sil.-Penn.*
Palaeophonoidea *(superfamily)* (1) *Sil.*
Palaeophonidae (1) *Sil.*
Mazonioidea *(superfamily)* (3) *Sil.-Penn.*
Dolichophonidae (2) *Sil.*
Mazoniidae (1) *Penn.*
Euscorpionina *(suborder)* (94) *Dev.-Rec.*
Palaeoscorpioidea *(superfamily)* (1) *Dev.*
Palaeoscorpiidae (1) *Dev.*
Archaeoctonoidea *(superfamily)* (2) *Carb.*
Archaeoctonidae (2) *Carb.*
Scorpionoidea *(superfamily)* (*14) *Carb.-Rec.*
Eoscorpiidae (12) *Carb.*
Buthidae (*1) *Oligo.-Rec.*
Scorpionidae (*1) *Oligo.-Rec.*
Cyclophthalmoidea *(superfamily)* (1) *Carb.*
Cyclophthalmidae (1) *Carb.*
Isobuthoidea *(superfamily)* (3) *Carb.*
Isobuthidae (3) *Carb.*
Centromachoidea *(superfamily)* (1) *Carb.*
Centromachidae (1) *Carb.*
Mesophonoidea *(superfamily)* (2) *Trias.*
Mesophonidae (2) *Trias.*
Incertae sedis (2) *Carb.*
Pseudoscorpionida *(order)* (239) *Oligo.-Rec.*
Chthoniina *(suborder)* (29) *Oligo.-Rec.*
Ditidae (*1) *Oligo.-Rec.*
Chthoniidae (*1) *Oligo.-Rec.*
Neobisiina *(suborder)* (66) *Oligo.-Rec.*
Neobisioidea *(superfamily)* (*1) *Oligo.-Rec.*
Neobisiidae (*1) *Oligo.-Rec.*
Garypoidea *(superfamily)* (*3) *Oligo.-Rec.*
Olpiidae (*1) *Oligo.-Rec.*
Garypidae (*2) *Oligo.-Rec.*
Feaelloidea *(superfamily)* (*1) *Oligo.-Rec.*
Pseudogarypidae (*1) *Oligo.-Rec.*
Cheliferina *(suborder)* (144) *Oligo.-Rec.*
Cheiridioidea *(superfamily)* (*1) *Oligo.-Rec.*
Cheiridiidae (*1) *Olig.-Rec.*
Cheliferoidea *(superfamily)* (*5) *Oligo.-Rec.*
Cheliferidae (*4) *Oligo.-Rec.*
Chernetidae (*1) *Oligo.-Rec.*
Phalangiida *(order)* (643) *Carb.-Rec.*
Cyphophthalmina *(suborder)* (10) *Rec.*
Palpatorina *(suborder)* (133) *Carb.-Rec.*
Troguloidea *(superfamily)* (*4) *Carb.-Rec.*
Eotrogulidae (1) *Carb.*
Nemastomatidae (*1) *Carb.-Rec.*
Nemastomoididae (1) *Carb.*
Ischyropsalidae (*1) *Oligo.-Rec.*
Phalangioidea *(superfamily)* (*6) *Oligo.-Rec.*
Phalangiidae (*6) *Oligo.-Rec.*
Laniatorina *(suborder)* (500) *Oligo.-Rec.*
Gonyleptidae *(superfamily)* (*1) *Oligo.-Rec.*
Gonyleptidae (*1) *Oligo.-Rec.*
Architarbida *(order)* (14) *Carb.*
Architarbidae (12) *Carb.*

Opiliotarbidae (1) *Carb.*
Heterotarbidae (1) *Carb.*
Acarida *(order)* (1,400) *Dev.-Rec.*
Notostigmatina *(suborder)* (3) *Rec.*
Holothyrina *(suborder)* (1) *Rec.*
Trombidiina *(suborder)* (550) *Dev.-Rec.*
Eupodidae (*1) *Dev.-Rec.*
Bdellidae (*1) *Oligo.-Rec.*
Erythraeidae (*2) *Oligo.-Rec.*
Trombidiidae (*1) *Oligo.-Rec.*
Anystidae (*1) *Oligo.-Rec.*
Cheyletidae (*1) *Oligo.-Rec.*
Tetranychidae (*1) *Oligo.-Rec.*
Parasitina *(suborder)* (345) *Oligo.-Rec.*
Parasitoidea *(superfamily)* (*1) *Oligo.-Rec.*
Phytoseiidae (*1) *Oligo.-Rec.*
Ixodoidea *(superfamily)* (*2) *Oligo.-Rec.*
Ixodidae (2) *Oligo.-Rec.*
Acarina *(suborder)* (455) *Oligo.-Rec.*
Acaroidea *(superfamily)* (*2) *Oligo.-Rec.*
Acaridae (2) *Oligo.-Rec.*
Oribatoidea *(superfamily)* (*48) *Oligo.-Rec.*
Belbidae (*3) *Oligo.-Rec.*
Camisiidae (*2) *Oligo.-Rec.*
Carabodidae (*8) *Oligo.-Rec.*
Ceratozetidae (*3) *Oligo.-Rec.*
Cymberemaeidae (*4) *Oligo.-Rec.*
Oribatidae (*10) *Oligo.-Rec.*
Galumnidae (*1) *Oligo.-Rec.*
Haplozetidae (*1) *Oligo.-Rec.*
Hypochthoniidae (*1) *Oligo.-Rec.*
Hermanniellidae (*1) *Oligo.-Rec.*
Liacaridae (*1) *Oligo.-Rec.*
Neoliodidae (*2) *Oligo.-Rec.*
Oribatellidae (*2) *Oligo.-Rec.*
Oribatulidae (*3) *Oligo.-Rec.*
Oripodidae (*1) *Oligo.-Rec.*
Parakalummidae (*1) *Oligo.-Rec.*
Phenopelopidae (*1) *Oligo.-Rec.*
Phthiracaridae (*2) *Oligo.-Rec.*
Tetrapodilina *(suborder)* (46) *Rec.*
Stethostomata *(subclass)* (12) *Carb.*
Haptopodida *(order)* (1) *Carb.*
Plesiosironidae (1) *Carb.*

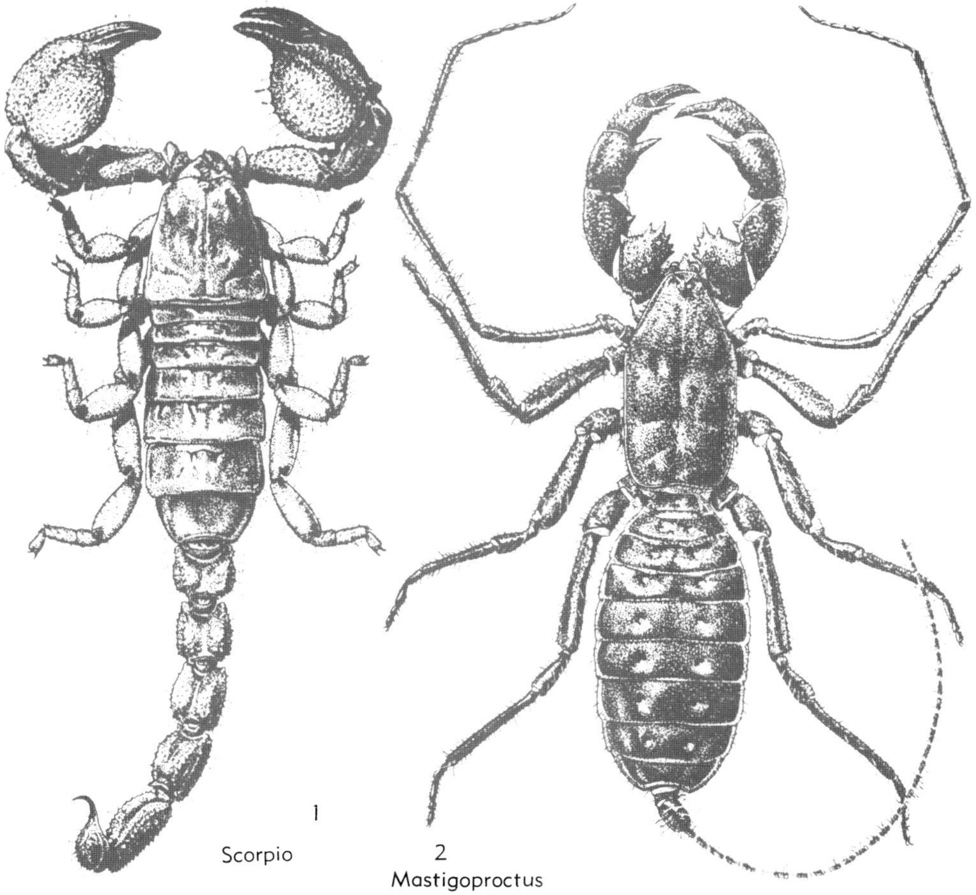

FIG. 33. Recent representatives of the orders Scorpionida and Thelyphonida. *1, Scorpio maurus,* a common scorpion, N.Afr., ×1.5. *2, Mastigoproctus giganteus,* a common whip scorpion, Mex., ×1.5 (by permission from *Traité de Zoologie,* tome 6, sous la direction du PR. GRASSÉ, Masson & Cie., Paris) (p. P75, P120).

Anthracomartida *(order)* (11) *Carb.*
Anthracomartidae (11) *Carb.*
Soluta *(subclass)* (24) *Dev.-Carb.*
Trigonotarbida *(order)* (24) *Dev.-Carb.*
Palaeocharinidae (2) *Dev.*
Anthracosironidae (1) *Carb.*
Eophrynidae (14) *Carb.*
Trigonotarbidae (1) *Carb.*
Trigonomartidae (6) *Carb.*
Caulogastra *(subclass)* (3,002) *Carb.-Rec.*
Latisterna *(superorder)* (5) *Jur.-Rec.*
Palpigradida *(order)* (5) *Jur.-Rec.*
Sternarthronidae (1) *Jur.*
Eukoeneniidae (4) *Rec.*
Camarostomata *(superorder)* (18) *Carb.-Rec.*
Thelyphonida *(order)* (12) *Carb.-Rec.*
Thelyphonidae (*2) *Carb.-Rec.*
Schizomida *(order)* (5) *?Plio.-Rec.*
Schizomidae (*) *?Plio.-Rec.*
Calcitronidae (2) *?Plio.*
Kustarachnida *(order)* (1) *Carb.*
Kustarachnidae (1) *Carb.*
Labellata *(superorder)* (2,840) *Carb.-Rec.*
Phrynichida *(order)* (21) *Carb.-Rec.*
Phrynichidae (*3) *Carb.-Rec.*
Araneida *(order)* (2,819) *Dev.-Rec.*
Liphistiina *(suborder)* (19) *Dev.-Rec.*
Liphistioidea *(superfamily)* (*11) *Dev.-Rec.*
Arthromygalidae (9) *Dev.-Rec.*
Arthrolycosidae (2) *Carb.*
Theraphosina *(suborder)* (370) *Oligo.-Rec.*
Theraphosoidea *(superfamily)* (*1) *Oligo.-Rec.*
Theraphosidae (1) *Oligo.-Rec.*
Atypoidea *(superfamily)* (*1) *Oligo.-Rec.*
Dipluridae (*1) *Oligo.-Rec.*
Incertae sedis (1) *Eoc.*
Hypochilina *(suborder)* (2) *Rec.*
Dipneumonina *(suborder)* (2,401) *Oligo.-Rec.*
Trionychi *(division)* (1,157) *Oligo.-Rec.*
Hersilioidea *(superfamily)* (*2) *Oligo.-Rec.*
Urocteidae (*1) *Oligo.-Rec.*
Hersiliidae (*1) *Oligo.-Rec.*
Scytodoidea *(superfamily)* (*1) *Oligo.-Rec.*
Pholcidae (*1) *Oligo.-Rec.*
Zodarioidea *(superfamily)* (*3) *Oligo.-Rec.*
Zodariidae (*1) *Oligo.-Rec.*
Spatiatoridae (2) *Oligo.*
Lycosoidea *(superfamily)* (*10) *Oligo.-Rec.*
Eresidae (*1) *Oligo.-Rec.*
Amaurobiidae (*2) *Oligo.-Rec.*
Psechridae (*1) *Oligo.-Rec.*
Agelenidae (*3) *Oligo.-Rec.*
Insecutoridae (1) *Oligo.*
Pisauridae (*1) *Oligo.-Rec.*
Lycosidae (*1) *Oligo.-Rec.*
Araneoidea *(superfamily)* (*34) *Oligo.-Rec.*
Archaeidae (*1) *Oligo.-Rec.*
Mimetidae (*1) *Oligo.-Rec.*
Arthrodictynidae (1) *Oligo.*
Dictynidae (*1) *Oligo.-Rec.*
Micryphantidae (*1) *Oligo.-Rec.*
Theridiidae (*7) *Oligo.-Rec.*
Adjutoridae (3) *Oligo.*
Ephalmatoridae (1) *Oligo.*
Linyphiidae (*9) *Oligo.-Rec.*
Araneidae (*11) *Oligo.-Rec.*
Archaeometoidea *(superfamily)* (3) *Carb.*
Archaeometidae (3) *Carb.*
Dionychi *(division)* (1,117) *?Carb., Oligo.-Rec.*
Zoropsoidea *(superfamily)* (*1) *Oligo.-Rec.*
Zoropsidae (*1) *Oligo.-Rec.*
Thomisoidea *(superfamily)* (*12) *Oligo.-Rec.*
Eusparassidae (*6) *Oligo.-Rec.*
Thomisidae (*6) *Oligo.-Rec.*
Clubionoidea *(superfamily)* (*11) *Oligo.-Rec.*
Drassodidae (*2) *Oligo.-Rec.*
Inceptoridae (1) *Oligo.*
Clubionidae (*2) *Oligo.-Rec.*
Parattidae (1) *Oligo.*
Salticoidea *(superfamily)* (*6) *Oligo.-Rec.*
Salticidae (*6) *Oligo.-Rec.*
Pyritaraneoidea *(superfamily)* (2) *Carb.*
Pyritaraneidae (2) *Carb.*
Quadrostiati *(division)* (127) *Oligo.-Rec.*
Dysderoidea *(superfamily)* (*5) *Oligo.-Rec.*
Segestriidae (*1) *Oligo.-Rec.*
Dysderidae (*3) *Oligo.-Rec.*
Oonopidae (*1) *Oligo.-Rec.*
Anyphaenoidea *(superfamily)* (*1) *Oligo.-Rec.*
Anyphaenidae (*1) *Oligo.-Rec.*
Argyronetoidea *(superfamily)* (*1) *Mio.-Rec.*
Argyronetidae (*1) *Mio.-Rec.*
Apneumonina *(suborder)* (16) *Rec.*
Rostrata *(superorder)* (135) *Carb.-Rec.*
Solpugida *(order)* (135) *Carb.-Rec.*
Galeodidae (*1) *Carb.-Rec.*
Podogona *(superorder)* (4) *Carb.-Rec.*
Ricinuleida *(order)* (4) *Carb.-Rec.*
Poliocheridae (*1) *Carb.*
Curculioididae (*1) *Carb.*

Class ARACHNIDA Lamarck, 1801

[*nom. correct. auctt.* (*pro.* Arachnidae LAMARCK, 1801, *emend.* STØRMER, 1944]

[=Embolobranchiata LANKESTER, 1905; Arachnida PETRUNKEVITCH, 1945]

[LAMARCK was the first to set aside LINNÉ's 7th order, Aptera, of the class Insecta as a separate class which he named Arachnidae, later accepting LATREILLE's modification of the name into Arachnides. For many years the class Arachnides still included all wingless insects, myriapods, crustaceans, horseshoe crabs, pycnogonids and arachnids. The present spelling "Arachnida" seems to have been used for the first time by NEWPORT in 1830, but the horseshoe crabs and the pycnogonids were retained in the class until the beginning of this century. Meanwhile, Tardigrada and Pentastomida were added to it by LEUCKART, and trilobites and eurypterids by LANKESTER. Only new evidence furnished by comparative anatomy and embryology finally made it possible to give the class Arachnida its present definition, adopted here.]

Air-breathing chelicerates with body divided into a cephalothorax and an abdomen, the former typically with 6 pairs of uniramous appendages. First pair (chelicerae) always preoral, 2- or 3-jointed. Second pair (pedipalpi) variously developed, but usually, typically 6-jointed. Correlated with the modification of structure, the function of the pedipalpi varies considerably from order to order. Third to 6th pairs typically 7-

jointed and usually developed as legs, although the 3rd pair is invariably modified as organs of chemotactile perception in some orders and the 5th pair as organs of copulation in the order Ricinuleida. Abdomen composed of a maximum of 12 segments and is either broadly joined to the cephalothorax or else the 1st abdominal segment is greatly constricted and forms a cone-shaped cylindrical pedicel (petiolus). Abdominal appendages present in the adult only in scorpions as a pair of combs (pectines) and in spiders as spinnerets.

Mouth parts vary considerably in structure and composition from order to order. They are formed by the chelicerae, upper lip (rostrum), lower lip (labium) and pedipalpal coxae or, as in scorpions, 1st and 2nd pedal coxae. Predigestion outside the body is the rule, with a preoral or a pharyngeal strainer permitting only fluid food to pass into the digestive system. The latter ends in an anus always situated on the last abdominal segment, which is closed either by a sphincter or an operculum.

Respiration by means of book lungs, tracheal tubes, or a combination of both, except in some cases of complete loss of organs of respiration in Acarida. Respiratory pigment always haemocyanin. Circulatory system composed of a dorsal heart, a complex arterial system and predetermined channels for the return of the blood to the heart through lateral ostia. Excretory system of 2 types of organs: coxal glands which are modified nephridia, and malpighian tubes which open into the digestive tract at the end of the mid-gut. Nervous system always with a dorsal brain characterized by a lack of a deutocerebrum, which is presumably fused with the protocerebrum. Ventral nervous chain of paired ganglia always exhibiting a trend to a forward displacement of abdominal ganglia and to a fusion and loss of some posterior abdominal ganglia. Eyes always of the simple type (ocelli). Organs of reproduction situated in the abdomen, with genital opening on the sternite of the embryologically 2nd abdominal segment, except in some Acarida in which it is secondarily displaced in the male. Sexes always separate. Females oviparous or ovoviviparous, rarely parthenogenetic.

Size varies from a fraction of a millimeter in some mites to 200 mm. in the largest scorpions. Habits terrestrial or secondarily aquatic, with complex inherited patterns of behavior. Most arachnids are carnivorous, but many mites are herbivorous. The great majority of arachnids are free-living, but many species of mites are temporarily or permanently parasitic. *Sil.-Rec.*

EXTERNAL CHARACTERS

CEPHALOTHORAX

The **head** is always fused with the 5 **thoracic segments,** forming a single unit or **tagma** always distinct from the **abdomen,** even when the latter is broadly joined with the last thoracic segment. Only when the segmentation of the abdomen also becomes obliterated by fusion, as in Acarida, the whole body forms a single unit and the limit between the **cephalothorax** and the abdomen can be determined only by the position of the 4th pair of legs, which belong to the last thoracic segment, and that of the genital opening which is on the 2nd abdominal segment. In all cases in which the abdomen is not fused with the cephalothorax, the dorsal surface of the latter is either covered with a single shield, the **carapace (peltidium),** which often shows its composition by the presence of a pair of **cephalothoracic sulci** outlining the limits between the head and the thorax, or else with several distinct plates. In the latter case, which is characteristic of Palpigradida, Schizomida and Solpugida and does not occur in other orders, one can always distinguish a single, large **propeltidium,** a paired (very rarely single) **mesopeltidium** and a single (rarely paired) **metapeltidium** (Figs. 83,*1b,2a;* 86,*1;* 112,*2a,3*).

The **eyes** of all arachnids are of the type of simple **ocelli,** with a corneal lens, the outer surface of which may be either convex or flat, a glass body which produces a new lens in each instar, and a retina which is either post- or prebacillar, depending upon whether the corresponding eyes receive their nerve supply from the 1st or the 2nd optic ganglia. The number and disposition of the eyes vary. One pair is often median in position, the others, of which there may be up to 5 pairs, as in some scorpions, are lateral. In many cases the eyes are elevated either on a common **eye tubercle** or on separate tubercles. In other cases they are sessile, i.e., their base is on a level with the surface of the carapace. The configuration of the **eye group** and the number of eyes furnish valuable systematic characters, but their value must not be exaggerated, as was often done in the past. Palpigradida, Anthracomartida and Ricinuleida have no eyes. In Schizomida the **median eyes** are wanting and the **lateral eyes** are rudimentary. Blind genera occur almost in all orders which normally possess well-developed eyes. **Compound eyes,** such as are found in Xiphosura, Eurypterida, Crustacea and Hexapoda, are never present in any Arachnida. The organs of scorpions belonging to the Triassic family Mesophonidae and those of the Palaeocharinidae (Devonian members of the order Trigonotarbida) claimed to be compound eyes are unlike true compound eyes and are probably sense organs of some unknown function.

The shape and proportions of the carapace are of importance in classification, but there are many cases in which the carapace of the male differs considerably from that of the female of the same species. A real "doublure" comparable to that in Xiphosura and Eurypterida is present only in the carapace of Thelyphonida (Fig. 84,*2*). In the order Ricinuleida a movable plate **(cucullus)** is always present at the anterior end of the carapace (Figs. 113-116). When flexed, the cucullus hides and protects the chelicerae.

VENTRAL SURFACE

The ventral surface of the cephalothorax is occupied by the **coxae** of the 6 pairs of appendages and the **thoracic sternites,** if such exist. The configuration of the coxosternal region is a very important taxonomic character of the highest order. In each case it presents a clear picture of the evolutionary changes which the cephalothorax of the particular order has undergone, changes involving partial or complete loss of thoracic sternites and displacement of the coxae from their original lateral position to the respective position which they now occupy. The extent of movability of the coxae is determined by the way of their articulation and the degree of their fusion with the ventral wall of the body. The movability of the coxae varies a great deal within the same order and even within the same species. Thus, in scorpions the anterior 2 pairs of coxae are movable individually, while the 3rd and 4th coxae are longitudinally fused together on each side of the body and possess jointly only very limited freedom of motion, being to all practical purposes immobile. The use of the degree of mobility of coxae has therefore a very limited systematic value.

ABDOMEN

The abdomen, which in the early embryological stages is always composed of 12 segments, may retain its segmentation in the adult or undergo various changes during development, masking or seriously changing its external appearance. Consequently external segmentation, even when present and well visible, does not necessarily represent the original segmentation. Theoretically, each abdominal segment should have a **tergite** and a **sternite** connected with each other laterally by a soft pleura and with the preceding and succeeding tergite and sternite by an intersegmental membrane. When the abdomen is distended, the tergites and sternites appear as independent plates, but in a normally contracted abdomen the intersegmental membrane of all segments becomes infolded under the posterior edge of the preceding tergite and sternite, so that the latter overlap the outer surface of the anterior edge of the succeeding tergite and sternite. In some segments, as in the 5 segments of the so-called postabdomen or "tail" of scorpions, the body wall of each segment forms a solid chitinous ring at the expense of the lateral pleurae, which are therefore wanting. In some cases, as in Anthracomartida and some Trigonotarbida, the last tergite is folded and forms the last plate on the ventral surface. A regular feature of the Anthracomartida is the fusion of the 2nd and 3rd tergites into a single plate, while the corresponding sternites remain separate (Fig. 66,*2a,b*). In the family Curculioididae of the order Ricinuleida, the 4 tergites visible as separate plates in the family Poliocheridae (Fig. 115,*1a*) are fused into a single plate (Fig. 116,*3b*). In Arachnida with a globular or ellipsoidal abdomen, the number of tergites visible in any position of the animal is always smaller than the total number (Figs. 97,*A,D,F*). The 1st tergite in Arachnida with a more or less flat abdomen is often concealed under the posterior edge of the carapace, as in Anthracomartida and Ricinuleida (Figs. 67, *1,5,7;* 68,*1,4*). Usually the tergites and sternites of Arachnida are entire, but in Anthracomartida the tergites are subdivided into 5 plates each and the sternites into 3 plates each. This is also the case in Architarbida and in one family of Ricinuleida. The tergites of some pseudoscorpions and of Haptopodida are subdivided into a pair of plates each. The 1st abdominal segment has completely disappeared in all Euscorpionina, Phalangiida and Ricinuleida. In the former, the first visible tergite is therefore the tergite of the 2nd embryonic segment, while in Ricinuleida it is the tergite of the 3rd segment, because the tergite of the 2nd segment, although present, is concealed under the carapace. All this shows clearly how misleading it would be, as it was done in the past, to number the *visible* tergites and sternites without taking into account their actual numerical order revealed by anatomical and embryological studies.

JUNCTURE BETWEEN CEPHALOTHORAX AND ABDOMEN

The juncture between the cephalothorax and the abdomen in Arachnida is of the greatest importance, but the outward appearance may be very misleading, as can be readily understood from what was just stated concerning the 1st *visible* tergite in Ricinuleida. In the case of the subclasses Latigastra and Stethostomata, the juncture is said to be "broad," which means that the carapace is connected with the abdomen by an intersegmental membrane of the full width of the posterior edge of the carapace and anterior edge of the abdomen. Because a broad connection is practically rigid, permitting but a very slight motion of the abdomen as a whole, the line of connection is usually more or less curved in a transverse

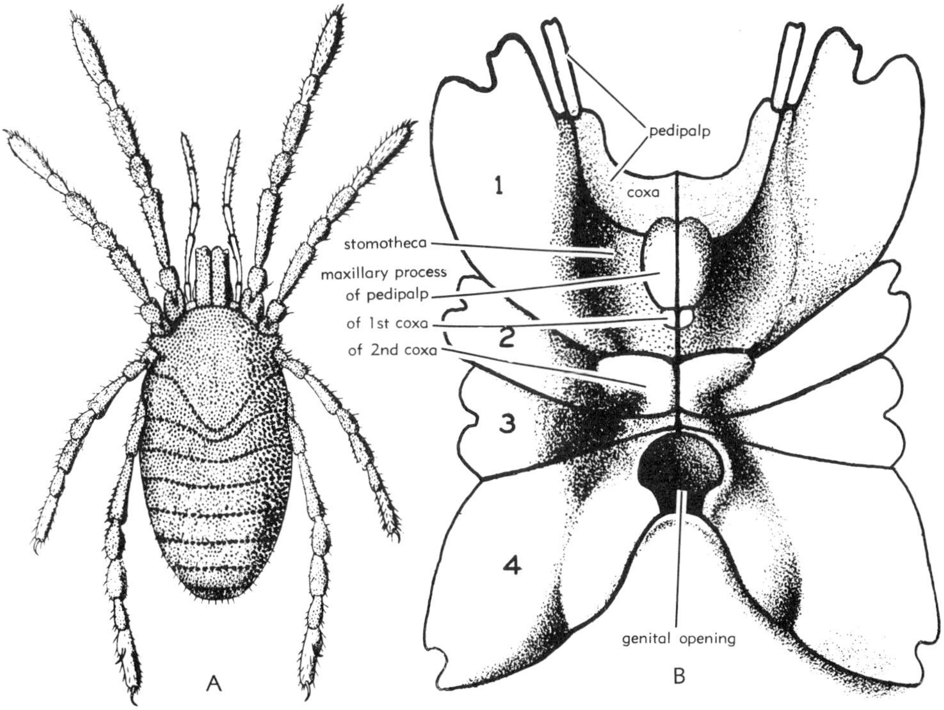

FIG. 34. Order Phalangiida; a Recent representative of the suborder Cyphophthalmina, *Siro duricorius*, Austria, ×18, and coxosternal region enlarged (68) (p. P85).

plane to the body, even when it has the appearance of a perfectly straight line when viewed from above. A similar connection is also found in the families Anthracosironidae and Trigonotarbidae in the order Trigonotarbida. In the family Eophrynidae of the same order, the connection is restricted to about the middle 3rd of the width of the carapace. In all Latigastra, Stethostomata and Soluta, the ventral connection between the cephalothorax and abdomen is more or less angular, because the anterior edge of the 1st sternite (or 2nd, when the 1st segment is lost) is triangular, as is also the diaphragm separating the cavity of the cephalothorax from that of the abdomen. In the subclass Caulogastra, the connection between the cephalothorax and the abdomen is considerably narrower than the width of the posterior edge of the carapace. In typical cases, such as presented by common spiders (Araneida), the 1st abdominal segment is reduced to a thin cylinder the anterior half of which is invaginated into the cavity of the cephalothorax with the entire connecting, cylindrical membrane. The dorsal portion of the wall of the 1st segment is of hard chitin. It is the tergite and is called **lorum pediculi.** The ventral wall remains usually soft, membranous, lacking a scleritized sternite. In less typical cases, such as presented by the whip scorpions (Thelyphonida), the dorsal tergite of the 1st segment is almost as wide as the posterior edge of the carapace, while the sternite is much narrower and quite short. In alcoholic speci-

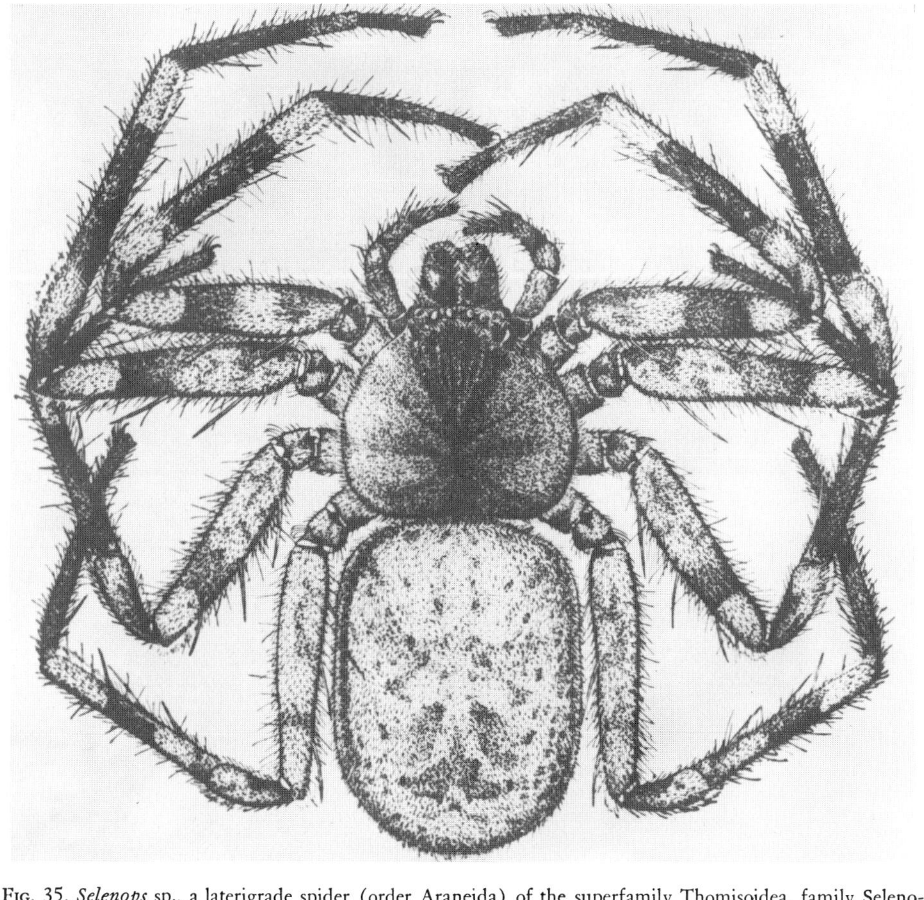

FIG. 35. *Selenops* sp., a laterigrade spider (order Araneida) of the superfamily Thomisoidea, family Selenopidae, from tropical Africa, ×6 (by permission from *Traité de Zoologie,* tome 6, sous la direction du PR. GRASSÉ, Masson & Cie., Paris) (p. P146).

mens of Recent species of caulogastric Arachnida, the abdomen breaks off easily, usually leaving the pedicel attached to the cephalothorax. In Arachnida with a broad juncture the abdomen remains firmly attached to the cephalothorax, from which it can be separated only by a sharp knife. But when, as in Anthracomartida, the abdomen is hard and its segments are firmly connected with each other, the soft membrane between the cephalothorax and the abdomen may disintegrate or break before fossilization is completed, with the result that the entire abdomen becomes severed in a single piece and may be the only portion of the animal preserved as a fossil (Figs. 34, 35).

CEPHALOTHORACIC APPENDAGES

The structure of the cephalothoracic appendages varies considerably from order to order and the terminology of their segments is often confusing, owing to the fact that different names have been given by specialists to the same segments in different orders of Arachnida. Moreover, the homology of the segments is not yet fully agreed upon. The greatest modifications occur in Acarida, and interested persons are referred to special articles on Recent representatives of that order.

The 1st pair of appendages in Arachnida was often termed antennae or mandibles in the past and even in the beginning of the present century. Now they are always called **chelicerae.** They are 3-jointed in Scorpionida, Phalangiida, Architarbida, Haptopodida, Anthracomartida and Palpigradida, but 2-jointed in all other orders. In both types they may be **retrovert** or **chelate,** but in the latter case one of the fingers of the chela is always immobile, being nothing but a rigid process of the segment proximal to the movable finger which forms the last segment (Fig. 112,*2c*). Phylogenetically and embryologically, the movable finger should be dorsal to the immobile one. But torsion is the cause of modification found in different orders and suborders. Thus, the movable finger is lateral in Scorpionida, Pseudoscorpionida and Palpigradida, medial in Haptopodida, ventral in Solpugida and Acarida.

The 2nd pair of appendages was sometimes called mandibles or maxillae in the past, but is now always termed **pedipalpi.** With the exception of the order Acarida, in which the number of pedipalpal segments varies from 7 in the Devonian *Protacarus* to 2 in some Recent species, the pedipalpi in all other orders are composed of 6 segments and may be **pediform** or **raptorial.** In the latter case they may be retrovert, as in Phrynichida, subchelate, as in some Thelyphonida, or chelate, as in Scorpionida and Pseudoscorpionida (Figs. 33, 36, 48). In Palpigradida (Fig. 82,*2d*) they are pediform and are used for walking, hence the name of the order in general use at present. In Araneida, they are usually much smaller than the legs and are used as sense organs by the young of both sexes and by adult females. Adult male spiders have their organs of copulation on the 6th joint of the pedipalpi. The basal joint, i.e., the coxa of the pedipalp, is usually participating in the process of ingestion, the exception being Scorpionida, Palpigradida and Solpugida, in all of which it has nothing to do with ingestion. In Schizomida, Thelyphonida, Kustarachnida and Ricinuleida, the pair of pedipalpal coxae are firmly grown together in the median line. Their common wall facing the body is deeply concave and covered with short, chitinous, fine spines. The convex **rostrum** (upper lip), covered with similar spines, fills out the concavity, the spines of the rostrum forming together with the spines of the coxae a perfect filtering apparatus. Into this concave filtering space (Fig. 84,*2*), called **"camarostome,"** the true mouth opens. In Araneida the pedipalpal coxae harbor so-called maxillary glands which produce digestive enzymes. For their accommodation these coxae are provided with maxillary lobes, conveniently termed **maxillae,** and either individual openings for each separate gland or a perforated disc on which all ducts open. The 6th, i.e., the terminal joint of the pedipalpi, is in many cases provided with a claw and, as a unique exception, is subdivided in Palpigradida, giving the impression that their pedipalp is composed of more than 6 segments.

The **legs,** of which there are never more than 4 pairs, are typically 7-jointed, but subsegmentation is common and this gives rise to some terminological confusion. In Araneida, 7 segments is the rule. They are called **coxa, trochanter, femur, patella, tibia,**

metatarsus and **tarsus.** Only the tarsus is subsegmented in a few species with slender and long legs. The trochanter serves as a pivot for the motion of the leg in any direction, thus physiologically corresponding to the hip joint of a vertebrate. The articulation between the femur and the patella is such that only flexion and extension are possible. This articulation corresponds physiologically to our knee. In some orders the coxa, trochanter and metatarsus may be subdivided into 2 segments each. The patella may be permanently fused with the tibia, leaving no trace of its existence. This composite segment is then called tibia. True homologies of the segments in the legs of various orders of Arachnida can be established only by a study of the type of their articulation, attachment of individual muscles and planes of possible motion. The tarsus is usually provided at its end with from 1 to 3 movable **claws**, which are either articulated directly with the tarsus or else with a terminal subsegment of the latter, called **onychium** or **posttarsus.** In legs which are modified as tactile organs, as in the case of the 1st pair of legs in Thelyphonida and especially in Phrynichida, claws are wanting. In legs of Arachnida accustomed to climbing on smooth surfaces, only 2 claws are present, but they are accompanied by some organ comparable to a sucker and called **arolium,** or by tufts of flattened hairs which function as a sucker owing to their peculiar structure permitting such close apposition to each other that the wall of the cup produced by them is as impermeable to air as if it were formed by a continuous membrane. The legs are often clothed with hair of various kinds, simple, clavate, plumose, etc. Movable spines are also present in many cases and their arrangement is often of systematic value. Thus, the spider family Mimetidae can be at once recognized by a row of special spines on their 1st and 2nd tibia and metatarsus, (Fig. 101,*3b*), while the most characteristic feature of the family Theridiidae is the presence of the so-called **comb** (Fig. 105,*5*) formed by a row of serrated bristles on the 4th tarsi.

ABDOMINAL APPENDAGES

Abdominal appendages are present only in scorpions and spiders, in the former in the shape of combs or pectines on the 3rd sternite (Figs. 37, 47), in spiders as **spinnerets** (mamillae) on that portion of the abdominal wall which belongs to the 4th and 5th segments.

ECDYSIS

In molting or ecdysis, not only the entire chitinous external skeleton is shed, but the chitinous lining of the book lungs, tracheal tubes, fore-gut and hind-gut is completely withdrawn from the inside of the corresponding organs. Such exuviae are common in amber, but some have been found even in the Carboniferous.

INTERNAL ANATOMY

Because internal organs are rarely preserved in fossil Arachnida they are here discussed only briefly. More detailed information may be found in various books listed at the end of this article.

DIGESTIVE SYSTEM

The digestive system follows more or less the same plan in all orders, but presents many interesting variations. Thus, in Araneida the filtering apparatus is located in the pharynx; in Scorpionida the filtering is accomplished in the channels formed by the adjoining maxillary lobes of the 1st and 2nd pairs of legs; in Schizomida, Thelyphonida and Ricinuleida by the camarostome. Phalangiida have no filter comparable to that in other Arachnida and are capable of ingesting larger particles. The esophagus is more or less similar in all orders. But a pumping gizzard is highly developed in several orders and very poorly in other orders. The mid-gut combines the functions of an organ producing enzymes and resorbing liquefied food. Usually it forms a system of blind, tubular sacs in the cephalothorax, for which I proposed the now generally accepted term **thoracenteron,** and a still more complicated system of branched ramifications in the abdomen. The abdominal system of branches used to be called "liver" in the past, but on my initiative is now called **chylenteron.** A large stercoral pouch in which the excrements are formed and temporarily stored, is usually present at the end of the median tube of the mid-gut, dorsal to it. The hind-gut is always short. The anus is at the end of the abdomen, ventral. In several orders a round anal operculum is present. In fossils, a clear impression of it is usually visible on the last tergite as well as on the last sternite, a circumstance which must be remembered when only one surface of the fossil is preserved. In many orders an operculum is lacking and the anus is closed by the contraction of a sphincter.

EXCRETORY SYSTEM

The excretory system consists of a pair of coxal glands which are modified nephridia, and a pair of malpighian tubes. The former are in the cephalothorax, the latter in the abdomen. The coxal glands open by 1 or 2 pairs of openings situated behind the coxae. The malpighian tubes open into the digestive system at the end of the mid-gut. The white crystals in the excreta of Arachnida are guanin.

CIRCULATORY SYSTEM

The circulatory system of arachnids may be very complex, as in scorpions and spiders, or subject to considerable reduction and even complete disappearance in some Acarida. Even in its highest development it is an "open" system, but the channels for the return of the blood to the heart are always predetermined. The heart, dorsal to the mid-gut in position, is restricted to the abdomen, except in Thelyphonida and Solpugida, in which a portion of it is in the cephalothorax. In fossil Arachnida, the heart has been found in only 2 specimens, one an amber spider, the other a Pennsylvanian architarbid. The blood in all Recent arachnids is propelled forward through the anterior aorta and backward through 3 posterior arteries. In scorpions the heart has 7 pairs of lateral ostia for the return of the blood; in spiders from maximum 5 pairs in Liphistiidae to 2 pairs in Quadrostiati and Apneumonina. All blood corpuscles are of the leucocyte type. The blood plasma is of a semitranslucent white color, turning rap-

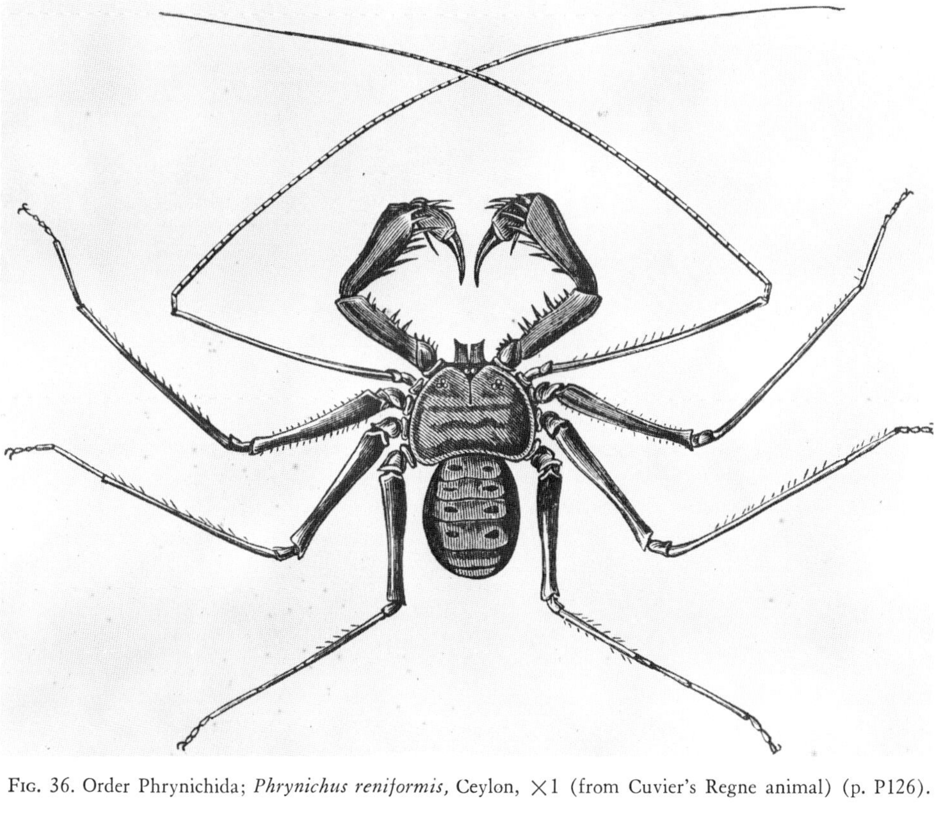

Fig. 36. Order Phrynichida; *Phrynichus reniformis*, Ceylon, ×1 (from Cuvier's Regne animal) (p. P126).

idly blue at exposure to air. The respiratory pigment of the blood is hemocyanin.

RESPIRATORY SYSTEM

The respiratory system consists of book lungs, tracheal tubes, or a combination of both types. It is lacking in many Acarida; in Araneida both types are present in the Dipneumonina, only book lungs in the Liphistiina, Theraphosina and Hypochilina, and only tracheal tubes in the Apneumonina. In all other orders either book lungs or tracheal tubes are present, but never both in the same order. Ventral evertible sacs, supposed to have respiratory function, are present in Palpigradida, some Phrynichida and Architarbida. The number and position of the respiratory stigmata varies from order to order, but is always a fixed one for each order.

NERVOUS SYSTEM

The main plan of the nervous system is remarkably uniform in all arachnids, but its detail structure varies considerably. The so-called deutocerebrum is always lacking, as in all Chelicerata. The chelicerae receive their nerves from the tritocerebrum, the pedipalpi from the 1st ventral pair of ganglia. The thoracic subesophageal ganglionic mass with a distinct pair of ganglia for each pair of thoracic appendages usually includes also several pairs of distinct abdominal ganglia, which have been drawn into the thorax from the abdomen during the embryological development by a process of contraction of the longitudinal connectives. The 1st pair of abdominal ganglia incorporated in the thoracic mass remains distinct even when the 1st abdominal segment is completely lost. In some orders, as for example in scorpions and whip scorpions, some of the abdominal ganglia remain in the abdomen even when they move forward out of their original position. In other orders, such as spiders, all abdominal ganglia are drawn into the cephalothorax in development and only their nerves remain in the abdomen. In all cases the maximum number of abdominal ganglia is 12, but with the loss of posterior segments the number of remaining ganglia may be smaller. A sympathetic nervous system has been described in several orders of Arachnida.

SENSE ORGANS

Only eyes, tactile and chemoreceptive organs are more or less well known. Although some arachnids produce sounds by organs of stridulation, no organs of hearing have yet been discovered. The eyes are always on the carapace and their number varies even within the same order. The maximum number of 6 pairs is found only in scorpions. Although presenting many differences in the structure of their lens, iris, vitreous body and retina, all arachnid eyes are of the type of simple ocelli. The corneal lens produced by the vitreous body may be disc-shaped in a transverse section, ellipsoidal or even triangular, with a flat or convex external surface. Some eyes have a retina with a tapetum which reflects light, making them luminous when a ray strikes them at night.

Tactile organs represented by sensory hairs of various types, are situated on the pedipalpi and legs. So-called trichobothria, at one time believed to be organs of hearing, are now known to be sensitive only to very slow motion of air or very light touch. They are very thin hairs arising from the center of a small membranous disc. They are more numerous in "hairy" species than in those devoid of hair. They are present, however, in all arachnids. The most common chemoreceptors are so-called slit organs found on various parts of the body as well as on the appendages. In spiders they are called lyriform organs and are found on the pedipalpi and legs close to the articulations. There are also various sense organs of unknown function, among them the combs of scorpions and the racket organs of solpugids (Fig. 112,*2b*).

REPRODUCTIVE SYSTEM

The reproductive system presents many variations. In some orders organs of copulation are wanting, in others they are well developed and peculiar to the different orders. Thus, the male copulatory organs of spiders are situated on the terminal segment of the pedipalpi; in Ricinuleida on the metatarsi of the 3rd pair of legs. Male Phalangiida have a penis on their 2nd abdominal sternite, their females possess an ovipositor in the corresponding place. In Acarida a penis is also present, but its position varies greatly. In scorpions a pair of so-called paraxial organs, connected with the vasa deferentia and capable of slight protrusion through the genital opening between the opercula, serve as organs of copulation. Correlated with the differences of structure of the reproductive organs the mating instincts also present great differences and reach the maximum of complexity in spiders.

GLOSSARY OF MORPHOLOGICAL TERMS

The following alphabetically arranged glossary of morphological terms applied to arachnids and used by specialists in the description of species and in keys for their identification, is designed to be of help to those who are not familiar with them. No distinction is made between different terms as to their systematic value because such distinction would be misleading. Alternative terms, used by different arachnologists but synonymous inasmuch as they refer to the same or homologous structures, are given in italics and their synonyms indicated. Obsolete terms, now out of use but common in publications of the past century, are marked with an asterisk (*).

abdomen. Complete portion of body following upon cephalothorax.

**antennae.* Obsolete term for chelicerae, used in the past to emphasize their homology with the antennae of crustaceans and insects.

apodeme. Invagination of the body wall for attachment of muscles.

arolium. Trumpet-shaped membranous structure at the end of tarsi in Pseudoscorpionida and some Phrynichida.

basitarsus. Leg segment preceding tarsus (protarsus, metatarsus).

beak. Hollow tube with mouth opening at its end in Solpugida (rostrum).

boss. Swelling at base of chelicerae for accommodation of muscles in spiders (condyle).

calamistrum. Row of special bristles on 4th metatarsi of spiders possessing a cribellum; used for drawing out a band of special silk from the latter (Fig. 103,*1b,2c*).

camarostome. Concave space formed by a depression in the common wall of fused pedipalpal coxae and the convex rostrum fitting into it; serves as filter of liquefied food before it reaches the mouth (found in Schizomida, Thelyphonida and Ricinuleida).

**capitulum.* Anterior portion of body in Acarida bearing mouth parts (gnathosoma).

carapace. Dorsal covering of cephalothorax in all arachnids, formed by fusion of cephalic and thoracic tergites; called entire when forming a single plate; when composed of several plates (as in Schizomida, Palpigradida and Solpugida) called segmented (peltidium).

caruncle. Sucker-like membrane at the end of tarsi in many Acarida.

cephalothoracic sulci. Pair of grooves separating the cephalic from thoracic portion of the carapace.

cephalothorax. Anterior portion of the body bearing 6 pairs of appendages (prosoma).

chela. Pincers of an appendage formed by a rigid process of the penultimate joint and a movable last joint.

chelicerae. Preoral appendages of all Chelicerata, corresponding to 2nd antennae of Crustacea, but modified for biting or piercing; composed of 3 or 2 segments (Figs. 70,*1d;* 82,*2c;* 112,*2c*).

chylenteron. Abdominal portion of enzyme-producing digestive diverticles of mid-gut.

clypeus. Portion of carapace between its anterior edge and the eyes.

comb. Term applied to 2 totally different structures. In Araneida, a row of serrated bristles on the 4th tarsi, found only in the family Theridiidae. In scorpions, a pair of abdominal appendages situated on the sternite following upon the genital opercula, present in all scorpions, but not in any other arachnids.

coxa. Basal segment of all cephalothoracic appendages, but name rarely used in the case of the chelicerae.

coxal glands. Modified nephridia situated in cephalothorax and opening to exterior by 1 or 2 pairs of lateral orifices behind the coxae.

cribellum. Single or paired perforated plate in a small group of spiders, corresponding to the anteromedial spinnerets of Liphistiina and serving as outlet for special silk glands.

cucullus. Term properly applied only to the movable plate, articulated to the front of the carapace in Ricinuleida; erroneously applied to part of carapace, anterior to the eyes in Pseudoscorpionida.

deutonymph. Second developmental stage in Acarida.

diaxial. Type of chelicerae in Araneida, in which the fangs move more or less in and out in a plane transverse to the longitudinal plane of symmetry of the body (labidognathous).

doublure. Ventral fold of anterior edge of carapace, present only in Thelyphonida.

drum. Small disclike sense organ of unknown function, situated on dorsal surface of tarsi in Araneida ("tarsal organ" of BLUMENTHAL).

empodium. Modified structure at end of tarsus in Acarida, taking place of the 3rd claw (pulvillus).

epigynum. Chitinous external structure in front of genital opening in female spiders of the suborders Dipneumonina and Apneumonina.

femur. Third segment of leg, forming the "hip" articulation with the preceding and the "knee" articulation with the following segment.

finger, immobile. Anterior, finger-like, rigid process of a chela.

finger, movable. Finger-like, movable last segment of a chela.

galea. Spinning tube on movable finger of chelicerae in Pseudoscorpionida.

gizzard. Last portion of the fore-gut developed as a pumping organ; its dorsal dilator muscle is attached to an apodeme visible on external surface of carapace and called thoracic groove.

gnathosoma. Anterior portion of body in Acarida, bearing the mouth parts; formerly called capitulum.

hand. Common appellation of pedipalpal chela; in restricted sense, the main part of the segment less its immobile finger, comparable to human palm.

hysterosoma. Section of body in Acarida, following upon 2nd pair of legs.

labidognathous. Alternative term for diaxial.

lorum pediculi. Dorsal sclerite of pedicel in Araneida.

malleoli. Racket organs of Solpugida on ventral surface of 4th pair of legs.

**mandible.* Obsolete term used in the past by some arachnologists for chelicerae and by others for pedipalpal coxae.

maxilla. Common name for coxa of pedipalp.

mesopeltidium. Single, more commonly paired, sclerites in arachnids with segmented carapace, situated immediately behind propeltidium.

metapeltidium. Usually single, rarely paired, last sclerite of a segmented carapace, following upon mesopeltidium.

metapodosoma. Section of body in Acarida, bearing the 3rd and 4th pairs of legs.

metatarsus. Typically 6th segment of a leg, following upon tibia and preceding tarsus (basitarsus, protarsus).

notogaster. Dorsal shield of hysterosoma in Acarida.

ocellus. Simple eye, as distinct from compound or facetted eye; only type of eyes found in arachnids.

onychium. Distal subsegment of tarsus carrying claws, found in some arachnids, but wanting in others (posttarsus).

operculum. Lid closing an opening, such as the anus or genital opening; usually disclike and flat, solid or composed of 2 parts, present only in some orders. Genital opercula in scorpions always paired, meeting in a median longitudinal line. Anal operculum of Anthracomartida composed of a larger anterior and a smaller posterior plate meeting in a transverse line.

opisthosoma. Posterior portion of body, following upon the 4th pair of legs (abdomen).

orthognathous. Type of chelicerae with fangs moving in a plane parallel to the plane of symmetry of the body (alternative term for paraxial).

palpi. Term applied either to pedipalpi including pedipalpal coxae, or more properly to the 5 segments following the coxa.

paraxial. Type of chelicerae in which the fangs move in a plane parallel to the plane of symmetry of the body, original, more primitive type (orthognathous, prognathous).

paracymbium. Basal apophysis of terminal segment of pedipalp in male spiders of the families Erigonidae and Linyphiidae.

patella. Fourth segment of a typical leg, following upon and forming the "knee" articulation with the femur; fused with tibia in some orders and then completely losing its identity.

pedicel. Greatly modified 1st segment of abdomen in Caulogastra, reaching its extreme development in spiders.

pedipalpi. Second pair of cephalothoracic appendages, subject to many variations in structure; largest and most conspicuous appendages in scorpions and pseudoscorpions, ending in a powerful chela; stout and conspicuous in whip scorpions and Phrynichida, but ending in a pointed joint; least conspicuous appendages in Architarbida. Terminal segment in male spiders bears at the time of maturity the organs of copulation.

peltidium. Latin name for carapace.

petiolus. Latin name for pedicel.

plagula sternalis. Ventral plate of pedicel, present in some genera of spiders.

postabdomen. Attenuated portion of abdomen in scorpions, composed of 5 segments and a telson modified as a poison gland (tail).

posttarsus. Distal subsegment of tarsus, bearing claws (onychium).

preabdomen. Anterior portion of abdomen in scorpions, composed of 7 segments, considerably wider than the tail.

pretarsus. Rarely used alternative of posttarsus.

prognathous. Alternative term for paraxial.

propeltidium. Anterior sclerite of a segmented carapace.

propodosoma. Section of body in Acarida, bearing the 1st and 2nd pairs of legs.

prosoma. Alternative term for cephalothorax.

protarsus. Alternative term for metatarsus.

proterosoma. Anterior section of body in Acarida, ending behind 2nd pair of legs.

protonymph. First postembryonic stage of Acarida.

pteromorph. Winglike outgrowth of body wall in Acarida.

pulvillus. Special structure at end of tarsus in Acarida (empodium).

quadrangle of eyes (or simply **quadrangle**). Rectangle formed by outer tangent lines to 4 median eyes of spiders.

racket organs. Special sense organs of Solpugida, situated on 4th pair of legs (malleoli).

rostrum. Term applied to totally different structures in different orders of Arachnida—in spiders, upper lip; in Solpugida, tubelike "beak"; in Eophrynidae, anterior spike of carapace, similar to the rostrum of a lobster.

sclerite. Hardened chitinous portion of any part of external skeleton.

scopula. Brush of short hair on any appendage; common on metatarsi and tarsi of many spiders.

serrula. Sawlike chitinous keel on aterior edge of maxillae in spiders; toothed keel on cheliceral fingers in pseudoscorpions.

somite. Body segment.

spinneret. Abdominal appendage of spiders, with spinning tubes at the end; maximum, 4 pairs, in Liphistiina; minimum, 1 pair, in a few genera of 4 different families (mamillae). In pseudoscorpions, a special spinning organ on the movable finger of chelicerae, called galea.

spiracle. Opening of tracheal tube or book lung (stigma).

sternite. Scleritized plate of ventral wall of a segment of the body.

sternum. Scleritized sternal plate of a single thoracic segment or more often of several thoracic segments, as in scorpions and spiders.

stigma. Opening of a book lung, or of a tracheal tube (spiracle).

stylophore. Fused base of chelicerae in some Acarida.

syngnathous. Same as diaxial.

tarsus. Last segment of leg, sometimes subsegmented.

tectopodium. Leaflike extension near base of coxa in Acarida.

telson. Dorsal, postanal extension of body, articulated to last abdominal segment; in scorpions, harbors poison glands and ends in a sting; in whip scorpions, developed as a multijointed flagellum; in Schizomida, rodlike in females, knoblike in males.

tergite. Hardened chitinous plate on the dorsal surface of a body segment.

thoracenteron. Thoracic diverticles of mid-gut.

thoracic groove. External indentation of dorsal cephalothoracic apodeme serving for attachment of dilator muscles of gizzard.

tibia. Fifth segment of typical leg or pedipalp, following upon patella which may be completely fused with it, in which case the resulting segment retains the name tibia.

trichobothrium. Sensory hair arising from the center of a disclike membrane on legs or pedipalpi and serving for perception of currents of air; number and distribution of trichobothria have value in taxonomy.

tritonymph. Third nymphal stage of Acarida.

trochanter. Second segment of pedipalp or leg, so articulated to coxa and femur as to permit motion of the entire leg in any direction; physiologically corresponds to vertebrate hip articulation.

ONTOGENETIC DEVELOPMENT

Most Arachnids are oviparous. Scorpions are ovoviviparous. Normal parthenogenesis occurs in some Acarida. Eggs are usually taken care of by the mother, either enclosed in a cocoon, as in the majority of spiders; or guarded inside a nest, as in Schizomida, until the embryological development is completed; or attached to the underside of a rock or loose bark and left to themselves. In almost all orders, including even scorpions in which development takes place inside the mother's body, the emerging 1st instar is unable to ingest food, because the development of the mid-gut is not completed at that time and the embryonic yolk, serving for nourishment, fills completely the mid-gut, which still lacks direct connection with the fore- and hind-gut. With few exceptions, ingestion of external food begins in the 2nd instar. Sexual maturity is reached after several molts, although in a few exceptional cases development is so abbreviated that adult males emerge from the eggs with fully developed reproductive organs, as has been shown for an Australian and an American species of spiders belonging to the family of orb weavers. In some families of spiders and in scorpions, the 1st instar is spent on the back of the mother. Except in the case of Acarida and Ricinuleida, in which the 1st instar differs from the following ones by having only 3 pairs of legs, the young of all other Arachnida possess 4 pairs of legs already in the 1st instar, though the proportions of body and appendages differ greatly from those in later stages. Because of their inability to ingest external food and lack of the sensation of hunger, young of the same batch of eggs keep together during the 1st instar. Dispersal coincides with the completion of digestion of the embryonic yolk, establishment of direct connection between the 3 sections of the digestive tract and the consequent awakening of the sensation of hunger. Young taken care of by their mother leave her and run away; young which spend their 1st instar in close association with each other, but without parental care, disperse similarly. Gregarious species are known only in a few spiders and some mites. In all carnivorous arachnids dispersal is a prerequisite check to cannibalism, which takes place invariably in confinement. Only in plant-eating mites is cannibalism nonexistent.

HABITAT AND GEOGRAPHICAL DISTRIBUTION

Most arachnids are free-living, terrestrial and carnivorous. Aquatic species occur only in a few genera of spiders and in several families of marine and fresh-water mites. Apart from the question of climate, food supply and extermination by enemies, geographical distribution is determined by the physically possible methods of transportation of eggs, young and adults in a viable condition. The most striking of such methods is the aerial transportation of spiderlings and small adult spiders on gossamer threads through dozens and hundreds of miles. Stronger winds carry to considerable distances dead leaves with attached egg cocoons of spiders and galls filled with mites. Parasitic Acarida depend partly upon the habits of their own nymphs and adults and partly upon the habits of their hosts.

Several orders of arachnids are found only in tropical and subtropical regions, the widest horizontal and vertical distribution being found in spiders and Acarida which exceed other orders in number of species. The smallest orders are Palpigradida, Schizomida and Ricinuleida. Very few species of Recent arachnids are cosmopolitan in their distribution, but many genera are represented in widely separated regions of the world. Thus, in the family Scorpionidae, the genus *Opisthophthalmus* is restricted to South Africa, *Opisthacanthus* is represented in Africa and America, and *Diplocentrus* only in America. Most families of spiders are represented on all continents, but many genera have a restricted distribution as have also 2 or 3 families. On the other hand, of the only 2 genera comprising the very clearly defined family Hypochilidae, the genus *Hypochilus* is found only in America, the genus *Ectatosticta* in China and Tasmania.

GEOLOGICAL RECORD AND EVOLUTION

As a class, the Arachnida were a richer group in the past than at present. Of the 16 known orders of Arachnida, only 11 survived, 5 became extinct in the late Paleozoic. At present the scorpions represent the oldest order because they are the only arachnids so far found in the Silurian. The first representatives of Acarida, Trigonotarbida and Araneida (the latter not quite certain) have been discovered in the Devonian. Presumably on account of their small size, no Acarida have been found in any later formation until the Oligocene (Baltic amber), but of the known 16 orders of Arachnida, 12 are represented in the Carboniferous. The only fossil representative of the order Palpigradida was found in the Jurassic; several genera of the order Pseudoscorpionida in the Oligocene (Baltic amber); and of the order Schizomida in the ?Pliocene (onyx marble). The fossil record of Arachnida is therefore very incomplete, with a big gap between the late Paleozoic and the Oligocene. A single fossil scorpion is known from the Devonian, none from the Permian, 2 genera from the Triassic and none again until the Oligocene. The order Phalangiida is represented by only 3 genera in the Carboniferous and none in later formations until the Oligocene. (Some have been reported and figured in the Lebanon Cretaceous, but not yet described.) The order Araneida, except for the doubtful representative described from the Devonian, is represented only in the Carboniferous and the Oligocene. But in their case we know at least definitely that fossil spiders have been finally found in the Cretaceous, because 3 collections of them have been placed in my hands for study.

Of the 5 extinct orders of Arachnida, Trigonotarbida are represented in the Devonian and were common in the Carboniferous, but have never been found in any later formation. The other 4 orders, namely Anthracomartida, Haptopodida, Kustarachnida and Architarbida have been found only in the Carboniferous. These 4 orders, while in some regards of great importance for understanding of the evolution of the orders of Arachnida, are themselves so distinct from all other orders that they can be regarded only as end products of macroevolution and not as intermediate stages. The unraveling of the process of evolution in Arachnida has to depend therefore on the tracing of the direction in evolutionary trends common to all of them and appearing also in the ontogeny of living representatives. In this respect, HAECKEL's biogenetic law, postulating that ontogeny is a recapitulation of phylogeny, is of great help, for it prevents making the mistake of assuming a reverse direction in certain trends. At the same time one must be on guard against the possibility that a trend which persisted through so many millennia and is still operative, may have remained latent at times and active at other times, being continuous in direction, yet discontinuous in operation, and therefore not necessarily synchronous with geological sequence.

Keeping these principles in mind we will not be astonished in finding that, for example, Recent scorpions do not represent the end of their macroevolutionary trend, consisting in a gradual loss of posterior thoracic sternites, but a middle stage which proved to be viable, while the extremes became extinct in the Carboniferous. Whether Recent scorpions still possess this trend in a latent stage or have completely lost it, we do not know and nothing but an experiment or a lucky monstrosity could give an answer to this question. But one fact remains unquestionable and that is that with the appearance of the first representatives of the Carboniferous superfamily Scorpionoidea this trend was arrested and all further morphological changes were of a smaller and different type.

Something similar may be observed in the displacement of the mouth opening backward from its original position at the end of the 1st cephalic segment. That process, as in all trends, is unidirectional. It reaches its maximum in Xiphosura and Stethostomata (among the Arachnida), but was arrested in all other Arachnida at a stage when the mouth reached the 1st postcheliceral segment, i.e., the pedipalpal segment. It is in that position in the Silurian protoscorpions, as well as in all Recent scorpions and other arachnids. In this single respect the Paleozoic Stethostomata are more advanced than now-living arachnids.

The most conspicuous trend, common to all Arthropoda and clearly operative in all Arachnida, is the one involving fusion of anterior abdominal segments with the cephalothorax and the gradual loss of posterior segments. Perhaps it would be more appropriate to treat them as 2 different trends, for they proceed in the opposite direction and are both, of course, irreversible. But the mechanics of both are essentially the same and consist either in a shortening of a segment without changing its transverse area or in a reduction of its transverse area by peripheral constriction without changing its length, until the segment disappears completely. At the anterior end of the abdomen, the 1st segment is the first to be affected. It disappears completely in scorpions, harvestmen and mites by the first method and in Ricinuleida by the second, i.e., by peripheral constriction. But in both cases, the 1st abdominal neuromere remains distinct and the same applies to the neuromeres of the following segments when they become fused with the cephalothorax, as in some harvestmen and mites.

Both methods of disappearance are also at work at the posterior end of the abdomen. But in all cases the last segment is the first to go, then the penultimate, then the antepenultimate and so on, the maximum involving 7 posterior segments in some spiders. In scorpions, 5 segments were involved, but, instead of being lost, they became attenuated and permanently retained in the form of a "tail." The gradual attenuation of these segments can be easily seen in the embryonic stages of any Recent scorpion and undoubtedly underwent a similar change in the Ordovician Eurypterida. In Phrynichida only the 12th segment became attenuated and remains as a single-jointed pygidium. In Thelyphonida and Ricinuleida, the 3 last segments underwent similar attenuation and these orders are characterized by the possession of a 3-jointed pygidium. In spiders one can see 12 abdominal segments in the early embryo and follow the attenuation of posterior segments forming an "evanescent" tail similar to that of scorpions, but then gradually lost, leaving only 1 to 3 anteriormost tergites which become the so-called anal tubercle of the adult spider. Holm, who gave an excellent description of this process, was misled by the similarity between the araneid evanescent postabdomen and that of scorpions and, assuming that scorpions are primitive arachnids in every respect because they were present in the Silurian, reached the erroneous conclusion that the spider *Segestria* is more primitive than a scorpion inasmuch as the scorpion has 5 postabdominal segments, whereas *Segestria* has 7. The reverse is, of course, the only possible interpretation. Scorpions are in several respects less primitive than some other Arachnida, as, for example, Palpigradida and segmented spiders of the suborder Liphistiina. In other respects they exhibit more primitive characters as, for example, in the presence of 7 pairs of cardiac ostia and 4 pairs of book lungs. Even *Liphistius* has only 5 pairs of

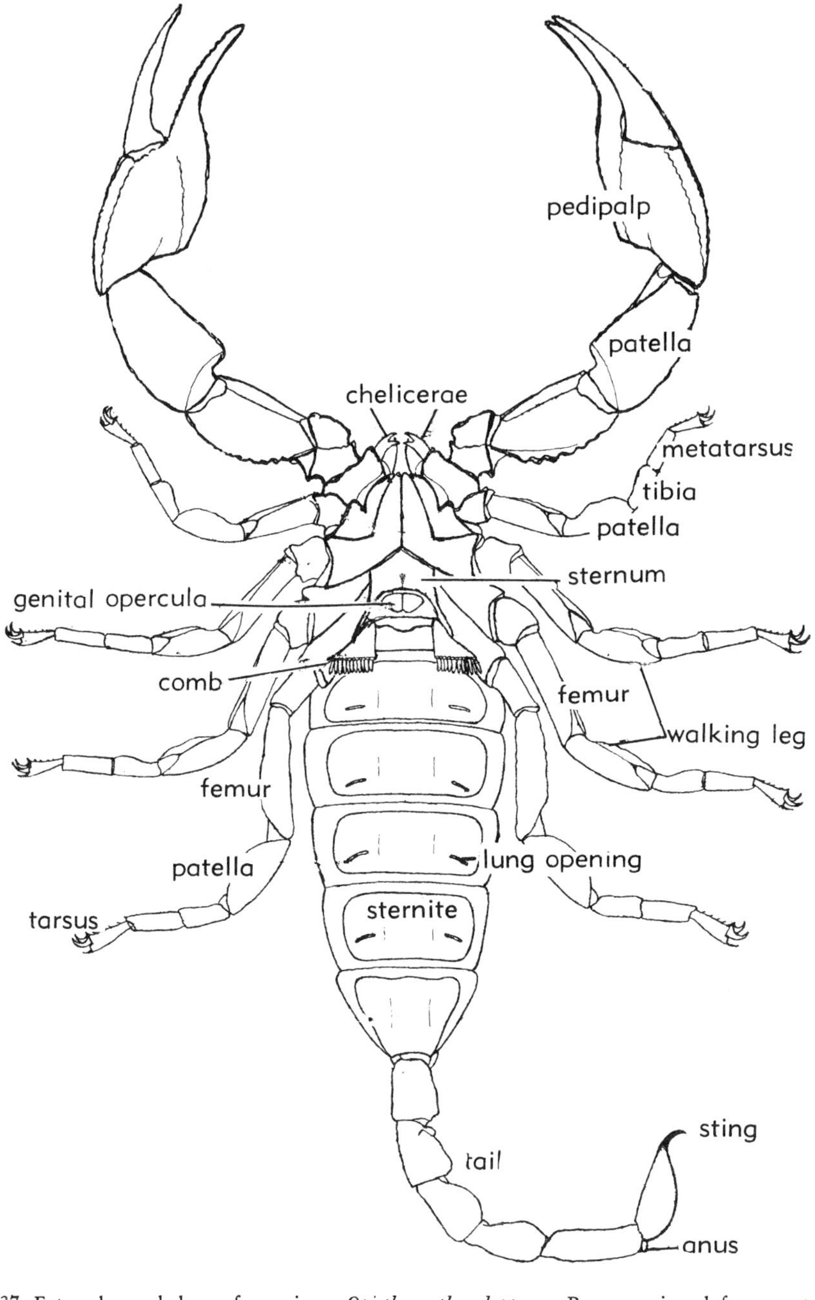

FIG. 37. External morphology of scorpions; *Opisthacanthus lepturus,* Panama, viewed from ventral side, ×2 (76).

ostia and 2 pairs of book lungs, while *Segestria* has 2 pairs of ostia and is in some respects one of the most advanced spiders.

The loss of segments may appear at first sight to be a trend involving only loss and not an increase in complexity of structure, which one associates with the idea of progressive evolution. But one must remember that Arthropoda developed from more primitive segmented animals, Recent relatives of which are the Annelida. In Annelida the segmentation of the body is still anomomeristic, the number of segments in many species increasing with age. The segregation of segments into tagmata of a definite number of segments was accomplished at a loss of segments and of the ability to increase the number of segments during life, but resulted in greater compactness of the body and in improved functional efficiency of its entire mechanism. This is certainly progressive evolution.

The tempo of evolution in Arachnida varies considerably from order to order. The most stable order seems to be the Thelyphonida. The Carboniferous *Prothelyphonus* and *Geralinura* present the same fundamental structures as the Recent genera. The macroevolution of the scorpions began in the Silurian and came to its end in the Carboniferous. Except for a belated straggler in the Triassic, all Recent scorpions can be readily traced to a single superfamily produced in the Carboniferous, along with several other, extinct superfamilies. Later formation of scorpion families involved only minor modifications. In Ricinuleida, the Recent family Ricinoididae has a different arrangement of coxae compared with the Carboniferous family Poliocheridae and produced by a trend similar to that which was responsible for the formation of Carboniferous superfamilies. But the paucity of material makes it impossible to say whether the family Ricinoididae was only recently produced or had its origin in the Carboniferous. The spiders began to split up into superfamilies already in the Carboniferous, but for some reason not yet clear continued to produce new families at least in the Oligocene. Only the Acarida exceeded the spiders in the lability of their organization and produced more than twice as many families for about a quarter of as many species.

One of the most interesting problems of relationship, and therefore of evolution, is presented by the harvestmen. This order Phalangiida is always divided into 3 suborders—Cyphophthalmina, Palpatorina and Laniatorina—on the basis of external characters which are clearly visible, but not in themselves of fundamental importance. However, important anatomical differences exist and, as Hansen & Sørensen have pointed out, Cyphophthalmina differ in many respects from the other 2 suborders, which have more features in common with each other. On the other hand Cyphophthalmina resemble Architarbida in many respects and it is not impossible that if we knew the internal anatomy of the latter, a still closer relationship between them could be established. Points of similarity between the Cyphophthalmina and some Acarida have also been discussed by various authors and it seems possible that Architarbida, Cyphophthalmina and Acarida represent 3 branches of a single ancestral group. For the present it is safer to consider them as separate orders.

Subclass LATIGASTRA Petrunkevitch, 1949

[Type: *Scorpio* Linné, 1758]

Arachnida with broad juncture between cephalothorax and abdomen. Basal joint of chelicerae anterior to pedipalpal coxae. Abdomen with distinct trend toward shortening of anterior segments, loss of 1st segment in Scorpionida and Phalangiida and fusion with cephalothorax in Phalangiida and Acarida. *Sil.-Rec.*

Order SCORPIONIDA Latreille, 1817

[*nom. correct.* Pearse, 1936 (*pro* Scorpionides Latreille, 1817)] [=Scorpiones Hemprich & Ehrenberg, 1826; Scorpiides C.L.Koch, 1837; Scorpiones *auctt.*] [Type: *Scorpio* Linné, 1758]

Abdomen segmented, clearly divided into a stout preabdomen (composed of 8 segments in Protoscorpionina, 7 segments in Euscorpionina) and a more slender postabdomen or "tail" composed in both suborders of 5 segments. At end of tail a dorsal telson, which has the shape of a bulb drawn out into a more or less curved sting and harbors within the bulb a pair of poison glands (Figs. 33, 37). Carapace entire, with a pair of median and 2 to 5 pairs of lateral eyes. Chelicerae 3-jointed, chelate (Fig. 44,*6*); their movable finger lateral. Pedipalpi 6-jointed, powerful, chelate; their movable finger ventrolateral (Fig. 37). Legs 7-jointed, with pair of claws and a spur at end, except in *Palaeophonus,* in which tarsi are simply pointed (the spur is often rated as a rudimentary 8th segment, "praetarsus" of BÖRNER, "posttarsus" of VACHON). First and 2nd coxae usually with maxillary glands harbored in maxillary processes. Single sternal plate followed by a pair of genital opercula, behind which are a pair of combs. Four pairs of book lungs with stigmata on 4 sternites following the combs. Anus terminal, at end of "tail." Ovoviviparous. *Sil.-Rec.*

DISCUSSION

Embryology of Recent scorpions shows that the 1st abdominal segment disappears, retaining in the adult only its neuromere, which becomes incorporated into the thoracic ganglionic mass as the 6th pair of its ganglia. The 1st visible tergite and the genital opercula on the 1st visible sternite in even just-born scorpions belong therefore to the 2nd embryonic segment. The original number (12) of abdominal segments is restored in the course of embryological development by subsegmentation of the 8th embryonic segment. The 1st segment does not disappear, but persists in the adult only in Protoscorpionina, which have consequently 8 preabdominal tergites and as many "tail" segments (5) as Recent scorpions. The corresponding 1st sternite is, however, not known in Protoscorpionina, because all 4 specimens, the only ones now known, present their dorsal surface with only some ventral lines impressed on it from below. All preabdominal tergites are connected with their sternites by a soft lateral pleural membrane. The last preabdominal segment is considerably narrowed posteriorly to the width of the "tail." The 5 segments of the latter have the shape of complete polygonal prisms with hard walls, lateral pleural membranes being absent. The intersegmental membranes of the tail are very short and the segments are so articulated that they never separate, unless forcibly torn apart. On the other hand, the tergites and sternites of the preabdomen, when the latter is distended by eggs or by any other cause, appear as individual plates well separated from each other by the soft intersegmental membrane. Granular or smooth keels are usually present on all tail segments and on the last preabdominal tergite.

The configuration of the coxosternal region (Figs. 38,*1c*; 40) is of suprafamilial value and is due to loss of certain thoracic sternites. All Recent families of scorpions have the same configuration of their coxosternal region as the Paleozoic family Eoscorpiidae. The coxa of the pedipalpi has never a maxillary lobe, nor does it harbor any maxillary gland. The size and shape of the pedipalpal "hand" varies greatly within each family and has therefore at best only generic value. The arrangement of granules on the hand, especially on the cutting edge of both fingers, is of a similar systematic value. The legs increase in length from the 1st to the 4th. In the superfamily Scorpionoidea, to which the Recent families also belong, the maxillary lobes of the 1st and 2nd pair of coxae harbor maxillary glands secreting digestive enzymes. The outer walls of the maxillary lobes of the 2nd pair, which are enclosed between the maxillary lobes of the 1st pair, and the medial walls of the latter are covered with small spines. The channel between the lobes functions as a filter allowing only fluid food to enter the mouth, drawn in by the suction of the powerful pumping pharynx. These coxae are independently movable, but the 3rd and 4th pedal coxae are permanently grown together, can move only jointly, and that motion is practically negligible. The proportions of the individual leg segments in Palaeophonidae are very different from those of all other scorpions, each segment being almost as wide as long, and their tarsi end in a point instead of in a pair of claws.

The mouth is situated between the bases of the pedipalpal coxae. In front of it is an anterior lip, behind it a posterior lip. The gizzard is scarcely different from the esophagus. The thoracenteron has only 2 pairs of diverticles, the chylenteron 5 pairs. The hind-gut is short and ends in an anus which is closed by a sphincter. The heart is restricted to the preabdomen and has 7 pairs of ostia. The openings of the coxal glands are at the base of the 3rd pedal coxae. Four abdominal neuromeres are incorporated in the thoracic ganglionic mass, 8 are in the abdomen and of these the 12th is fused with the 11th. The reproductive glands are in both sexes of the fenestrated type. The male has a pair of copulatory organs (so-called paraxial organs, connected with the vasa deferentia) which can be protruded through the genital opening. The latter is closed by a pair of opercula in both sexes. Secondary sexual characters are not conspicuous. Males are often more slender and have a longer tail than the females. The newly born young climb on the back of their mother and stay there until their first skin is shed, when they disperse. The size of adult scorpions varies from 13 to 200 mm.

Suborder PROTOSCORPIONINA Petrunkevitch, 1949

[*nom. correct.* PETR., herein (*pro* Protoscorpiones PETR., 1949)] [Type: *Palaeophonus* THORELL & LINDSTRÖM, 1885]

Scorpions with 1st abdominal segment persisting in the adult and preabdomen with 8 tergites (43). *Sil.-U.Carb.*

Superfamily PALAEOPHONOIDEA Thorell & Lindström, 1885

[*nom. transl.* PETR., herein (*ex* Palaeophonoidae THORELL-L., 1885)] [=Apoxypodes THORELL-L., 1885 (suborder)]

Legs with segments almost as wide as or even wider than long. Tarsi terminating in a point, without claws. *Sil.*

Family PALAEOPHONIDAE Thorell & Lindström, 1885

[*nom correct.* FRITSCH, 1904 (*pro* Palaeophonoidae THORELL-L., 1885); *emend.* PETR., 1953]

First abdominal tergite concealed under carapace, its anterior edge indicated by a transverse furrow. First pair of coxae wide apart, without maxillary lobes (Fig. 38,*1c*). *Sil.*

Palaeophonus THORELL & LINDSTRÖM, 1884 [*nom. correct.* THORELL-L., 1885 (*pro Paleophoneus* LINDSTRÖM, 1884)[1]] [**Palaeophoneus nuncius* THORELL-L., 1884]. Carapace subquadrate, with concave anterior edge. Hand, to base of fingers, almost as wide as long, with strongly convex ventral edge (45). *Sil.*, Eu.—FIG. 38,*1*. **P. nuncius* THORELL-L., Gotl.; *1a*, specimen as preserved in matrix, ×1.9; *1b*, distal half of 2nd leg, showing tibial spur and pointed tarsus, ×7; *1c*, outline of carapace with coxae and genital opercula superposed, ×3.2 (76).—FIGS. 38,*2*; 39*B*. *P. caledonicus* HUNTER, Scot.; 38,*2*, dorsal surface of specimen, showing eyes, ×3; 39*B*, photograph of holotype, ×7 (76*).

[1] Owing to an unfortunate *lapsus calami* the name of the genus was first misspelled with *-eus* instead of *-us* at the end, and the names of its 2 authors printed not immediately following the name of the species, but in a footnote on the same page 985 of the Comptes Rendus Acad. Sci. Paris, 1884, in the abstract of the letter written by LINDSTRÖM to MILNE-EDWARDS, informing him of the important discovery of a Silurian scorpion. That the spelling of the name was a *lapsus calami* and its subsequent correction by the authors themselves not an attempt to *change* the name, but to restore its original spelling is quite clear from the statements of the authors in their joint paper of 1885 (*On a Silurian Scorpion from Gotland:* K.Svensk. Akad., Band 21, No. 9). They say (1885) on p. 7: "Owing to various hindrances no opportunity was found to make this discovery known before November the 12th 1884, when it was announced at the evening meeting of the Royal Swedish Academy of Sciences. A photograph, double the size of the original, was sent abroad to several museums and zoologists, after we had agreed jointly to undertake this memoir and had named the animal *Palaeophonus nuncius*. This new species was first mentioned in print in the 'Comptes Rendus de l'Académie des Sciences,' Paris, 1884, Dec. 1, p. 984, in an article 'Sur un Scorpion du terrain Silurien de Suède.' And (1885) on p. 9 the authors say (in footnote 2): "The name of the genus should be written, as it is here, *Palaeophonus*, not *Palaeophoneus*." The name is spelled correctly in the Glasgow *Herald* of December 19, 1884, and on the original label accompanying the specimen in the Stockholm Museum. It is therefore quite wrong to assign the authorship of the name to LINDSTRÖM alone and to revert to the first, incorrectly spelled name, merely on the purely formal basis of priority of its publication and in disregard of the authors' own explanation, as the Editor of the *Treatise* suggested in his correspondence with me concerning this subject. LINDSTRÖM himself acknowledged the joint authorship in the above letter; and neither he, nor THORELL, nor anybody else has ever used the wrong spelling.

(Prof. PETRUNKEVITCH's foregoing note presents adequate statement of THORELL & LINDSTRÖM's intentions, making clear at the same time that evidence of a *lapsus* is not contained in the original publication of 1884. Accordingly, the only legal way to validate the universally accepted spelling, which is that here used, is through application to ICZN to place *Palaeophonus* THORELL & LINDSTRÖM, 1884, on the "Official List of Generic Names in Zoology" and to place the name *nuncius* THORELL & LINDSTRÖM, 1884, as published in the combination *Palaeophonus nuncius*, on the "Official List of Specific Names in Zoology"; also to place the name *Palaeophoneus* THORELL & LINDSTRÖM, 1884, on the "Official List of Rejected and Invalid Generic Names in Zoology" and to place the name *nuncius* THORELL & LINDSTRÖM, 1884, as published in the combination *Palaeophoneus nuncius*, on the "Official List of Rejected and Invalid Specific Names in Zoology." If this seems cumbersome, it is only needful to point out that until this is done, under the Rules nomenclatural instability relating to these names will continue to exist, being not removable by lengthened years of usage or expressions of individual opinion.—EDITOR)

(Subsequently to notes given above, Prof. PETRUNKEVITCH has been informed by Mr. FRANCIS HEMMING, Secretary of ICZN, that action recommended by the Editor should be taken; hence application has been made to designate *Palaeophonus* THORELL & LINDSTRÖM and *Palaeophonus nuncius* THORELL & LINDSTRÖM as *nomina conservanda* by use of the plenary powers of ICZN.—EDITOR)

Superfamily MAZONIOIDEA Petrunkevitch, 1913

[*nom. transl.* PETR., herein (*ex* Mazoniidae PETR., 1913)]

Legs with segments distinctly longer than wide, their proportions similar to those of Euscorpionina. Tarsi not preserved, but presumably ending in a pair of claws (38). *Sil.-U.Carb.*

Family DOLICHOPHONIDAE Petrunkevitch, 1953

First abdominal tergite concealed under carapace. Coxosternal region and armature of legs unknown. *Sil.*

Dolichophonus PETRUNKEVITCH, 1949 [**Palaeophonus loudonensis* LAURIE, 1889]. Carapace much longer than wide, with concave anterior edge. Eyes lacking. Hand longer than fingers (43,45). *Sil.(Wenlock.),* Scot.—FIG. 38,*4*. **D. loudonensis* (LAURIE); dorsal surface of holotype, ×1 (76).

Proscorpius WHITFIELD, 1885 [**Palaeophonus osborni* WHITF., 1885]. Carapace as wide as long, convex in front, with flat ocular tubercle, but no eyes. Hand shorter than fingers (45). *U.Sil.,* N. Am.—FIGS. 38,*3*; 39*A*. **P. osborni* (WHITF.), N.Y.; 38,*3*, dorsal surface, ×2.3; 39*A*, photograph of holotype ×2.8 (76).

Family MAZONIIDAE Petrunkevitch, 1913

[Correctly spelled on p. 35; misspelled Mazonidae on p. 54]

First abdominal tergite fully visible. *Penn.*

Mazonia MEEK & WORTHEN, 1868 [**M. woodiana*]. Carapace about as long as wide, with a pointed projection in front and a pair of eyes on an oval tubercle close behind anterior end (36). *Penn.,* N.Am.—FIG. 38,*5*. **M. woodiana,* Ill.; ×1 (76).

Suborder EUSCORPIONINA Petrunkevitch, 1949

[*nom. correct.* PETR., herein (*pro* Euscorpiones PETR., 1949] [Type: *Scorpio* LINNÉ, 1758] [=Dionychopodes THORELL, 1885 (rejected because used in different arachnid orders at various times)]

Scorpions in which the 1st abdominal segment, present in the early embryo, disappears in the course of development and becomes reduced to its neuromere incorporated in the thoracic ganglionic mass as its 6th pair of ganglia. First tergite of even 1st instar is really that of 2nd embryonic segment. Leg segments distinctly longer than wide, tarsi with a pair of claws. First coxae always with maxillary lobes, 2nd coxae with maxillary lobes in all Recent and most fossil genera, but lacking in some fossil ones. Maxillary lobes of 2nd pair contiguous along median line and wedged in between those of 1st coxae (45). *Dev.-Rec.*

DISCUSSION

The subdivision of Euscorpionina into families, as recently proposed by me and used here for subdivision into superfamilies, is based on differences in the configuration of the coxosternal region. These differences are caused by the disappearance of a portion of the sternum and the shifting into the vacated place of 1 or 2 pairs of coxae. Only in the case of the Palaeoscorpioidea a different character had to be used for separation from the other superfamilies because their coxosternal region is unknown. In the Archaeoctonoidea (Fig. 40,*2*) the anterior quarter of the sternum disappeared and the 1st coxae occupy the vacant place, while the next 3 pairs are in their original position. In the Scorpionoidea (Fig. 40,*1*) the sternum lost its anterior half and the 1st and 2nd pair of coxae occupy the vacant place. In the Mesophonoidea (Fig. 40,*6*) the arrangement of the coxae is essentially the same as in the Scorpionoidea, but only the 1st coxae have maxillary lobes. In the Cyclophthalmoidea (Fig. 40,*3*) the sternum lost its anterior ¾, the vacated space is occupied by 3 pairs of coxae and only the 4th pair is in its original position. In the Isobuthoidea (Fig. 40,*5*) the space vacated by the disappearance of the anterior ¾ of the sternum is occupied by the 2 anterior pairs of coxae and by the last quarter of the sternum, which moved forward, followed by the genital opercula which occupied its place. In the Centromachoidea (Fig. 40,*4*) the remnant of the sternum and the genital opercula moved still one segment further, with the result that the 3rd and 4th pairs of coxae are abutting against the genital opercula.

Superfamily PALAEOSCORPIOIDEA Lehmann, 1944

[*nom. transl.* PETRUNKEVITCH, herein (*ex* Palaeoscorpiidae, *nom. correct.* PETR., herein, *pro* Palaeoscorpionidae LEHMANN, 1944)]

Preabdomen with median longitudinal fold continuous with tail and resembling it. *Dev.*

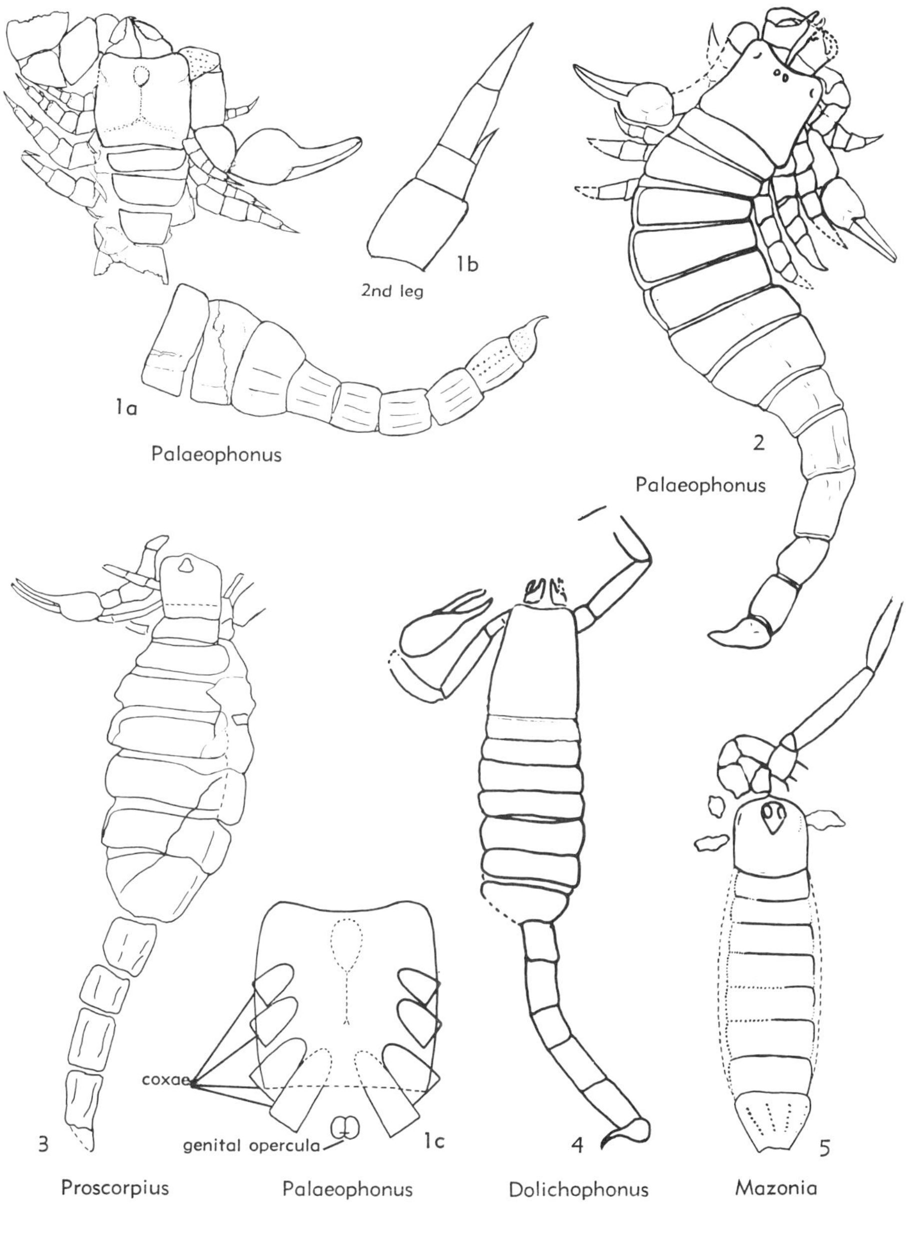

FIG. 38. Order Scorpionida, suborder Protoscorpionina (76). *1, Palaeophonus nuncius,* Sil., Swed.; *1a,* ×1.9; *1b,* 2nd leg showing pointed tarsus, ×7; *1c,* coxosternal region with outline of carapace, ×3.2. *2, Palaeophonus caledonicus,* Sil., Eng.; ×3. *3, Protoscorpius osborni,* Sil., N.Y.; ×2.3. *4, Dolichophonus loudonensis,* Sil., Scot.; ×1.5, *Mazonia woodiana,* Penn., Ill.; ×1 (p. P69-P70).

FIG. 39. Order Scorpionida, suborder Protoscorpionina. *A, Proscorpius osborni,* Sil., N.Y.; ×2.8. *B, Palaeophonus caledonicus,* Sil., Scot.; ×7. Photographs of holotypes (*76) (p. P69-P70).

Family PALAEOSCORPIIDAE Lehmann, 1944

[*nom. correct.* PETRUNKEVITCH, herein (*pro* Palaeoscorpionidae LEHMANN, 1944)]

Median eyes situated behind middle of carapace, lateral eyes not known. Tarsi with 2 claws. Coxosternal region unknown (23, 45). *Dev.*

Palaeoscorpius LEHMANN, 1944 [**P. devonicus*] (23). *Dev.*, Ger.

Superfamily ARCHAEOCTONOIDEA Petrunkevitch, 1949

[*nom. transl.* PETR., herein (*ex* Archaeoctonidae PETR., 1949)]

First pair of coxae meeting in median line in front of sternum, the other 3 pairs abutting against sternum. *Carb.*

Family ARCHAEOCTONIDAE Petrunkevitch, 1949

[*emend.* PETR., 1953]

Only 1st pair of coxae with maxillary lobes (43, 45) (Fig. 40,*2*). *Carb.*

Archaeoctonus POCOCK, 1911 [**Eoscorpius glaber* PEACH, 1883]. Sternum with slightly converging sides, truncated at both ends. Carapace as wide in front as behind with a slight anterior median projection, much wider than long. Eyes seemingly wanting. Hand shorter than fingers, which curve toward each other (45). *Carb.*, Eu.—FIG. 41,*1*. **A. glaber* (PEACH), Scot.; ventral view, ×1.4 (76).

Eoctonus PETRUNKEVITCH, 1913 [*emend.* PETR., 1953] [**E. miniatus*]. Carapace wider than long, with distinctly converging sides. Pair of eyes ¼ length of carapace behind anterior edge (34). *Penn.*, N.Am.—FIG. 41,*2*. **E. miniatus*, Ill.; dorsal side, ×4.6 (76).

Superfamily SCORPIONOIDEA Leach, 1815

[*nom. transl.* PETRUNKEVITCH, herein (*ex* "family" Scorpionides LEACH, 1815[1])]

First and 2nd pairs of coxae with well-developed maxillary lobes (Fig. 40,*1*), those of 2nd pair meeting in median line and wedged in between maxillary lobes of 1st pair; 3d and 4th pairs of coxae abutting against sternum (45). *Carb.-Rec.*

[1] The name Scorpionides was proposed by LATREILLE as name for an order. LEACH was the first to use it as a *family* name, with *Scorpio* LINNÉ as type, but like LATREILLE's order, LEACH's family still included pseudoscorpions besides scorpions.

Family EOSCORPIIDAE Scudder, 1884

[*nom. correct.* PETRUNKEVITCH, herein (*pro* Eoscorpioidae SCUDDER, 1884)] [=Eoscorpionidae *auctt.*]

Coxae of 3d and 4th pairs of legs not much longer than those of 1st and 2nd pairs and probably not yet grown together along their line of contact, retaining independent motion. *Carb.*

Eoscorpius MEEK & WORTHEN, 1868 [**E. carbonarius*]. Carapace rectangular, median eyes situated at considerable distance from anterior edge (38). *U.Carb.*, N.Am.-Eu.—FIG. 42. **E. carbonarius*, Ill.; dorsal side, ×1.4 (73).

Alloscorpius PETRUNKEVITCH, 1949 [**Eoscorpius granulosus* PETR., 1913]. Carapace rectangular, median eyes not far behind anterior edge (45). *U.Carb.*, N.Am.-Eu.—FIG. 43,*1*. **A. granulosus* (PETR.), Ill.; dorsal side, ×0.7 (76).

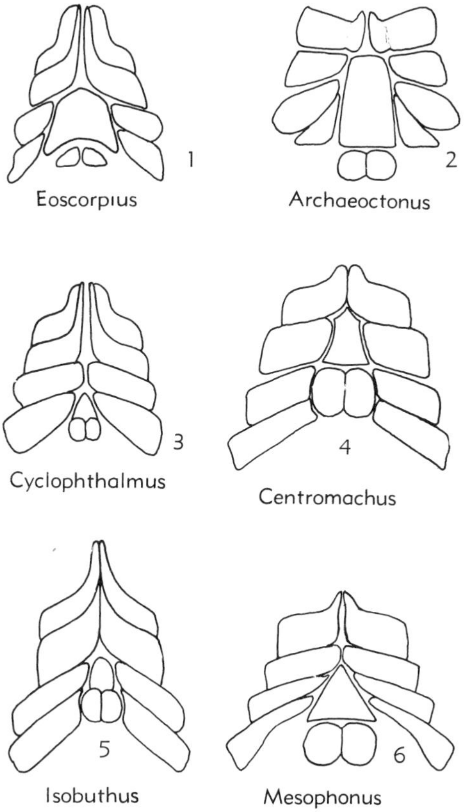

FIG. 40. Coxosternal region of fossil scorpions belonging to 6 families of the suborder Euscorpionina: *1*, Eoscorpiidae; *2*, Archaeoctonidae; *3*, Cyclophthalmidae; *4*, Centromachidae; *5*, Isobuthidae; *6*, Mesophonidae (76) (p. P73-P78).

Trigonoscorpio PETRUNKEVITCH, 1913 [**T. americanus*]. Carapace triangular, with truncated apex; median eyes close to anterior edge. Hand slender, with long fingers. Tail slender (34). *U.Carb.*, N. Am.-Eu.—FIG. 43,*2*. **T. americanus*, Ill.; dorsal side, ×2 (76).

Buthiscorpius PETRUNKEVITCH, 1953 [**Anthracoscorpio buthiformis* POCOCK, 1911]. Carapace with convex converging sides and convex anterior edge, median eyes on tubercle about ⅓ from anterior edge. Fingers considerably longer than hand (45). *U.Carb.*, Eu.—FIG. 43,*3*. **B. buthiformis* (POCOCK), Eng.; carapace, ×6.5 (76).

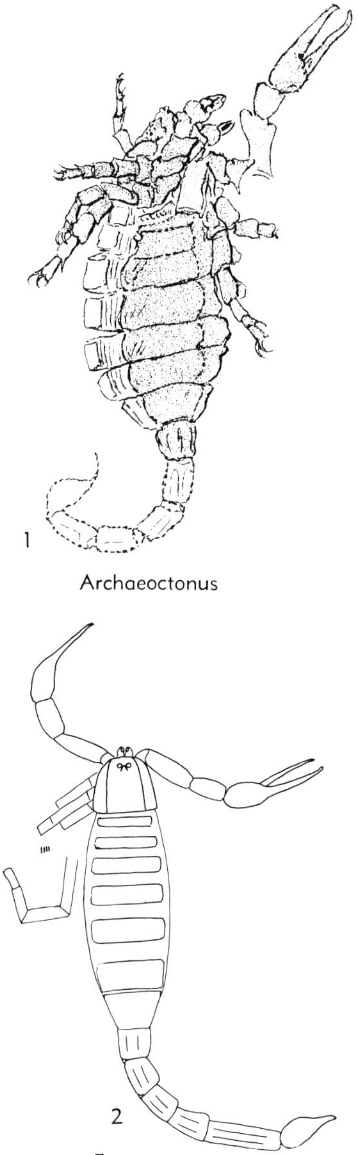

FIG. 41. Family Archaeoctonidae: *1, Archaeoctonus glaber*, U.Carb., Scot., ×1.4; *2, Eoctonus miniatus*, Penn., Ill., ×4.6 (76) (p. P73).

FIG. 42. Family Eoscorpiidae: *Eoscorpius carbonarius*, Penn., Ill., ×1.4 (80) (p. P73).

Anthracoscorpio KUŠTA, 1888 (**Cyclophthalmus juvenis* KUŠTA, 1885] [=*Lichnoscorpius* PETRUNKEVITCH, 1949]. Carapace semicircular, with slightly concave posterior edge; median eyes sessile, situated in advance of middle (45). *Carb.*, Eng.-Czech.—FIG. 43,*4*. **A. juvenis* (KUŠTA), Czech.; carapace, ×20 (76).

Lichnophthalmus PETRUNKEVITCH, 1949 [**L. pulcher*]. Carapace trapezoidal, wider than long, with cephalic portion distinctly elevated and posterior edge margined; eyes ellipsoidal, on tubercle not far from anterior edge (43). *Carb.*, Eng.—FIG. 43,*7*. **L. pulcher;* carapace, ×2.7 (76).

Typhlopisthacanthus PETRUNKEVITCH, 1949 (**Palaeopisthacanthus mazonensis* PETR., 1913]. Tail very short and slender. Carapace much longer than wide, flat, with converging sides. No eyes (43). *Carb.*, N.-Am.-Eu.—FIG. 43,*5*. **P. mazonensis* (PETR.), Ill.; dorsal side, ×2 (76).

Palaeopisthacanthus PETRUNKEVITCH, 1913 [**P. schucherti*]. Tail very short and slender. Carapace wider than long, trapezoidal; median eyes on tubercle ⅓ from anterior edge (38). *Carb.*, N. Am.—FIG. 43,*6*. **P. schucherti*, Ill.; dorsal side, ×2.7 (76).

Compsoscorpius PETRUNKEVITCH, 1949 [**C. elegans*]. Carapace subtriangular, with rounded front; sides and posterior edge margined; median eyes on large longitudinally elliptic tubercle situated on well-developed cephalic portion; 3 pairs of lateral eyes close to edge, slightly anterior to median ones. Hand slender, with long curved fingers (43). *Carb.*, Eu.—FIG. 44,*1*. **C. elegans*, Eng.; carapace, ×5.5 (76).

Typhloscorpius PETRUNKEVITCH, 1949 [**T. distinctus*]. Carapace wider than long, with cephalic portion strikingly elevated, posterior edge straight, margined; front almost as wide as posterior edge. Other parts unknown (43). *Carb.*, Eu.—FIG. 44,*2*. **T. distinctus*, Eng.; carapace, ×3.3 (76).

Europhthalmus PETRUNKEVITCH, 1949 [**E. longimanus*]. Carapace elliptic, with straight posterior edge; median eyes close to anterior edge. Pedipalpi ⅔ length of body (43). *Carb.*, Eu.—FIG. 45,*1*. **E. longimanus*, Eng.; dorsal side, ×1.5 (76).

Garnettius PETRUNKEVITCH, 1953 [**Mazonia hungerfordi* ELIAS, 1936]. Carapace rectangular, with pointed anterior projection; median eyes on tubercle close to anterior edge. Hand stout, shorter than fingers, which are also stout; 3rd leg with dorsal spurs on tibia and metatarsus (43). *U.Carb.*, N. Am.—FIG. 44,*8*. **G. hungerfordi* (ELIAS), Kans.; ×0.5 (76).

Family SCORPIONIDAE Leach, 1815

[*nom. transl.* POCOCK, 1893 (*ex* Scorpionini PETERS, 1861, *nom. correct. pro* Scorpionides LEACH, 1815, *non* Scorpionides C.L. KOCH, 1837)]

Third and 4th pairs of coxae grown together along line of contact and much longer than 1st and 2nd pairs. Sternum pentagonal. Single spur in intersegmental membrane between metatarsus and tarsus of 1st and 2nd pairs of legs, no such spur on 3rd and 4th pairs. Three pairs of lateral eyes. *Oligo.-Rec.*

Scorpio LINNÉ, 1758 (**S. maurus* = *S. europaeus* LINNÉ 1758, *non* 1754). Median eyes in middle of carapace. Last preabdominal sternite and ventral

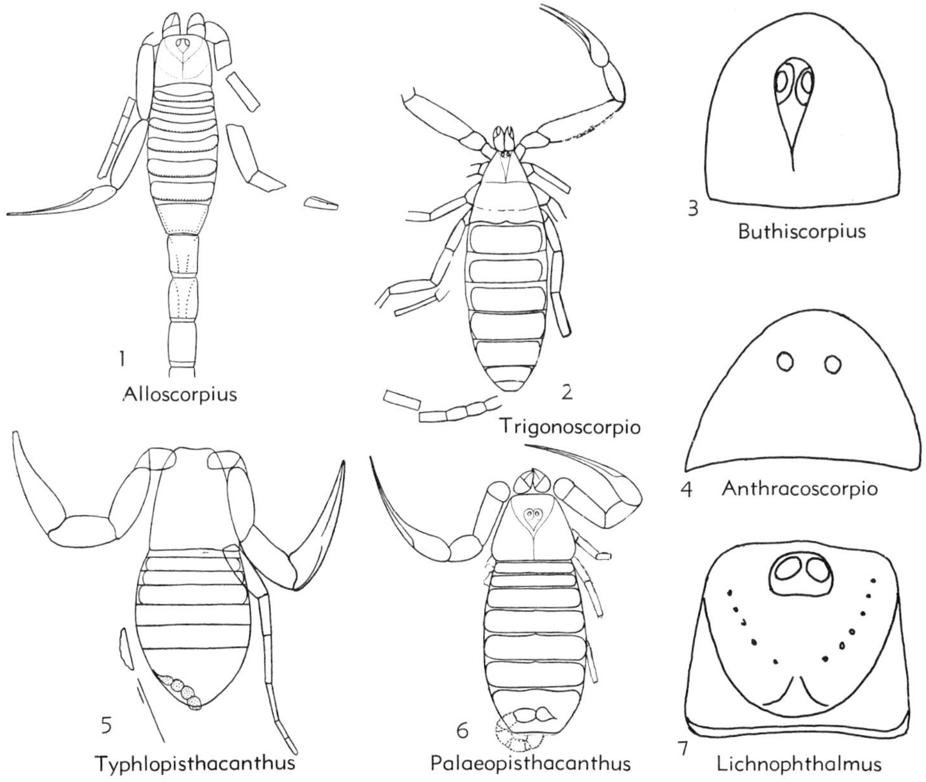

FIG. 43. Family Eoscorpiidae: *1, Alloscorpius granulosus*, Penn., Ill., ×0.7; *2, Trigonoscorpio americanus*, Penn., Ill., ×2; *3, Buthiscorpius buthiformis*, U.Carb., Eng., carapace, ×6.5; *4, Anthracoscorpio juvenis*, U.Carb., Czech., carapace, ×20; *5, Typhlopisthacanthus mazonensis*, Penn., Ill., ×2; *6, Palaeopisthacanthus schucherti*, Penn., Ill., ×2.7; *7, Lichnophthalmus pulcher*, U.Carb., Eng., carapace, ×2.7 (76) (p. P73-P74).

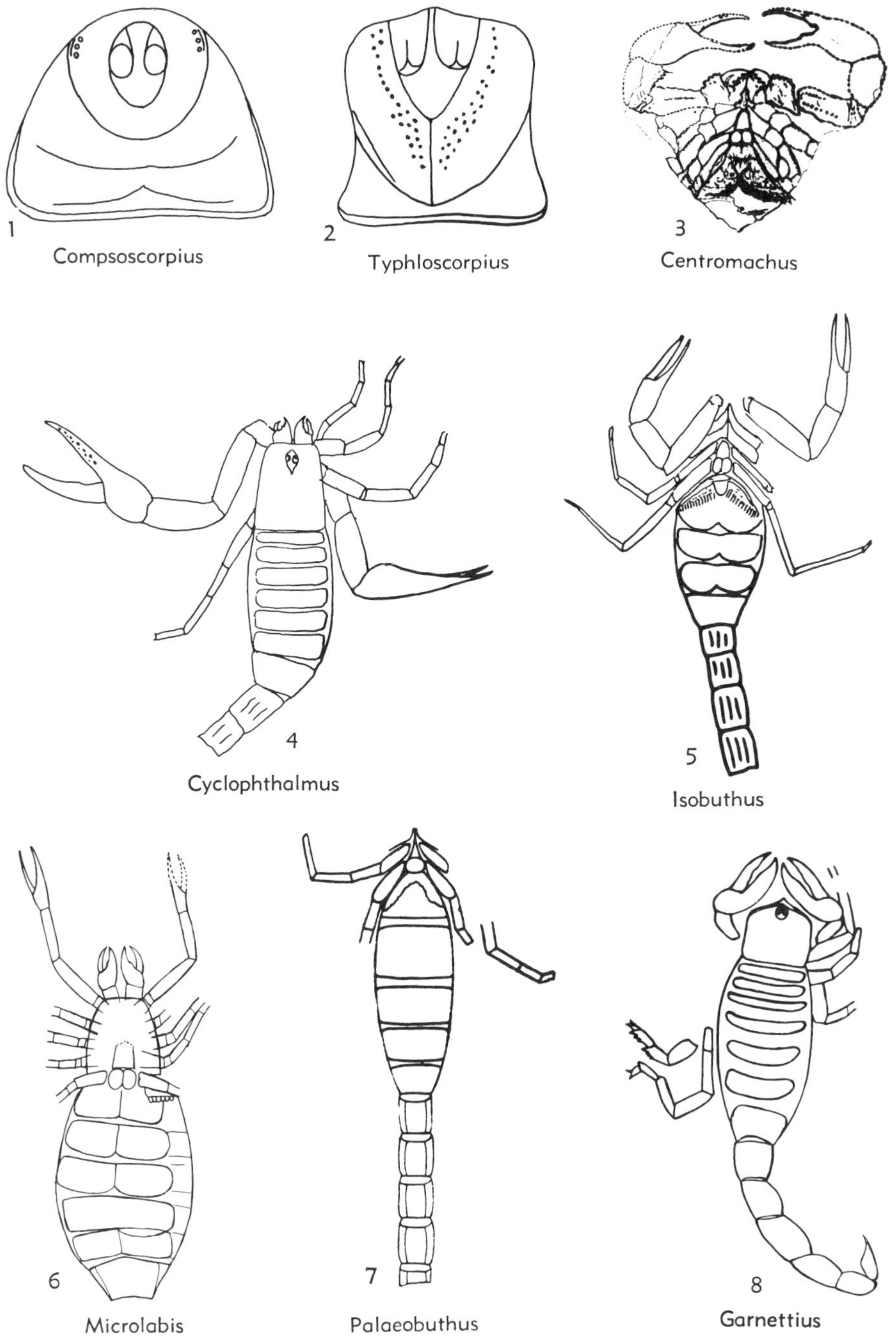

FIG. 44. Eoscorpiidae: *1, Compsoscorpius elegans,* U.Carb., Eng., carapace, ×5.5; *2, Typhloscorpius distinctus,* U.Carb., Eng., carapace, ×3.3; *7, Garnettius hungerfordi,* Penn., Kans., ×0.5. Centromachidae: *3, Centromachus euglyptus,* U.Carb., Scot., ×2. Cyclophthalmidae: *4, Cyclophthalmus senior,* U.Carb., Czech., ×0.6. Isobuthidae: *5, Isobuthus kralupensis,* U.Carb., Czech., ×0.4; *6, Microlabis sternbergii,* U.Carb., Czech., ×1.5 (76) (p. P75-P78).

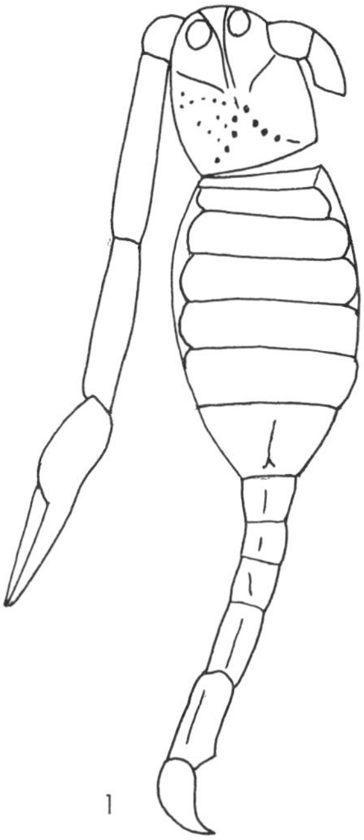

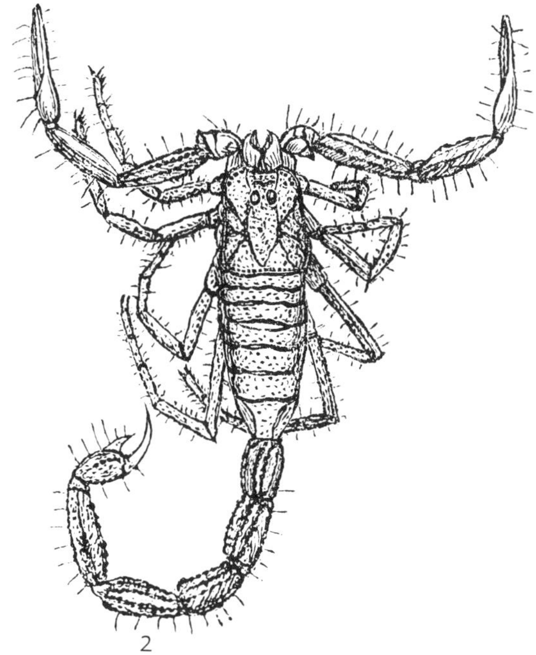

FIG. 45. Eoscorpiidae: *1, Europhthalmus longimanus,* U.Carb., Eng., ×1.5 (76). Buthidae: *2, Tityus eogenus,* Oligo., Balt., ×7 (74) (p. P75-P77).

surface of 1st tail segment with 4 keels (31). *Oligo.-Rec.* (*S. schweiggeri* HOLL, *Oligo.,* Balt.; *S. zeuneri* HADZI, *Mio.,* Ger.).—FIG. 33,*1*. **S. maurus, Rec.,* N.Afr.; dorsal view, male, ×1.5 (*75*a*).

Family BUTHIDAE C. L. Koch, 1837

[*nom. correct.* SIMON, 1864 (*pro* Buthides C.L. KOCH, 1837)]

Third and 4th coxae as in Scorpionidae. Sternum triangular. Two spurs in intersegmental membrane between metatarsus and tarsus. Three or 5 pairs of lateral eyes. *Oligo.-Rec.*

Tityus C.L. KOCH, 1836 [*Scorpio bahiensis* PERTY, 1833]. *Oligo.-Rec.*—FIG. 45,*2*. *T. eogenus* MENGE, *Oligo.,* Balt.; dorsal side, ×7 (74).

Superfamily CYCLOPHTHALMOIDEA Petrunkevitch, 1913

[*nom. transl.* PETR., herein (*ex* Cyclophthalmidae PETR., 1913)]

Three pairs of coxae in front of sternum, 4th pair abutting against sternum; 1st and 2nd pairs with maxillary lobes (34) (Fig. 40,*3*). *Carb.*

Family CYCLOPHTHALMIDAE Petrunkevitch, 1913

Characters of superfamily. *Carb.*

Cyclophthalmus CORDA, 1835 [**C. senior*]. Carapace much longer than wide, with slightly converging sides, anterior edge slightly concave, posterior edge straight; median eyes on tubercle, not far from anterior edge. Hand shorter than fingers (34,35). *Carb.,* Eu.—FIG. 44,*4*. **C. senior,* Czech.; dorsal side, ×0.6 (76).

Superfamily ISOBUTHOIDEA Petrunkevitch, 1913

[*nom. transl.* PETR., herein (*ex* Isobuthidae PETR., 1913]

Two pairs of coxae in front of sternum, 3rd pair abutting against sternum, 4th pair against genital opercula (34, 45) (Fig. 40, *5*). *Carb.*

Family ISOBUTHIDAE Petrunkevitch, 1913

Characters of superfamily. *Carb.*

Isobuthus FRITSCH, 1904 [**Cyclophthalmus kralupensis* THORELL, 1885] [=*Eobuthus, Feistmantelia* FRITSCH, 1904]. Abdominal sternites distinctly bilobed. Carapace as wide as long. Hand longer than wide, with long fingers. Comb with more than 20 teeth (34, 45). *U.Carb.,* Eu.-N.Am.—FIG. 44,*5*. **I. kralupensis* (THORELL), Czech.; ventral side, ×0.4 (76).—FIG. 46. *I. rakovnicensis* (FRITSCH); specimen and counterpart, ×0.85 (76).

Microlabis CORDA, 1839 [**M. sternbergii*]. Abdominal sternites with slight median indentation and ridge. Carapace much narrower in front than behind. Sternum with parallel sides, ?pentagonal. Comb with less than 10 teeth (45). *Carb.,* Eu.—FIG. 44,*6*. **M. sternbergii,* Czech.; ventral side, ×1.5 (76).

Palaeobuthus PETRUNKEVITCH, 1915 [**P. distinctus*]. Abdominal sternites with straight posterior edge without indentation or median ridge. Sternum triangular (34, 45). *Penn.,* N.Am.—FIG. 44,7. **P. distinctus,* Ill.; ventral side, ×1.35 (76).

Superfamily CENTROMACHOIDEA Petrunkevitch, 1953

[*nom. transl.* PETR., herein (*ex* Centromachidae PETR., 1953)]

First pair of coxae in front of sternum, 2nd pair abutting against sternum, 3rd and 4th pairs abutting against genital opercula (45) Fig. 40,*4*). *Carb.*

Family CENTROMACHIDAE Petrunkevitch, 1953

Sternum pentagonal. Other characters of superfamily. *Carb.*

Centromachus THORELL, 1885 [**Eoscorpius euglyptus* PEACH, 1883]. Comb massive, with many intermediate plates and about 40 teeth. Hand stout, with short fingers curving toward each other (45). *Carb.,* Eu.—FIG. 44,*3*. **C. euglyptus* (PEACH), Scot.; ventral side of cephalothorax, ×2 (76).

Superfamily MESOPHONOIDEA Wills, 1910

[*nom. transl.* PETRUNKEVITCH, herein (*ex* Mesophonidae WILLS, 1910)]

Two pairs of coxae in front of sternum, but only 1st pair with maxillary lobes; 3rd and 4th pairs abutting against sternum (Fig. 40,*6*). Stigmata of book lungs at posterior edge of sternites (not on the sternites themselves), with a row of thorns on both lips (45, 61). *Trias.*

Family MESOPHONIDAE Wills, 1910

Combs with 8 to 17 teeth. Other characters of superfamily. *Trias.*

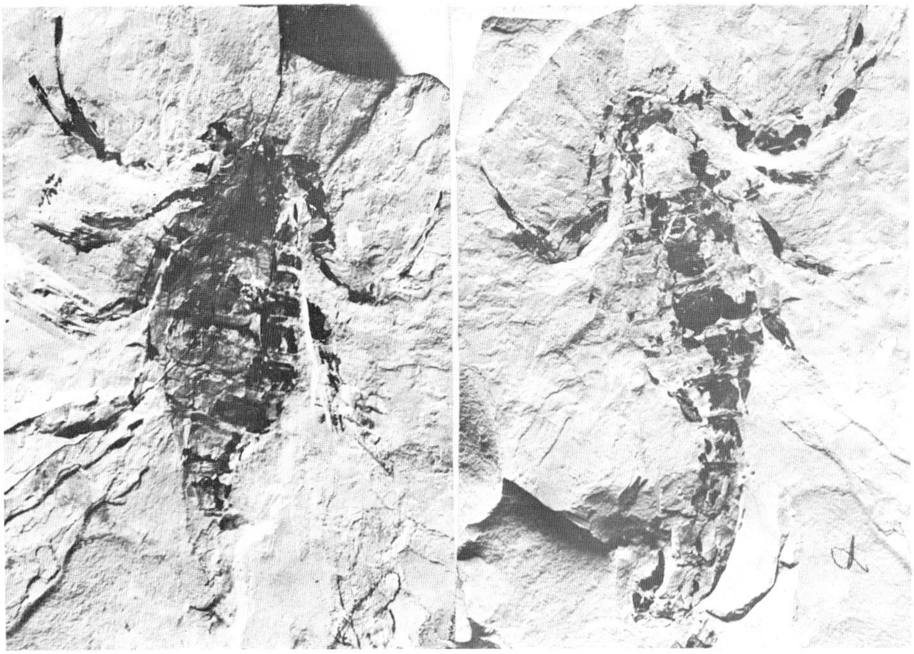

FIG. 46. Isobuthidae: *Isobuthus rakovnicensis,* U.Carb., Czech., photographs of both pieces, ×0.85 (*76).

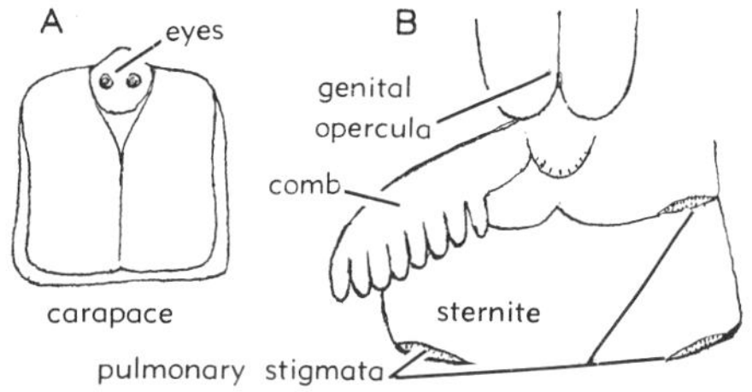

FIG. 47. Mesophonidae: *Mesophonus gracilis,* U. Trias., Eng.; *A,* carapace of male, ×7.7; *B,* genital opercula, comb and stigmata, ×7.7 (82).

Mesophonus WILLS, 1910 [**M. perornatus*]. Carapace rectangular, with anterior median process; pair of median eyes almost in line with anterior edge (Fig. 47*A*). Carapace presumably dimorphic: in presumptive male squarish and without organs described as "compound facetted eyes," in presumptive female transversely oblong and having such "eyes." Hand narrow, with long slender fingers. Sternum pentagonal (45, 61). *U.Trias.,* Eu. —FIG. 47. *M. gracilis* WILLS, L.Keuper, Eng.; *A,* carapace of male; *B,* genital opercula, right comb, sternite and stigmata, ×7.7 (82).

Spongiotarsus WILLS, 1947 [**S. pustulosus*]. Sternum an equilateral triangle. Cuticle of segments with polygonal texture and pustulose surface (61). *Trias.,* Eng.

Genera Incertae Sedis

Lissoscorpionides BODE, 1951 [**L. schmidti*]. Tail short and slender. Poorly known (?related to *Palaeopisthacanthus* or *Typhlopisthacanthus*). *L. Jur.,* Ger.

Palaeomachus POCOCK, 1911 (**Eoscorpius anglicus* WOODWARD, 1876). Only a pedipalp known. *U. Carb.,* Eng.

Order PSEUDOSCORPIONIDA Latreille, 1817

[*nom. correct.* PETRUNKEVITCH, herein (*pro* Pseudoscorpiones LATREILLE, 1817)] [=*Chernetes* SIMON, 1879; Chelonethi THORELL, 1882] [Type: *Chelifer* GEOFFROY, 1763]

Latigastra with segmented abdomen composed of 12 segments, but the 12th rudimentary and so small that it is easily overlooked and the end of the abdomen looks evenly rounded. Carapace entire. Coxosternal region without true sternum, coxae contiguous in median line or with a "pseudosternum" or with rudimentary sternal plates. Chelicerae 2-jointed, chelate, with movable finger lateral in position, supplied with a serrula and a galea (spinneret). Coxa of pedipalp with maxillary lobe; pedipalpi 6-jointed, chelate, powerful. Movable finger of hand ventral in position. Legs 6- or 7-jointed, ending in a pretarsus which forms 7th or 8th joint and is armed with a pair of claws and an arolium. Gizzard and malpighian tubes wanting. Respiration by tracheal tubes with pair of stigmata on the 1st abdominal sternite and another pair on 3rd sternite. Poison glands, if present, situated in the hand and opening either on one of the fingers, or, in case of a pair of glands, on both fingers. Male reproductive organs with an atrium (*uterus masculinus externus* of German authors) and a pair of eversible "ram's-horn organs," but without organs of copulation. Female with a single median ovary and a pair of tubular sperm receptacles. Ventral nervous system concentrated in the cephalothorax. Oviparous (3, 7, 52, 56). *Oligo.-Rec.*

DISCUSSION

The pseudoscorpions are small arachnids which resemble scorpions on account of their powerful chelate pedipalpi. They can be easily distinguished from scorpions by the absence of a "tail." Fossil pseudoscorpions can be recognized by the presence of 11 plainly visible abdominal tergites whereas a scorpion with a broken-off tail has only 7 tergites. Both tergites and sternites may be entire or divided by a median longitudinal line, those of the 12th segment (between which the slitlike anus opens) being very narrow and having the appearance of a pygidium when protruded. The carapace is always entire, but may be impressed transversely by 1 or 2 furrows; commonly it bears 1 or 2 pairs of eyes, but some genera are blind (Fig. 48).

The mouth, situated between the pedipalpal coxae, has an anterior and a posterior lip. The lips are covered with fine hair which serves as a filter. The lips and the pedipalpal coxae form together a prebuccal cone. The pharynx is a powerful pumping organ. The esophagus opens directly into the mid-gut without a gizzard. The mid-gut has an extensive system of diverticles forming a chylenteron. It has a stercoral pouch at its end. The hind-gut is a short rectum.

The heart is limited to the abdomen and has a maximum of 3 pairs of ostia in some genera and a minimum of a single pair in others. The tracheal tubes are highly branched, the 1st pair supplies the cephalothorax, the 2nd pair the abdomen. The eggs develop in a ventral brood pouch on the abdomen.

About 1,000 species are known, ranging in size from 0.8 to 6 mm. Although several fossils from the Carboniferous and one from the Jurassic were classed by their discoverers as representatives of the order Pseudoscorpionida, such identification proved to be erroneous. The Carboniferous species are scorpions, the Jurassic form is a malacostracan. True pseudoscorpions are fairly common in Oligocene and Miocene deposits; of 19 families recognized in living forms, 9 are represented by fossils described from Baltic amber (Oligocene). They include 12 genera and 18 species (Figs. 49, 50).

Suborder CHTHONIINA Beier, 1932

[*nom. correct.* PETRUNKEVITCH, herein (*pro* Chthoniinea BEIER, 1932)] [=Chthonioidea CHAMBERLIN, 1929] [Type: *Chthonius* C.L. KOCH, 1843]

First and 2nd pairs of legs without basitarsus, 3rd and 4th pairs with basitarsus (3). *Oligo.-Rec.*

Family CHTHONIIDAE Hansen, 1894

Spiracles transverse. Epigynal area nonsclerotic. *Oligo.-Rec.*

Chthonius C. L. KOCH, 1843 [**Obisium orthodactylum* LEACH, 1815]. Spurs present on 1st, 2nd, and 3rd coxae. *Oligo.-Rec.* [*C. mengei* BEIER, Oligo., Balt.].

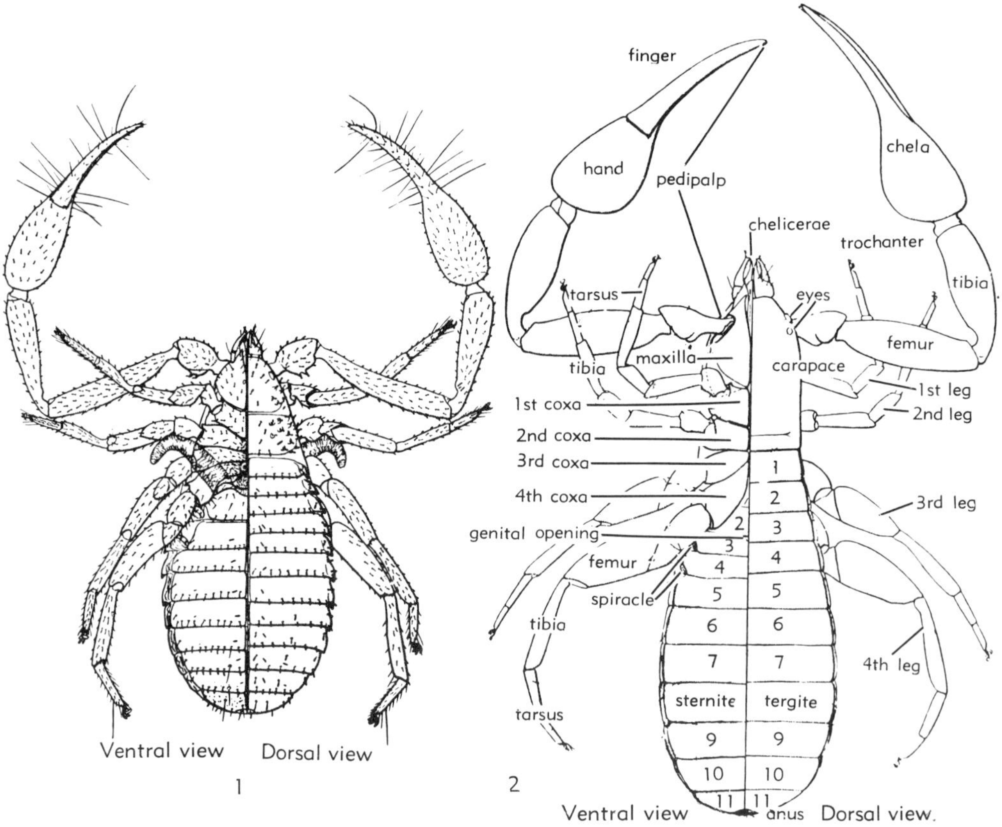

FIG. 48. External morphology of Recent pseudoscorpions, ventral (left half) and dorsal (right half): *1, Chelifer cancroides; 2, Hesperolpium slevini* (66).

Family DITHIDAE Chamberlin, 1931

Spiracles obliquely oriented. Epigynal area sclerotized. *Oligo.-Rec.*

Heterolophus TÖMÖSVARY, 1884 [**H. guttiger*]. Carapace smooth. Two pairs of eyes of equal size. Pedipalpal fingers with teeth. *Oligo.-Rec.*—FIG. 49,*2*. *H. kochii* (MENGE), Oligo., Balt.; dorsal side ×20 (74).

Suborder NEOBISIINA Beier, 1932

[*nom. correct.* PETRUNKEVITCH, herein (*pro* Neobisinea BEIER, 1932)] [Type: *Neobisium* CHAMBERLIN, 1930]

Basitarsus present in all legs. *Oligo.-Rec.*

Superfamily NEOBISIOIDEA Beier, 1932

[*nom. correct.* PETRUNKEVITCH, herein (*pro* Neobisiides BEIER, 1932)]

Movable finger of chelicerae with teeth. *Oligo.-Rec.*

Family NEOBISIIDAE Chamberlin, 1930

Poison gland only in immobile finger of pedipalp. Opisthosomal pleura granular. *Oligo.-Rec.*

Neobisium CHAMBERLIN, 1930 [**Obisium muscorum* LEACH]. *Oligo.-Rec.*—FIG. 49,*1*. *N. rathkii* (KOCH & BERENDT), Oligo., Balt.; dorsal side, ×18 (71).

Superfamily GARYPOIDEA Beier, 1932

[*nom. correct.* PETRUNKEVITCH, herein (*pro* Garypides BEIER, 1932)]

Movable finger of chelicerae smooth. *Oligo.-Rec.*

Family OLPIIDAE Chamberlin, 1930

Poison glands in both fingers of pedipalp. Opisthosomal pleura smooth. *Oligo.-Rec.*

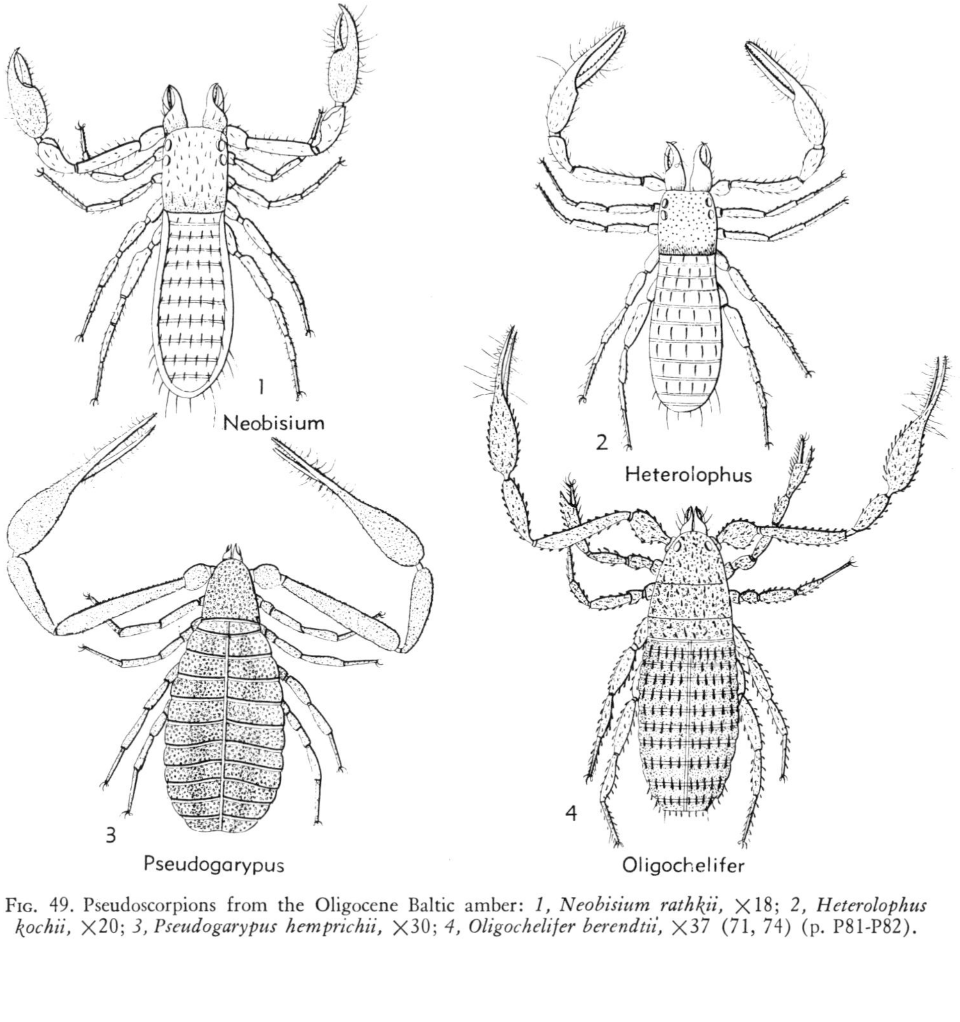

FIG. 49. Pseudoscorpions from the Oligocene Baltic amber: *1, Neobisium rathkii,* ×18; *2, Heterolophus kochii,* ×20; *3, Pseudogarypus hemprichii,* ×30; *4, Oligochelifer berendtii,* ×37 (71, 74) (p. P81-P82).

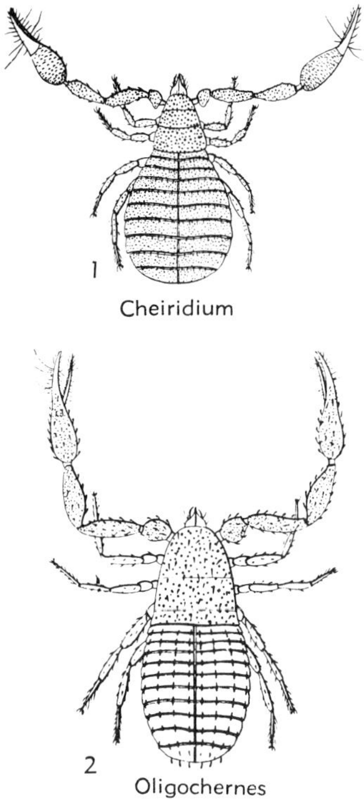

FIG. 50. Pseudoscorpions from the Oligocene Baltic amber: *1, Cheiridium hartmanni,* ×33; *2, Oligochernes weigandi,* ×25 (74) (p. P82).

Garypinus DADAY, 1888 [**Olpium dimidiatum* L. KOCH, 1873]. *Oligo.-Rec.* [*G. electri* BEIER, Oligo., Balt.].

Family GARYPIDAE Hansen, 1894

Poison glands as in Olpiidae. Opisthosomal pleura granular. *Oligo.-Rec.*

Garypus L. KOCH, 1873 [**Obisium beauvoisii* SAVIGNY, 1829]. *Mio.-Rec.* [*G. burmiticus* COCKERELL, Mio., Burma].

Geogarypus CHAMBERLIN, 1930 [**Garypus minor* L. KOCH, 1873]. *Oligo.-Rec.* [*G. macrodactylus* BEIER, Oligo., Balt.].

Superfamily FEAELLOIDEA Beier, 1932

[*nom. correct.* PETRUNKEVITCH, herein (*pro* Feaellides BEIER, 1932)]

All tarsi with a single segment. *Oligo.-Rec.*

Family PSEUDOGARYPIDAE Chamberlin, 1923

Two pairs of eyes. *Oligo.-Rec.*

Pseudogarypus ELLINGSEN, 1909 [**Garypus bicornis* BANKS, 1895]. *Oligo.-Rec.*—FIG. 49,*3*. *P. hemprichii* (KOCH & BERENDT), Oligo., Balt.; dorsal side, ×30 (71).

Suborder CHELIFERINA Hagen, 1878

[*nom. correct.* PETRUNKEVITCH, herein (*pro* Cheliferinea BEIER, 1932)]
[Type: *Chelifer* GEOFFROY, 1763]

All legs without basitarsus. One pair of eyes or none. *Oligo.-Rec.*

Superfamily CHEIRIDIOIDEA Chamberlin, 1931

Femoral articulations of all legs of same morphological type, vestigial, functionless. *Oligo.-Rec.*

Family CHEIRIDIIDAE Chamberlin, 1924

All legs 5-jointed. Poison glands only in immobile fingers of pedipalpi. *Oligo.-Rec.*

Cheiridium MENGE, 1855 [**Chelifer muscorum* LEACH, 1817]. *Oligo.-Rec.*—FIG. 50,*1*. *C. hartmanni* (MENGE), Oligo., Balt.; dorsal side, ×33 (74).

Superfamily CHELIFEROIDEA Chamberlin, 1931

[*nom. transl.* CHAMBERLIN, 1931 (*ex* Cheliferidae STECKER, 1874)]

Femoral articulations heteromorph, of one type in legs 1 and 2, of another type in legs 3 and 4. *Oligo.-Rec.*

Family CHERNETIDAE Menge, 1855

All legs 6-jointed. Poison glands only in movable fingers of pedipalpi. *Oligo.-Rec.*

Oligochernes BEIER, 1937 [**O. bachofeni*]. *Oligo.*, Balt.—FIG. 50,*2*. *O. weigandi* (MENGE); ×25 (74).

Family CHELIFERIDAE Stecker, 1874

[*nom. transl.* HAGEN, 1879 (*ex* Cheliferinae STECKER, 1874)]

Poison glands in both fingers of pedipalpi. *Oligo.-Rec.*

Electrochelifer BEIER, 1937 [**E. mengei*]. *Oligo.*, Balt.

Oligochelifer BEIER, 1937 [**Dichela berendtii* MENGE, 1854]. *Oligo.*, Eu.—FIG. 49,*4*. **O. berendtii* (MENGE), Balt.; ×37.

Oligowithius BEIER, 1937 [**Obisium sieboldi* MENGE, 1854]. *Oligo.*, Balt.

Pycnochelifer BEIER, 1937 [**Chelifer kleemanni* KOCH & BERENDT, 1854]. *Oligo.*, Balt.

Order PHALANGIIDA Perty, 1833

[*nom. correct.* PETRUNKEVITCH, herein (*pro* Phalangida PERTY, 1833, *nom. transl.* PERTY *ex* family Phalangides LEACH, 1815)] [=Opiliones SUNDEVALL, 1833] [The name Opiliones was proposed as a substitute for Phalangides because it limited the order to arachnids of the same group, whereas the name Phalangides did not originally have such a restricted meaning either in its French vernacular form used by LATREILLE or in its Latin form used by both LATREILLE and LAMARCK. LEACH was the first to restrict the name to a "family." As the name of an order, Phalangida was in common use up to 1909 and has not only priority over Opiliones, but also the advantage that the type of the order is *Phalangium* LINNÉ]

Latigastra with compact body 1 to 22 mm. in length, usually with slender legs which are in some species 200 mm. long. First abdominal segment lost, its neuromere incorporated in the thoracic ganglionic mass. Maximum number of abdominal segments 10 (if anal operculum is counted as 10th tergite). Carapace entire, but the last 2 thoracic tergites, corresponding to the segments bearing the 3rd and 4th pairs of legs, may be clearly outlined by transverse furrows. Three to 8 abdominal tergites usually fused with carapace, forming dorsal scutum, but the tergites may remain recognizable by transverse furrows between them. Configuration of coxosternal region subject to several structural variations. All coxae movable in Phalangioidea and some other groups; immobile in Nemastomatidae and Trogulidae. First pair of coxae movable, other pairs immobile and coalesced in Laniatorina. Sternum may be present or wanting. Chelicerae 3-jointed, chelate. Pedipalpi 6-jointed, pediform or raptorial, with or without a claw at end. Legs 7-jointed, tarsi often subsegmented and flexible, those of 1st and 2nd pairs always with a single claw, those of 3rd and 4th pairs either with a single claw or with 2 claws. Coxae of pedipalpi and of 1st pair of legs with maxillary lobes, those of 2nd legs usually without lobes, but in some groups with lobes (Fig. 51).

Mouth opens either directly to the outside or into a stomotheca formed by the inner surface of the 1st coxae. Pharynx hexagonal, lacking an apparatus for filtering

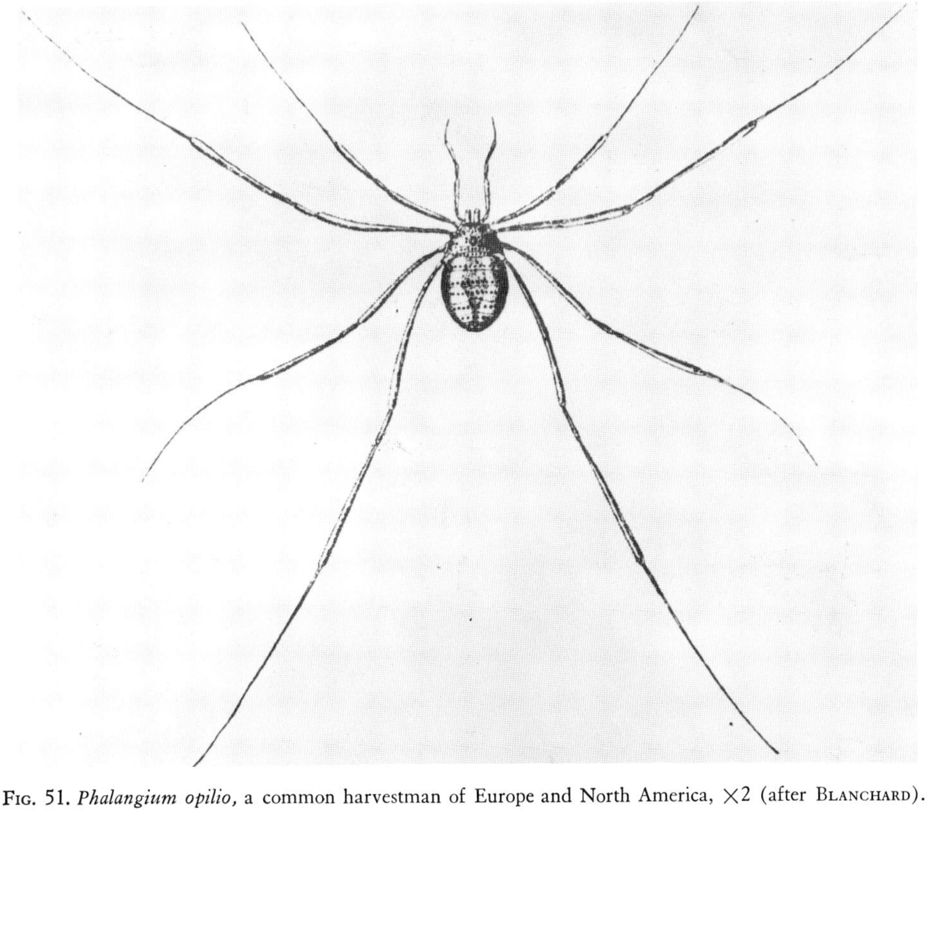

FIG. 51. *Phalangium opilio,* a common harvestman of Europe and North America, ×2 (after BLANCHARD).

food, so that coarse particles are regularly ingested. Mid-gut with spacious diverticles occupying major portion of body cavity. Hind-gut quite short, anus with an operculum. Respiration by means of a pair of highly ramified tracheal tubes with a pair of spiracles on 2nd abdominal sternite, behind 4th pair of coxae. A pair of coxal glands with openings behind 3rd pair of coxae. Heart with only 2 pairs of ostia. Arterial system lacking. A pair of odoriferous glands in the cephalothorax, with openings close to 2nd or 4th coxae, dorsal to them. Pair of median eyes, usually elevated on a tubercle, but sessile and wide apart in Cyphophthalmina. Reproductive organs of male with testes fused into a single U-shaped organ, with protrusible penis. Females with a U-shaped ovary and protrusible ovipositor. Oviparous (4, 21, 35, 51). *Penn.-Rec.*

DISCUSSION

Recent Phalangiida are divided into 3 suborders—Cyphophthalmina, Palpatorina and Laniatorina—with more than 2,350 species. Owing to poor preservation, the few known Paleozoic Phalangiida cannot be assigned unreservedly to any of the Recent suborders,

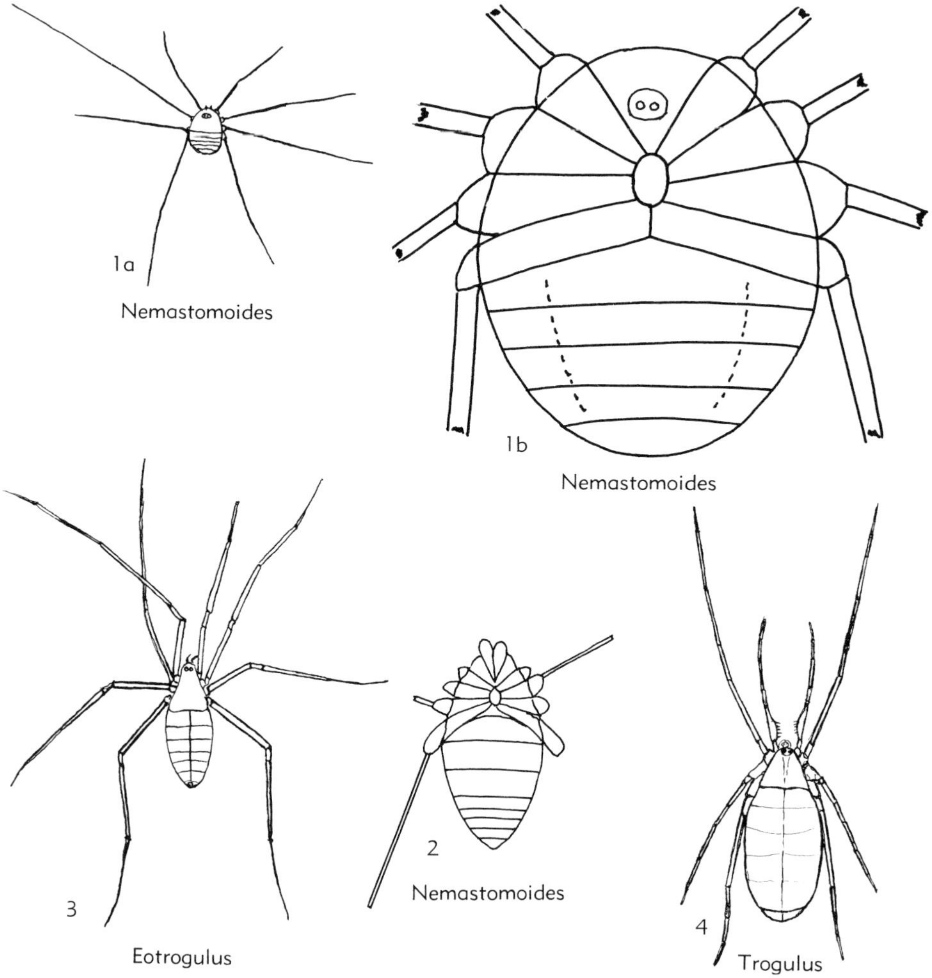

FIG. 52. Phalangiida. *1, Nemastomoides elaveris,* U.Carb., Fr.; *1a,* dorsal side, ×2.2; *1b,* ventral side of body, ×20. *2, Nemastomoides longipes,* Penn., Ill.; ventral side, ×2.7. *3, Eotrogulus fayoli,* U.Carb., Fr., ×3. *4, Trogulus gypseus,* representative of the suborder Cyphophthalmina, Rec., Morocco, ×1.8 (76) (p. P85).

but their general appearance favors placing them in the Palpatorina. A Jurassic fossil, *Hasseltides primigenius* (WEYENBERGH), which formerly (45) was considered to be a phalangiid, recently has been shown by VAN REGTEREN ALTENA, on the basis of other specimens, to be a distorted incomplete crinoid (*Saccocoma*). Most Tertiary Phalangiida belong to the Palpatorina but at least one representative of the Laniatorina has been found.

Suborder CYPHOPHTHALMINA Simon, 1879

[*nom. correct.* PEARSE, 1936 (*pro* Cyphophthalmi SIMON, 1879)] [=Anepignathi THORELL, 1882] [Type: *Siro* LATREILLE, 1796 (=*Cyphophthalmus* JOSEPH, 1868)]

Openings of stink glands at end of a pair of lateral conic protuberances of cephalothorax. Genital opening without an operculum. *Rec.*

Suborder PALPATORINA Thorell, 1876

[*nom. correct.* PEARSE, 1936 (*pro.* Palpatores THORELL, 1876)] [=Plagiostethi SIMON, 1879] [Type: *Phalangium* LINNÉ, 1758]

Openings of stink glands not on protuberances. Pedipalpi slender. All tarsi with a single, simple claw. *Carb.-Rec.*

Superfamily TROGULOIDEA Sundevall, 1833

[*nom. transl.* PETRUNKEVITCH, herein (*ex* Trogulides SUNDEVALL, 1833)] [=Dyspnoi HANSEN & SÖRENSEN, 1904]

Terminal joint of pedipalpi shorter than tibia. *Carb.-Rec.*

Family EOTROGULIDAE Petrunkevitch, nov.

Coxae free, arranged in 2 nearly parallel rows. *Carb.*

Eotrogulus THEVENIN, 1901 [**E. fayoli*]. Carapace longer than wide, subtriangular, with rounded apex and slightly concave sides. Pair of sessile eyes close to anterior end. Abdominal tergites bisected by median line. Legs not exceeding 2.5 times body length. *Carb.*, Fr.—FIG. 52,*3*. **E. fayoli;* dorsal side, ×3 (76).

Family TROGULIDAE Sundevall, 1833

[*nom. correct.* SIMON, 1879 (*pro* Trogulides SUNDEVALL, 1833)]

Carapace forms a bifurcated "hood" in front of the eyes. *Rec.*

Trogulus LATREILLE, 1802 [**Acarus nepaeformis* SCOPOLI, 1763]. *Rec.*—FIG. 52,*4*. *T. gypseus* SIMON, Rec., Morocco; dorsal side, ×1.8 (76).

Family NEMASTOMATIDAE Simon, 1872

Sternum fused with arculi genitales separating posterior coxae. Dorsal scutum composed of carapace and 5 abdominal tergites grown together. *Carb.-Rec.*

Nemastoma C.L. KOCH, 1836 [**Phalangium lugubre* MÜLLER, 1776]. Abdominal tergites of scutum firmly grown together, *Oligo.-Rec.* [*N. denticulatum* KOCH & BERENDT, Oligo., Balt.].

Family NEMASTOMOIDIDAE Petrunkevitch, nov.

Sternum oval, separated from 1st abdominal sternite by 4th coxae meeting in median line. *Carb.*

Nemastomoides THEVENIN, 1901 [**N. elaveris*] [=*Protopilio* PETR., 1913]. Carapace wider than long, with evenly rounded front and sides. Eyes on tubercle considerably in front of middle. *U. Carb.*, Eu.-N.Am.—FIG. 52,*1*. **N. elaveris,* Fr. (Commentry); *1a,* dorsal side, ×2.2; *1b,* ventral side, ×20 (76).—FIG. 52,*2*. *N. longipes* (PETR.), Ill.; ventral side, ×2.7 (76).

Family ISCHYROPSALIDAE Simon, 1879

Carapace separated from abdominal scutum by a transverse line. *Oligo.-Rec.*

Sabacon SIMON, 1879 [**S. paradoxus*]. All femora lacking false articulations. All coxae movable. *Oligo.-Rec.* [*S. bachofeni* RÖWER, Oligo., Balt.].

Superfamily PHALANGIOIDEA Thorell, 1876

[*nom. transl.* PETRUNKEVITCH, herein (*ex* Phalangiidae SIMON, 1879, *nom correct. pro* Phalangioidae THORELL, 1876] [=Eupnoi HANSEN & SÖRENSEN, 1904]

Terminal joint of pedipalpi longer than tibia. Maxillary lobe of 2nd coxae long and slender. *Oligo.-Rec.*

Family PHALANGIIDAE Thorell, 1876

[*nom. correct.* SIMON, 1879 (*pro* Phalangioidae THORELL, 1876)]

Body soft. Eyes on tubercle in middle of carapace. Tarsi with numerous segments, flexible. *Oligo.-Rec.*

Caddo BANKS, 1892 [**C. agilis*]. Eyes very large, occupying almost entire carapace. *Oligo.-Rec.*—FIG. 53,*1*. *C. dentipalpus* (KOCH & BERENDT); oblique dorsal view, ends of legs omitted, ×7 (71).

Cheiromachus MENGE, 1854 [**C. coriaceus*]. Trochanter of pedipalp short, cone-shaped, wider in front than at base; femur very stout, dorsally convex, laterally compressed, with fairly long ventral thorns; patella much thinner and half as long as femur, wider in front than at base; tibia

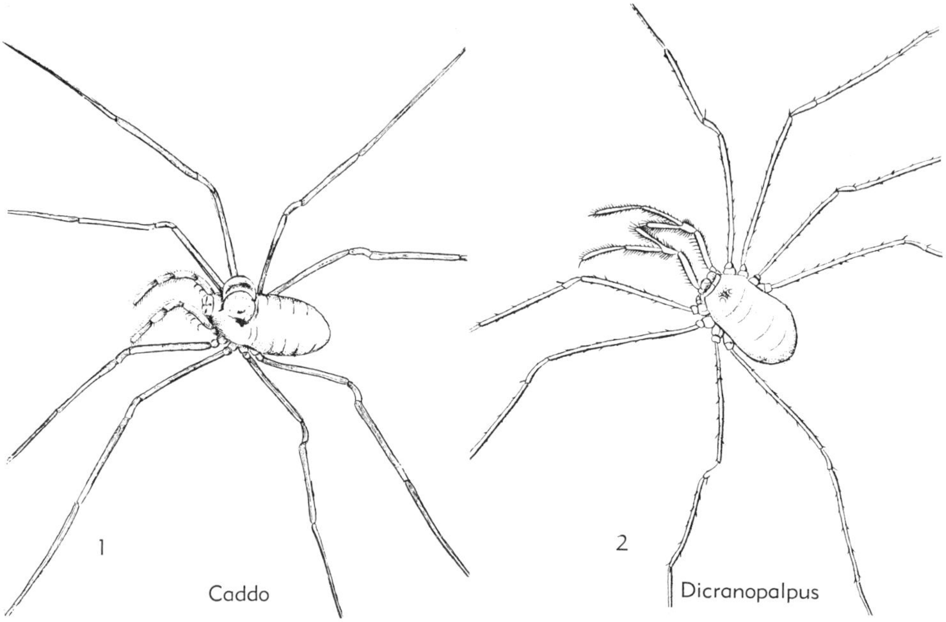

FIG. 53. Phalangiida from the Oligocene Baltic amber: *1, Caddo dentipalpus,* ×7; *2, Dicranopalpus ramiger,* ×5.4 (71) (p. P85-P86).

short, spindle-shaped; terminal joint cylindrical, short. Ocular tubercle with strong thorns. *Oligo.,* Balt.

Dicranopalpus DOLESCHALL, 1852 [*D. *gasteinensis*]. Patella of pedipalp with long apophysis. *Oligo.-Rec.*—FIG. 53,2. *D. ramiger* (KOCH & BERENDT), Oligo., Balt.; dorsal side, ends of legs omitted, ×5.4 (71).

Liobunum C.L. KOCH (**Phalangium rotundum* LATREILLE, 1798). Palpal claw denticulate. Femur of 1st leg longer than body. *Oligo.-Rec.* [*L. longipes* MENGE, Oligo., Balt.].

Opilio HERBST, 1798 [**Phalangium parietinum* DEGEER, 1778]. Middle joint of chelicerae without dorsal spur in male. *Oligo.-Rec.* [*O. ovalis* KOCH & BERENDT, Oligo., Balt.]

Phalangium LINNÉ, 1758 [**P. opilio*]. Middle joint of pedipalp with dorsal spur in male. *Oligo.-Rec.* FIG. 51. **P. opilio,* Rec., Eu.-N.Am.(cosmop.); ×2 (76).—FIG. 54. *P. oculatum* PETRUNKEVITCH, Oligo., Colo.; ×6.5 (76).

Suborder LANIATORINA Thorell, 1876

[*nom. correct.* PEARSE, 1936 (*pro* Laniatores THORELL, 1876)] [=Mecostethi SIMON, 1879] [Type: *Gonyleptes* KIRBY, 1819]

Pedipalpi raptorial, powerful. Tarsi of 1st and 2nd legs with single claw, those of 3rd and 4th with 2 claws or a trident claw (4). *Oligo.-Rec.*

Family GONYLEPTIDAE Sundevall, 1833

Third and 4th tarsi with 2 claws. Carapace fused with 5 abdominal tergites into a solid scutum. Fourth coxae very large and immobile. *Oligo.-Rec.*

Gonyleptes KIRBY, 1819 [**G. horridus*]. First tarsus 6-jointed. Femur of pedipalp with spine. Ocular tubercle with pair of thorns. *Oligo.-Rec.* [*G. nemastomoides* KOCH & BERENDT, Oligo., Balt.].

Order ARCHITARBIDA Petrunkevitch, 1945

[*nom. correct.* PETR., herein (*pro* Architarbi PETR., 1945)] [=Phalangiotarbi HAASE, 1890] [Type: *Architarbus* SCUDDER, 1868]

Latigastra with segmented abdomen, 5 or 6 anterior tergites of which are distinctly abbreviated. Carapace entire, with either a pair of sessile eyes (Fig. 56,*8*) or 3 pairs of eyes on a flat, trifoliate tubercle (Figs. 55, 58, 59) or without eyes. Abdomen composed of not more than 11, more commonly of 10 segments. Abbreviated tergites thickened and longitudinally bisected by a median line (Fig. 55,*7a*). Owing to fusion, number of sternites always smaller than that of tergites, but not less than 7 nor more than 9, and always divided by a pair of longitudinal lines (Fig. 55,*6*). Last sternite with circular anal operculum. Lateral pleural membrane not segmented. Except in Heterotarbidae, coxosternal region has typically only 4 pairs of more or less triangular coxae radiating from a small sternum. Pedipalpal coxae hidden from view by 1st pedal coxae, above which they are situated; they become occasionally partly visible when the 1st coxae are not in full contact with each other, as usual (Figs. 55,*6;* 58,*4b*). Laterally, on each side of sternum, the 4 pedal coxae are in contact with each other. The triangular space between the 4th coxae is oc-

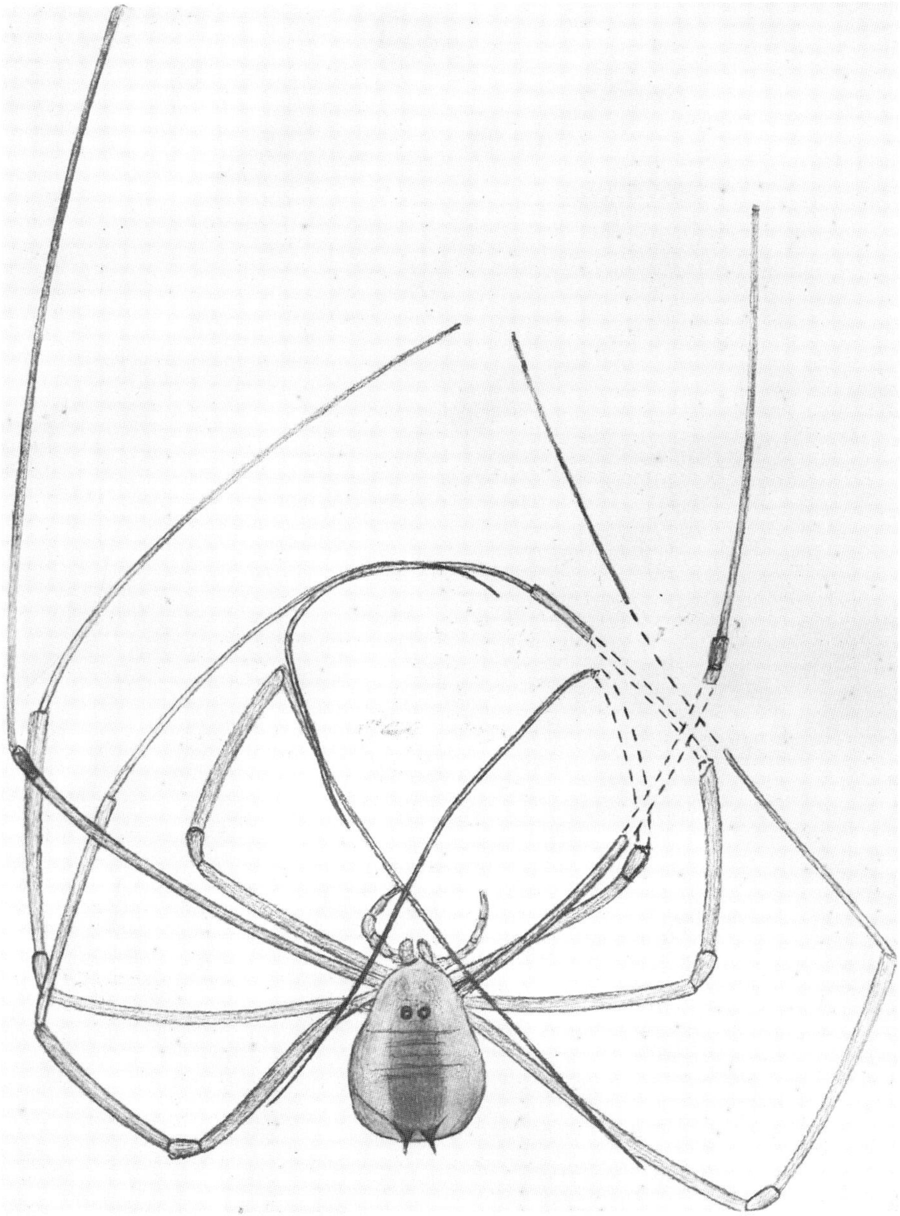

FIG. 54. *Phalangium oculatum,* Oligo., Colo. (Florissant), ×6.5 (76) (p. P86).

cupied by the 1st and 2nd abdominal sternites. Sternum small, consisting of 3 (Fig. 55,*6*) or 4 sclerites (Fig. 58,*4b*), or completely wanting (Fig. 56,*2b*). Chelicerae 3-jointed, chelate, small. Pedipalpi 6-jointed, slender, short. Legs 7-jointed, except in Opiliotarbidae in which trochanters of 3rd and 4th pairs of legs are 2-jointed, making these legs 8-jointed.

Mouth concealed, presumably opening

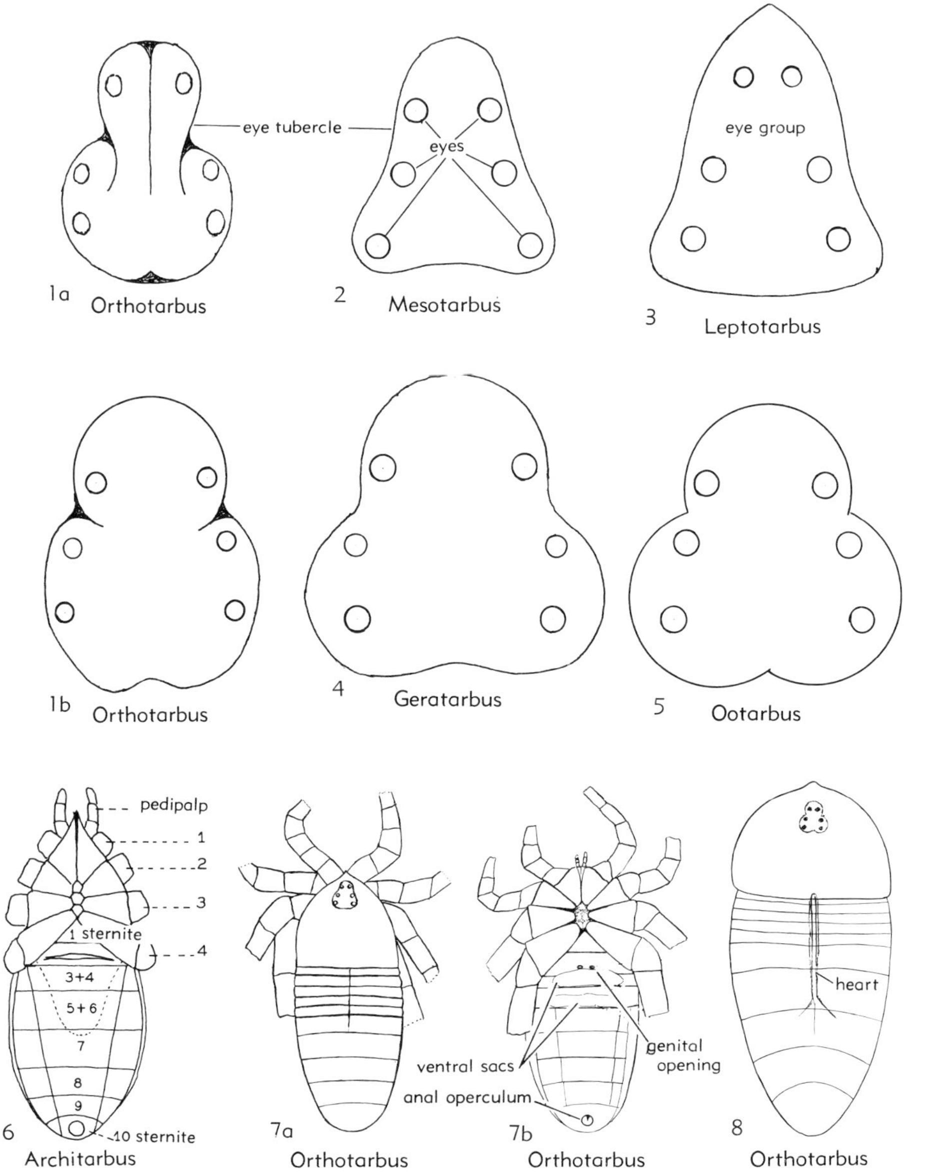

FIG. 55. External morphology of Architarbida. *1-5,* Eye tubercle with eyes; *1, Orthotarbus robustus,* Penn., Ill., *1a,* type, ×27, *1b,* paratype, ×30; *2, Mesotarbus intermedius,* U.Carb., Eng., ×27; *3, Leptotarbus torpedo,* U.Carb., Eng., ×29; *4, Geratarbus lacoei,* Penn., Ill., ×50; *5, Ootarbus pulcher,* Penn., Ill., ×33; *6, Architarbus minor,* Penn., Ill., ventral side of body showing segmentation and coxosternal region, ×2.8; *7a,b, Orthotarbus minutus,* Penn., Ill., dorsal and ventral side, ×3.5; *8, Orthotarbus robustus,* specimen with preserved heart, ×2.4 (76) (p. P90-P92).

into a stomotheca as in Recent Cyphophthalmina of the order Phalangiida, formed by medial walls of 1st coxae. A pair of round genital openings, usually present on 2nd abdominal sternite, probably are openings of sperm receptacles (Fig. 55,7*b*). Book lungs wanting, but 1 to 4 ventral sacs are present, one sac per segment on 3rd to 6th abdominal sternites, which presumably are organs of respiration. The presence of

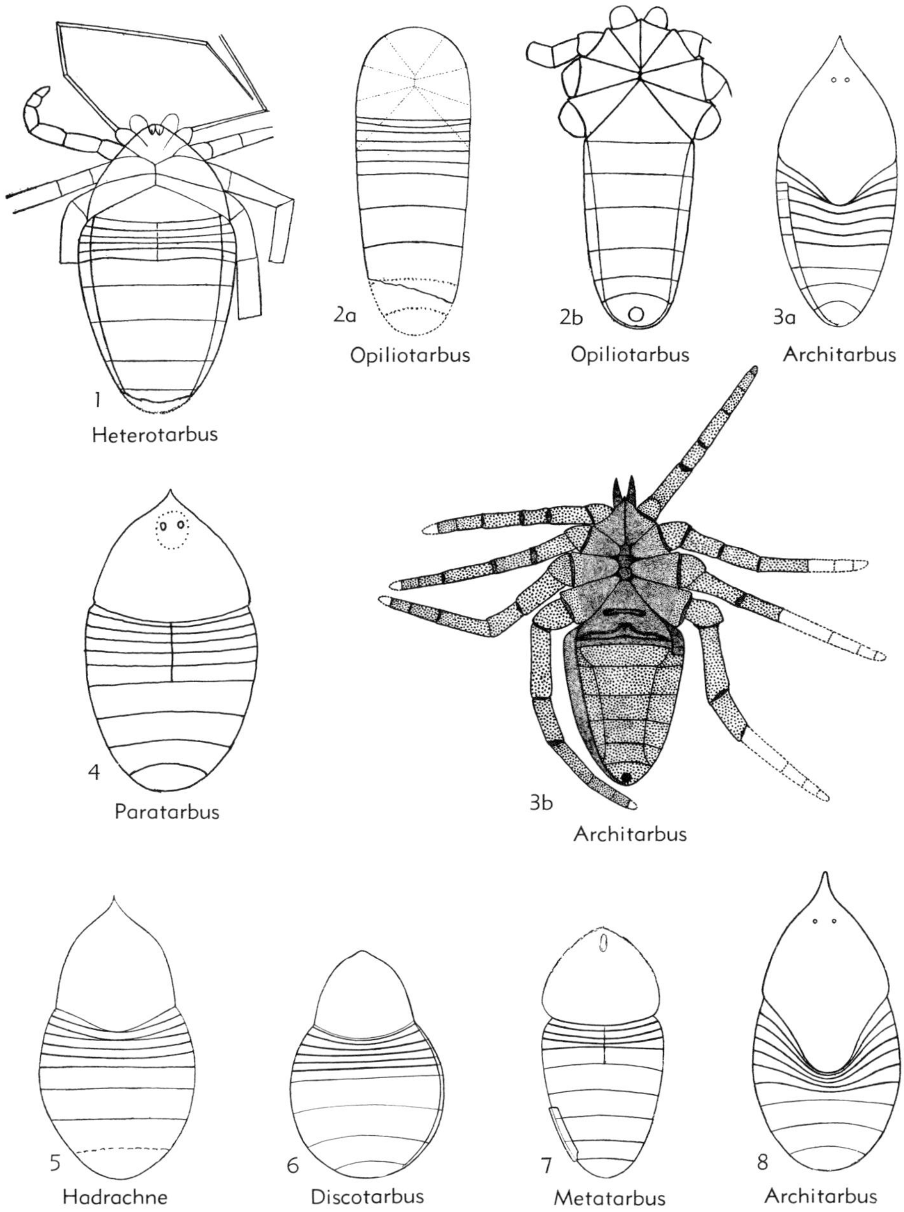

FIG. 56. Architarbida. *1, Heterotarbus ovatus,* Penn., Ill., ×2.7; *2, Opiliotarbus elongatus,* Penn., Ill.; *2a,b,* dorsal and ventral side, ×2.7; *3, Architarbus rotundatus,* Penn., Ill.; *3a,b,* dorsal and ventral side, ×1.65; *4, Paratarbus carbonarius,* Penn., Ill., ×2.1; *5, Hadrachne horribilis,* Penn., Ill., ×1; *6, Discotarbus deplanatus,* Penn., Ill., ×2.7; *7, Metatarbus triangularis,* Penn., Ill., ×2; *8, Architarbus minor,* Penn., Ill., ×2.7 (76) (p. P90-P93).

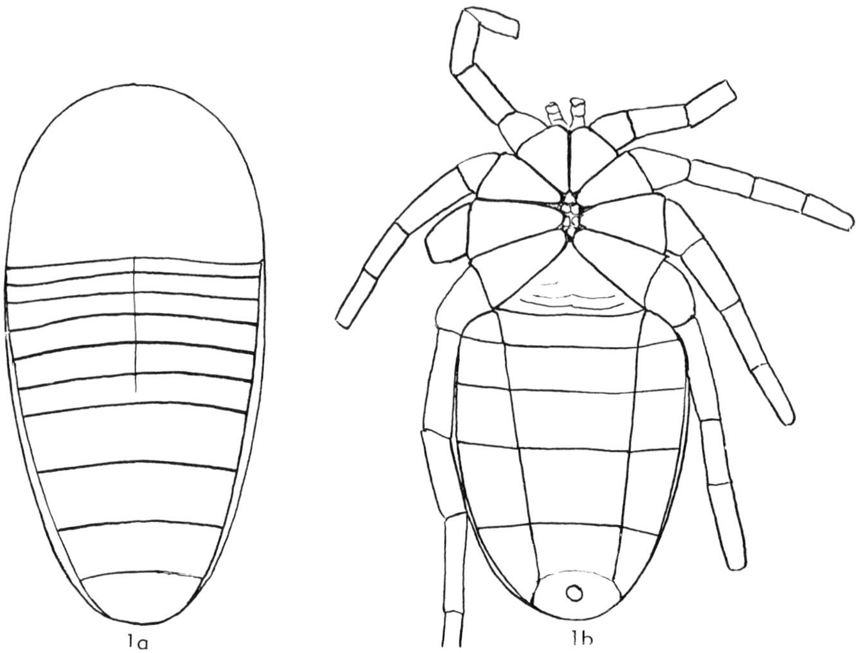

FIG. 57. *Phalangiotarbus subovalis,* U.Carb., Eng.; *1a,b,* dorsal and ventral sides of lectotype, ×9 (76).

tracheal tubes cannot be demonstrated, but is not excluded, because tracheal tubes cannot be seen in fossils which undoubtedly possessed them. Heart, enclosed in its pericardium and ending in 3 posterior arteries, is preserved in one specimen (Fig. 55,*1c*) and seems to be limited to abdomen. Nothing known of other internal organs. (14, 34, 43, 45.) *Carb.*

Family ARCHITARBIDAE Karsch, 1882

[=Phalangiotarbidae HAASE, 1890]

Abdomen with 10 tergites and 8 or 9 sternites. All coxae triangular, trochanters single-jointed, legs similar, 7-jointed (45). *Carb.*

Architarbus SCUDDER, 1868 [**A. rotundatus*] [=*Geraphrynus* SCUDDER, 1884]. Carapace rhomboidal, pointed in front, rounded behind, longer than wide, with pair of sessile eyes; sides of posterior portion of carapace concave; abbreviated abdominal tergites sigmoidal. Order of legs 4321 (34, 38, 43, 45). *Penn.,* N.Am.—FIG. 56,*3*. **A. rotundatus,* Ill.;*3a,* dorsal side, ×1.65; *3b,* ventral side, ×1.65 (76).—FIGS. 55,*6;* 56,*8*. *A minor* PETR., Ill.; 55,*6,* ventral side showing position of posterior edge of carapace by dotted line reaching beyond 6th sternite, ×2.8; 56,*8,* dorsal side, ×2.7 (76).

Paratarbus PETRUNKEVITCH, 1945 [**P. carbonarius*]. Carapace wider than long, subtriangular, with pointed anterior end and convex posterior edge. Pair of oval eyes on low, longitudinally elliptic elevation (38). *Penn.,* N.Am.—FIG. 56,*4*. **P. carbonarius,* Ill.; dorsal side, ×2.1 (76).

Hadrachne MELANDER, 1903 [**H. horribilis*]. Carapace slightly longer than wide, pointed in front, with convex posterior edge. No eyes (38). *Penn.,* N.Am.—FIG. 56,*5*. **H. horribilis,* Ill.; dorsal side, ×1 (76).

Discotarbus PETRUNKEVITCH, 1913 [**D. deplanatus*]. Carapace subtriangular, flat, slightly wider than long, with sides convex. Abdomen flat, disc-shaped, wider than carapace (34, 36). *Penn.,* N. Am.—FIG. 56,*6*. **D. deplanatus,* Ill.; dorsal side, ×2.7 (76).

Phalangiotarbus HAASE, 1890 [**Architarbus subovalis* WOODWARD, 1872]. Carapace fairly semicircular, with straight posterior edge. ?Eyes lacking. *Carb.,* Eu.—FIG. 57,*1*. **P. subovalis* (WOODW.), Eng.; *1a,* dorsal side; *1b,* ventral side, ×9 (76).

Goniotarbus PETRUNKEVITCH, 1945 [**Geraphrynus tuberculatus* POCOCK, 1911]. Carapace with 3 pairs of eyes on flat, trifoliate tubercle; posterior edge convex, slightly sinuous, sides convex (43, 45). *U.Carb.,* Eu.—FIG. 58, *4*. **G. tuberculatus* (POCOCK), Eng.; *4a,* dorsal surface, ×3.6; *4b,* ventral surface, ×4.6 (76).—FIG. 58,*5*. *G. angulatus* (POCOCK), Eng.; dorsal, ×3.5 (76).

Mesotarbus PETRUNKEVITCH, 1949 [**M. inter-*

medius]. Carapace longer than wide, with 3 pairs of eyes on flat tubercle (43, 45). *U.Carb.*, Eu.—FIGS. 55,*2;* 59,*2*. **M. intermedius*, Eng.; 55,*2*, ocular tubercle with eyes, ×27; 59,*2*, dorsal side, ×4.7 (76).—FIG. 58,*2*. *M. angustus* (POCOCK), Eng.; dorsal side, ×5.4 (76).—FIG. 59,*5*. *M. eggintoni* (POCOCK), Eng.; dorsal side, ×5.4 (76).—FIG. 58,*3*. *M. hindi* (POCOCK), Eng.; dorsal side, ×5.5 (76).

Leptotarbus PETRUNKEVITCH, 1945 [**Geraphrynus torpedo* POCOCK, 1911]. Carapace much longer than wide, pointed in front; posterior edge slightly convex. Three pairs of eyes on flat tubercle. Abdomen narrow (38, 43, 45). *U.Carb.*, Eu.—FIGS. 55,*3;* 59,*3*. **L. torpedo* (POCOCK), Eng.; 55,*3*, ocular tubercle with eyes, ×29; 59,*3a*, dorsal side, ×7; 59,*3b*, ventral side, ×7 (76).

Orthotarbus PETRUNKEVITCH, 1945 [**Geratarbus minutus* PETR., 1913]. Carapace wider than long, with straight posterior edge. Three pairs of eyes on flat tubercle (38, 43, 45). *Carb.*, N.Am.-Eu.—FIG. 55,*7*. **O. minutus* (PETR.), Ill.; *7a*, dorsal

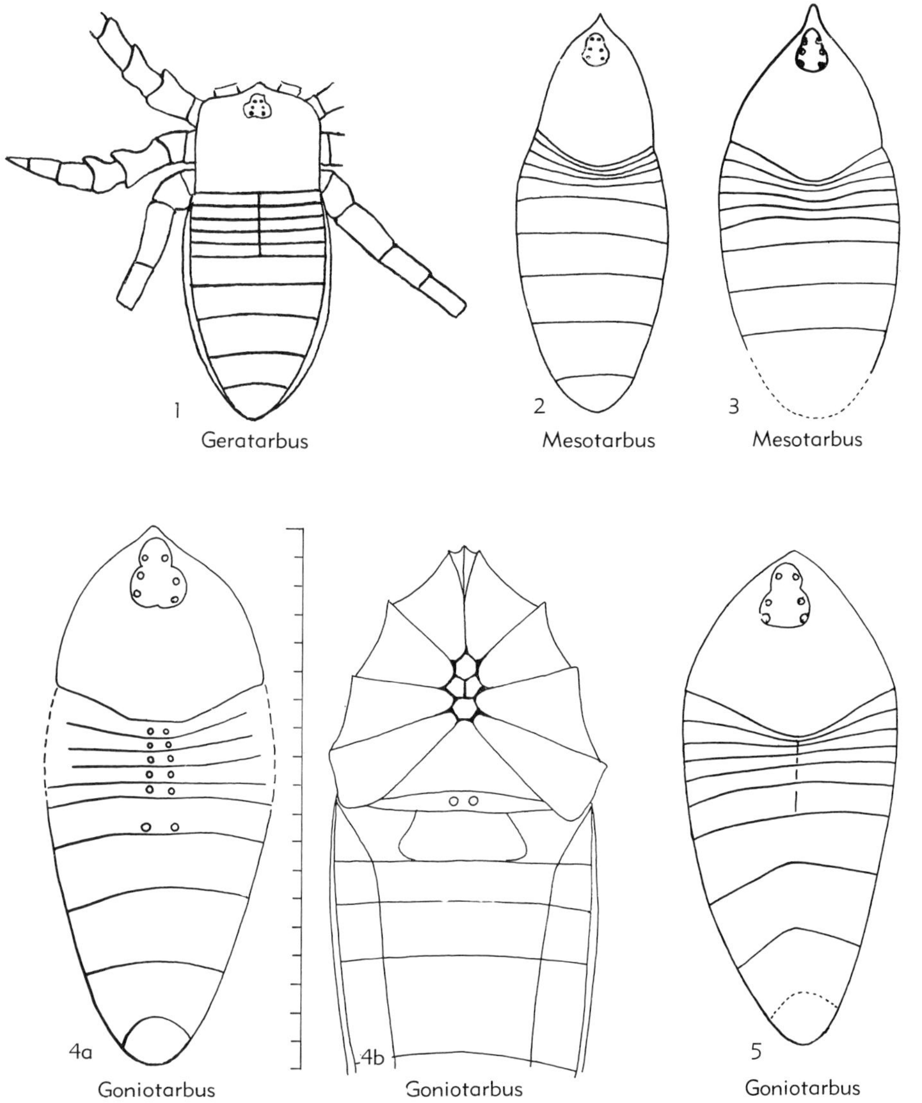

FIG. 58. Architarbida. *1, Geratarbus lacoei*, Penn., Ill., ×3; *2, Mesotarbus angustus*, U.Carb., Eng., ×5.4; *3, Mesotarbus hindi*, U.Carb., Eng., ×5.5; *4, Goniotarbus tuberculatus*, U.Carb., Eng.; *4a*, dorsal side, ×3.6; *4b*, ventral side, ×4.6; *5, Goniotarbus angulatus*, U.Carb., Eng., ×3.5 (76) (p. P90-P92).

side, ×3.5; *7,* ventral side, ×3.5.—FIG. 55,*1,8*. *O. robustus* PETR., Ill.; *1a,* ocular tubercle of type, ×27; *1b,* of paratype, ×30; *8,* dorsal side of another paratype, showing heart, ×2.4; 60*A,B,* photographs of holotype showing dorsal and ventral sides, ×4 (76).—FIG. 59,*1*. *O. nyranensis* PETR., Czech.; ventral side, ×3.7 (76).

Geratarbus SCUDDER, 1890 [**G. lacoei*]. Carapace rectangular, wider than long, with 3 pairs of eyes on flat tubercle (38, 45). *Carb.,* N.Am.-Eu.—FIGS. 55,*4;* 58,*1*. **G. lacoei,* Ill.; 55,*4,* ocular tubercle, ×50 (76); 58,*1,* dorsal side of holotype, ×3 (76).

Ootarbus PETRUNKEVITCH, 1945 [**O. pulcher*]. Carapace almost semicircular with straight posterior edge; 3 pairs of eyes on flat tubercle (38). *Penn.,* N.Am.—FIGS. 55,*5;* 59,*4*. **O. pulcher,* Ill.; 55,*5,* ocular tubercle, ×33; 59,*4,* dorsal side, ×3 (76).

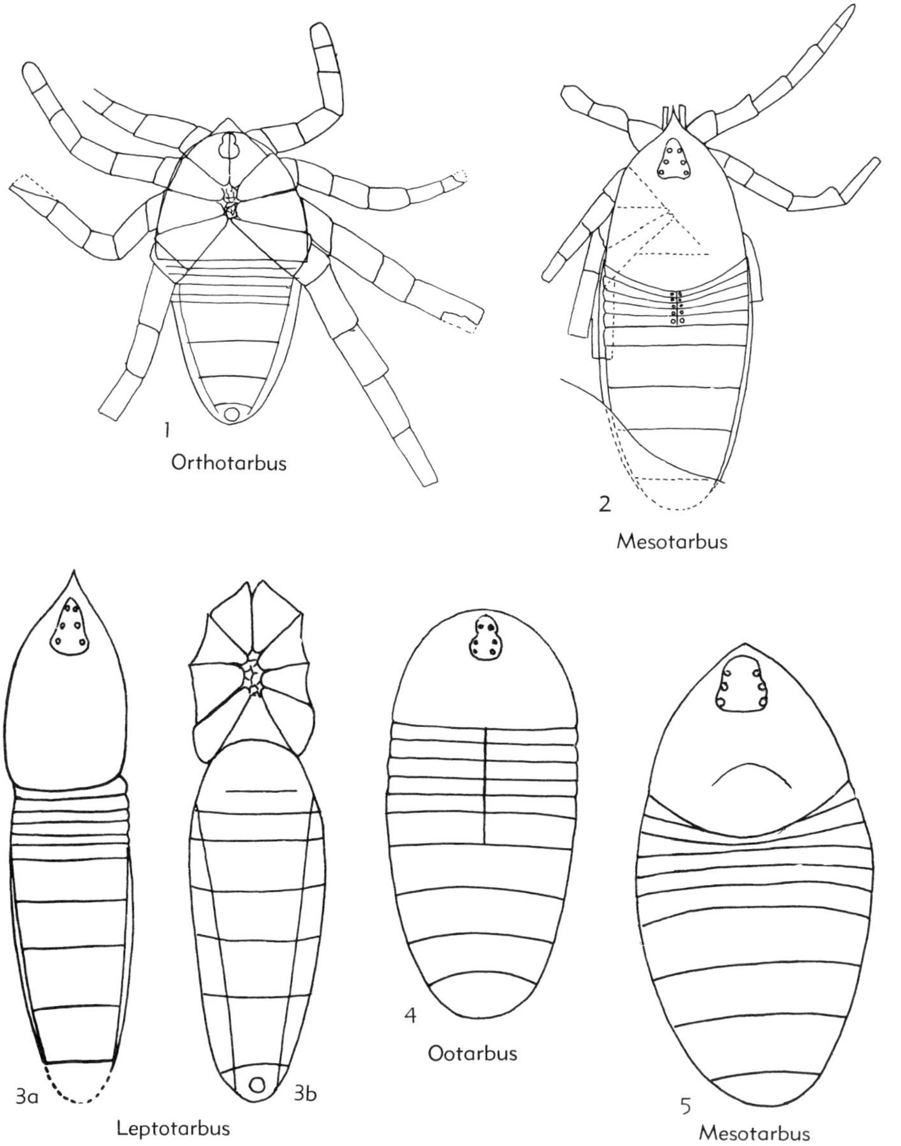

FIG. 59. Architarbida. *1, Orthotarbus nyranensis,* U.Carb., Czech., ×3.7; *2, Mesotarbus intermedius,* U. Carb., Eng., ×4.7; *3, Leptotarbus torpedo,* U.Carb., Eng.; *3a,b,* dorsal and ventral sides, ×7; *4, Ootarbus pulcher,* Penn., Ill., ×3; *5, Mesotarbus eggintoni,* U.Carb., Eng., ×5.4 (76) (p. P91-P92).

FIG. 60. *Orthotarbus robustus*, Penn., Ill.; *A,B*, photographs of dorsal and ventral sides, ×4 (76).

Metatarbus PETRUNKEVITCH, 1913 [**M. triangularis*]. Carapace subtriangular, wider than long, all sides convex. Ocular tubercle without eyes (34). *Penn.*, N.Am.—FIG. 56,*7*. **M. triangularis*, Ill.; dorsal side, ×2 (76).

Family OPILIOTARBIDAE Petrunkevitch, 1945

Coxae triangular, trochanters of 1st and 2nd pairs of legs single-jointed, those of 3rd and 4th pairs 2-jointed. Abdomen with 11 tergites. *Carb.*

Opiliotarbus POCOCK, 1900 [**Architarbus elongatus* SCUDDER, 1890]. Carapace almost semicircular, with straight posterior edge as in *Ootarbus*, but without ocular tubercle or eyes. Sternum wanting, all coxae meet in median line (38, 45). *Penn.*, N. Am.—FIG. 56,*2*. **O. elongatus* (SCUDDER), Ill.; *2a*, dorsal side, ×2.7; *2b*, ventral side, ×2.7 (76).

Family HETEROTARBIDAE Petrunkevitch, 1913

Coxae of 1st and 2nd pairs of legs small and separated, those of 3rd and 4th pairs much longer, cylindrical, contiguous in median line by their proximal ends. Trochanters single-jointed. First legs very long and slender (34, 45). *Carb.*

Heterotarbus PETRUNKEVITCH, 1913 [**H. ovatus*]. Carapace subtriangular, much wider than long, with convex sides and posterior edge. No eyes (34). *Penn.*, N.Am.—FIG. 56,*1*. **H. ovatus*, Ill.; dorsal side, with chelicerae and coxae superimposed on carapace, ×2.7 (76).

Order ACARIDA Nitzsch, 1818

[*nom. correct.* ENGELMANN, 1860 (*ex* Acari SUNDEVALL,[1] 1833, *ex* Acarina NITZSCH, 1818] [=Monomerostomata LEACH, 1815; Acarides VAN HEYDEN, 1826; Acarenses DUGÈS, 1834; Acari *auctt.*] [Type: *Acarus* LINNÉ, 1758]

Latigastra in which the abdominal segmentation is usually lost, so that the body shows no external segmentation (Figs. 61, 62) or may be divided into a proterosoma and a hysterosoma, with a division line between the 2nd and 3rd pairs of legs, the

[1] NITZSCH was the first to use the name Acarina as an order. BRÜNNICH used the name Acari in 1772 merely as the plural of the genus *Acarus*, the only genus recognized by him. LATREILLE used the name Acaridiae in 1806 as a family name and included it in only 5 genera. LEACH substituted for his order Monomerostomata the name Acari in 1819, but raised it to the rank of a class.

former usually subdivided into a gnathosoma (formerly called capitulum) and a propodosoma. Gnathosoma bears chelicerae and pedipalpi, both presenting numerous modifications of structure. Chelicerae are of 2 main types which may be found in different representatives of the same suborder or an entire suborder may have the same type of structure. The chelate type occurs in all suborders. It is the only type in Holothyrina, Notostigmatina and Tetrapodilina, in which they are 2-jointed, with movable finger either ventral or lateral in position and used for grasping. Chelicerae of the piercing type are modified as stylets of different structure. Pedipalpi are 6-jointed, but like chelicerae, subject to modifications. Typical number of legs is, of course, 4 pairs, but some have 3 pairs, some, as Tetrapodilina, have 2 pairs, and a few Acarida have even a single pair. Number of leg segments varies from 2 to 7, but usual number is 6. Tarsi end in a pair of claws often accompanied by "caruncle" or arolium. In most Acarida body is soft, some have a dorsal scutum covering hysterosoma. Mouth ventral to chelicerae and protected by a labrum from above and by an extension of the ventral body wall of the gnathosoma from below.

Digestive system presents several variations. Usually a well-developed fore-gut with pharynx and esophagus is followed by a complicated mid-gut with diverticles and ends in a hind-gut which opens to outside by an anus. The latter is ventral in most cases, but dorsal in others, situated near end of abdomen. Excretory system consists of a pair of malpighian tubes and an excretory vesicle. Most species have no heart. When one is present, it is small, almost rudimentary, with a single pair of ostia. Respiration accomplished either by means of tracheal tubes, or is cutaneous when organs of respiration are wanting. Position of tracheal spiracles is of greatest importance in classi-

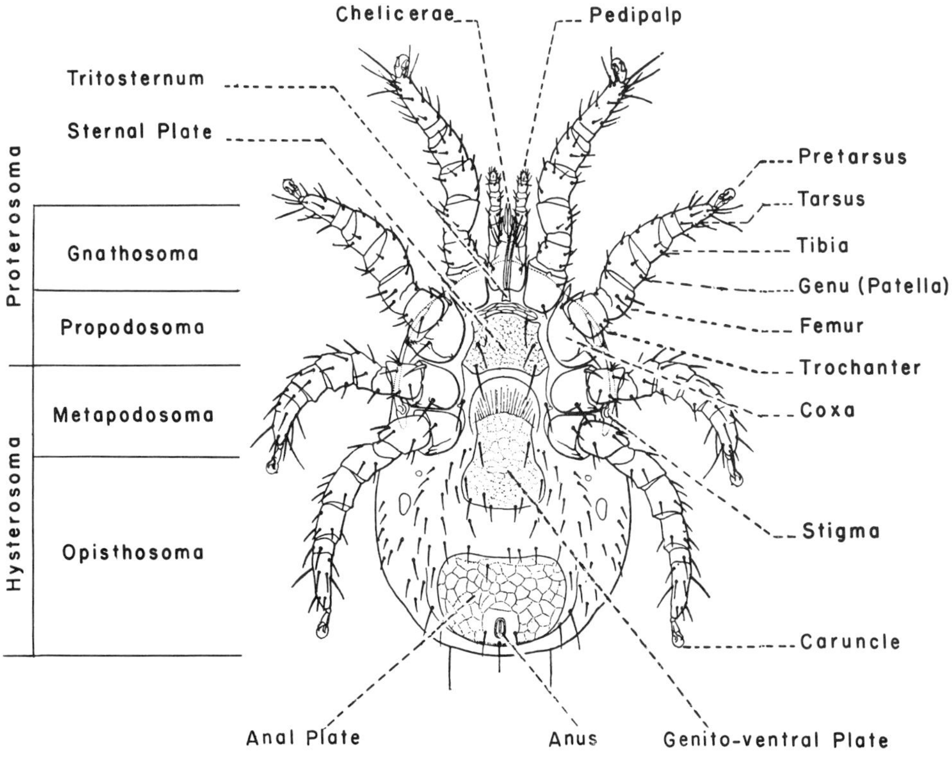

FIG. 61. External morphology of Acarida. *Myonyssus decumani,* a mite parasitic on rats, ventral side (81).

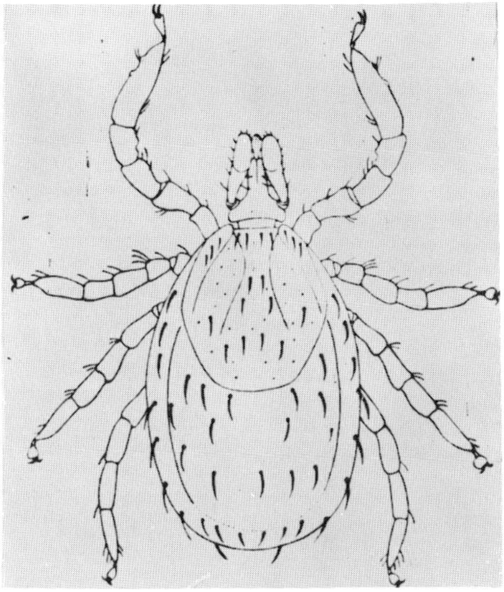

FIG. 62. *Ixodes ricinus,* common European tick, female, ×8 (after Nuttal).

fication. Notostigmatina have 4 pairs of spiracles on dorsal surface of opisthosoma. Parasitoidea (Mesostigmata) have a single pair on metapodosoma. Oribatoidea have *areae porosae* in various regions of their body. Nervous system highly concentrated. Eyes present in some species, but majority are blind. Pair of highly developed salivary glands open into mouth cavity; in some species they are modified as silk glands.

Sexes are separate. Genital openings vary considerably in position. Embryologically these openings belong to the 2nd abdominal sternite in all Arachnida. In some Acarida this position has been secondarily displaced. Males of the parasitic family Demodicidae present an extreme in this respect. In them the genital opening is dorsal in position, situated above either 1st, 2nd or 3rd pair of legs, while in the female it occupies its normal position on the ventral side between the hind coxae. Male copulatory apparatus consists of a penis. Some females have an ovipositor, but eggs are usually laid directly from the genital opening. Development is often complicated. From the egg emerges a 6-legged larva which on molting first turns into a protonymph, then a deutonymph and a tritonymph before reaching adult stage. In some species, as for example in *Pediculoides ventricosus,* all these stages are passed within the egg. The development in some itch mites is so rapid, that a new generation is produced in 10 to 14 days.

Acarida is the 2nd largest order of Arachnida, one of the richest in variations of structure and habit. They are the only order in which plant-eating species are found. Some lead completely parasitic existence on vertebrate and invertebrate hosts. Others are temporary parasites. Several families are strictly aquatic, fresh-water and marine. Some are polymorphic. Many are only a fraction of a millimeter in size, some on the limit of unaided human vision. The largest female ticks gorged with blood reach 25 mm. in size, but their normal, undistended length is no more than 5 or 6 mm. On account of their small size, fossil Acarida are best known from the Baltic amber. The oldest fossil acarid was fairly common in the Devonian. *Dev.-Rec.*

Recent Acarida are divided into 5 suborders, 162 families and nearly 1400 genera.

Suborder NOTOSTIGMATINA With, 1903

[*nom. correct.* PEARSE, 1947 (*pro* Notostigmata WITH, 1903)] [Type: *Opilioacarus* WITH, 1903]

Four pairs of dorsolateral stigmata. Two pairs of eyes. *Rec.*

Suborder HOLOTHYRINA Reuter, 1909

[*nom. correct.* PETRUNKEVITCH, herein (*pro* Holothyroidea REUTER, 1909)] [Type: *Holothyrus* GERVAIS, 1842]

Two pairs of lateral stigmata. No eyes. *Rec.*

Suborder TROMBIDIINA Dugès, 1839

[*nom. correct.* PETRUNKEVITCH, herein (*pro* Trombidiei DUGÈS, 1839)] [=Trombidiformes REUTER, 1909] [Type: *Trombidium* FABRICIUS, 1775]

A pair of stigmata on or near gnathosoma. *Dev.-Rec.*

Superfamily TROMBIDIOIDEA Leach, 1815

[*nom. transl.* PETRUNKEVITCH, herein (*ex* Trombidides LEACH, 1815)] [=Prostigmata KRAMER, 1877]

Stigmata at base of chelicerae. *Dev.-Rec.*

Family EUPODIDAE C.L.Koch, 1842

Body soft, with short gnathosoma. Chelae modified. *Dev.-Rec.*

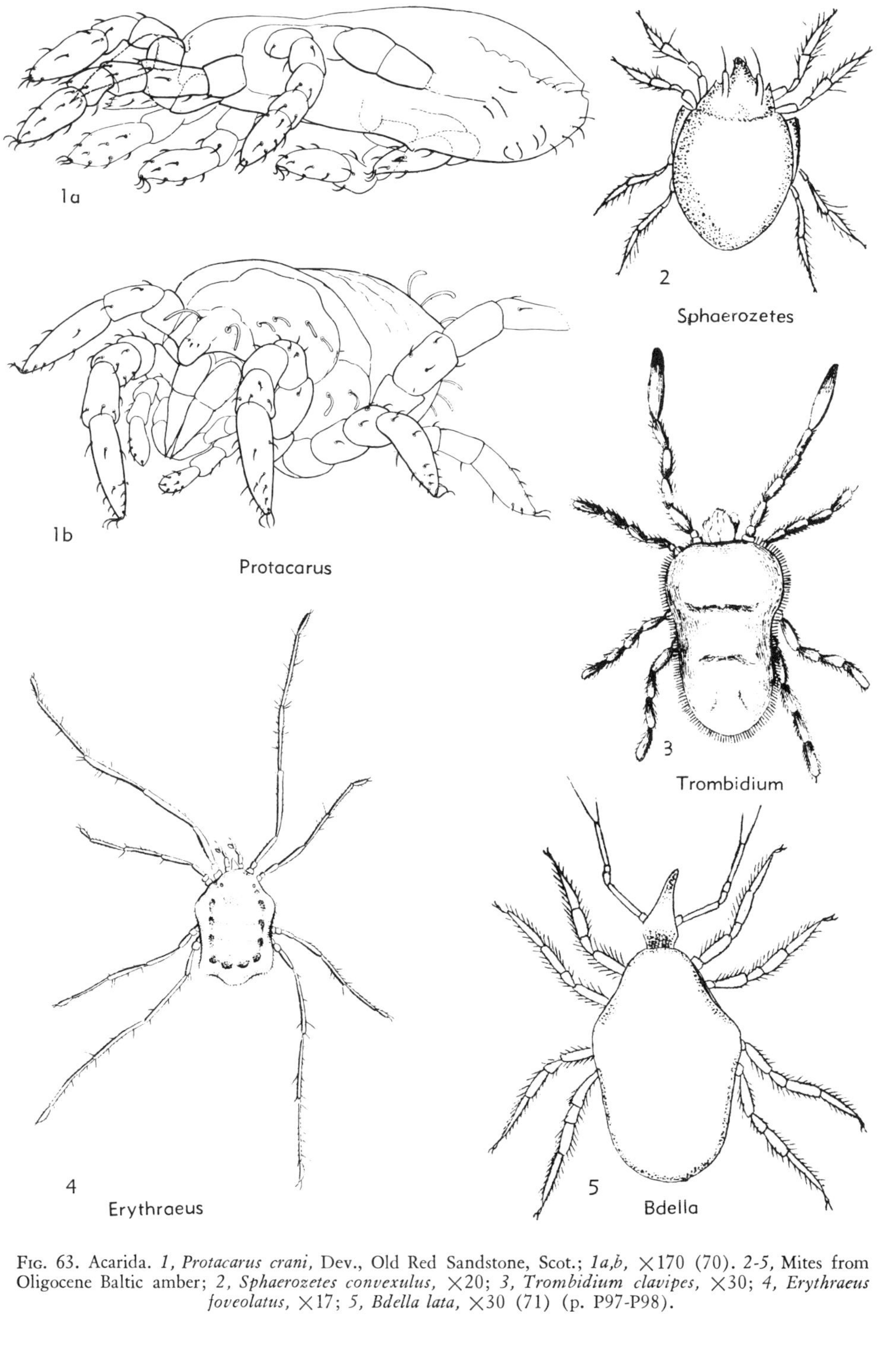

FIG. 63. Acarida. *1, Protacarus crani,* Dev., Old Red Sandstone, Scot.; *1a,b,* ×170 (70). *2-5,* Mites from Oligocene Baltic amber; *2, Sphaerozetes convexulus,* ×20; *3, Trombidium clavipes,* ×30; *4, Erythraeus foveolatus,* ×17; *5, Bdella lata,* ×30 (71) (p. P97-P98).

Protacarus HIRST, 1923 [**P. crani*]. Pedipalpi 7-jointed, with 3 terminal hairs shaped like claws. *Dev.*, Scot.—FIG. 63,*1*. **P. crani; 1a*, side view, ×270; *1b*, holotype, ×170 (70).

Family BDELLIDAE Dugès, 1834

A suture between proterosoma and hysterosoma. Tarsi with pretarsus, 2 claws and pulvillus. Two pairs of eyes. *Oligo.-Rec.*

Bdella LATREILLE, 1795 [**Acarus longicornis* LINNÉ, 1758]. *Oligo.-Rec.*—FIG. 63,*5*. *B. lata* KOCH & BERENDT, Oligo., Balt.; ×30 (71).

Family ERYTHRAEIDAE Oudemans, 1802

Chelicerae stylet-like, single-jointed. Legs long and slender. One or 2 pairs of eyes. *Oligo.-Rec.*

Erythraeus LATREILLE, 1806 [**Acarus phalangioides* DEGEER, 1778] [=*Rhyncholophus* DUGÈS, 1834] *Oligo.-Rec.* [7 species in Baltic amber].—FIG. 63,*4*. *E. foveolatus* (KOCH & BERENDT), Oligo., Balt.; dorsal view, ×17 (71).

Arythaena MENGE, 1854 [**A. troguloides*]. *Oligo.*, Balt.

Family TROMBIDIIDAE Leach, 1815

[*nom. correct.* FLACH, 1864 (*pro* Trombidides LEACH, 1815]

Palpal tibia with apical claw. Chelicerae with bladelike 2nd joint. *Oligo.-Rec.*

Trombidium FABRICIUS, 1775 [**Acarus holosericeus* LINNÉ, 1758]. *Oligo.-Rec.* [6 species in Baltic amber].—FIG. 63,*3*. *T. clavipes* KOCH & BERENDT, Oligo., Balt.; ×30 (71).

Family ANYSTIDAE Oudemans, 1902

Body short and broad. Palpal tibia with 3 claws. Two pairs of eyes. *Oligo.-Rec.*

Anystis VAN HEYDEN, 1826 [**Trombidium cornigerum* HERMANN] [=*Actineda* C.L. KOCH, 1836]. *Oligo.-Rec.* [*A. venustula* (KOCH & BERENDT), Oligo., Balt.].

Family CHEYLETIDAE Leach, 1815

[*nom. correct.* BANKS, 1904 (*pro* Cheyletides LEACH, 1815)]

Body with 1 or 2 dorsal shields. Chelicerae stylet-like. Tarsi with 2 claws and empodium. *Mio.-Rec.*

Cheyletus LATREILLE, 1796 [**Acarus eruditus* SCHRANK, 1781]. *Mio.-Rec.* [*C. burmiticus* COCKERELL, 1917, Mio., Burmese amber].

Family TETRANYCHIDAE Donnadieu, 1875

Plant-feeders with soft skin. Chelicerae fused at base forming a stylophore which is used as a piercing organ. One pair of eyes. *Oligo.-Rec.*

Tetranychus DUFOUR, 1832 [**T. lintearius*]. *Oligo.-Rec.* [*T. gibbus* KOCH & BERENDT, Oligo., Balt.].

Suborder PARASITINA Reuter, 1909

[*nom. correct.* PEARSE, 1936 (*pro* Parasitiformes REUTER, 1909)] [=Mesostigmata + Metastigmata CANESTRINI, 1891] [Type: *Parasitus* LATREILLE, 1795]

One pair of stigmata, lateral or posterior to legs. *Oligo.-Rec.*

Superfamily PARASITOIDEA Oudemans, 1901

[*nom. transl.* BANKS, 1915 (*ex* Parasitidae OUDEMANS, 1901)] [=Mesostigmata CANESTRINI, 1891]

Stigmata lateral to legs. *Oligo.-Rec.*

Family PHYTOSEIIDAE Berlese, 1916

No constriction between podosoma and opisthosoma. Dorsal plate entire. Coxae movable. *Oligo.-Rec.*

Seius C.L. KOCH, 1836 [**S. togatus*]. *Oligo.-Rec.* (*S. bdelloides* KOCH & BERENDT, Oligo., Balt.).

Superfamily IXODOIDEA Leach, 1815

[*nom. transl.* BANKS, 1904 (*ex* Ixodides LEACH, 1815)]

Stigmata behind legs, with stigmal plates. Body sclerotized. *Oligo.-Rec.*

Family IXODIDAE Leach, 1815

[*nom. correct.* MURRAY, 1877 (*pro* Ixodides LEACH, 1815)]

Dorsal scutum present. Hypostome modified as piercing organ. *Oligo.-Rec.*

Ixodes LATREILLE, 1795 [**Acarus ricinus* LINNÉ, 1758]. *Oligo.-Rec.* (*I. tertiarius* SCUDDER, Oligo., Wyo.).—FIG. 62,*1*. *I. ricinus,* Rec., Eu.; female, dorsal side, ×8 (62).

Dermacentor C.L. KOCH, 1844 [**Acarus reticulatus* FABRICIUS]. *Pleisto.-Rec.,* Eu.; *Pleisto.,* Galicia.

Suborder ACARINA Leach, 1815

[*nom. correct.* NITZSCH, 1818 (*pro* Acarides LEACH, 1815] [=Sarcoptiformes REUTER, 1909] [Type: *Acarus* LINNÉ, 1758]

Oligo.-Rec.

Superfamily ACAROIDEA Latreille, 1802

[*nom. transl.* EWING & NESBITT, 1942 (*ex* Acaridiae LATREILLE, 1802)]

Tarsi with caruncles. *Oligo.-Rec.*

Family ACARIDAE Latreille, 1802

[*nom. correct.* EWING & NESBITT, 1942 (*pro* Acaridiae LATREILLE, 1802)] [=Tyroglyphidae DONNADIEU, 1868]

Transverse groove between protero- and hysterosoma. Five pairs of setae on propodosoma. *Oligo.-Rec.*

Acarus LINNÉ, 1758 [**A. siro*]. *Oligo.-Rec.* [*A. rhombeus* KOCH & BERENDT, Oligo., Balt.].

Tyroglyphites PAMPALONI, 1902 [**T. miocenicus*]. *M.Mio.*, Sicily.

Superfamily ORIBATOIDEA Dugès, 1834

[*nom. correct.* BANKS, 1904 (*pro* Oribatei DUGÈS, 1834)]

Body wall sclerotized. Tarsi without caruncles. *Oligo.-Rec.*

Family BELBIDAE Willmann, 1931

Genital and anal openings on common ventral plate. Chelicerae chelate. Legs long. *Oligo.-Rec.*

Damaeus C.L. KOCH, 1836 [**D. auritus*]. *Oligo.-Rec.* [*D. genadensis* SELLNICK, Oligo., Balt.].

Gymnodamaeus KULCZYNSKI, 1902 [**Damaeus bicostatus* C.L. KOCH, 1835]. *Oligo.-Rec.* [*G. kulczynskii* SELLNICK, Oligo., Balt.].

Belbites PAMPALONI, 1902 [**B. disodilia*]. *Mio.*, Sicily.

Family CAMISIIDAE Sellnick, 1928

Genital and anal openings contiguous; anal opening with a pair of opercular plates. *Oligo.-Rec.*

Camisia VAN HEYDEN, 1826 [**Notaspis segnis* HERMANN, 1804]. *Oligo.-Rec.* [*C. horrida fossilis* SELLNICK, Oligo., Balt.].

Nothrus C.L. KOCH, 1836 [**N. palustris*]. *Oligo.-Rec.* [*N. illantus* SELLNICK, Oligo., Balt.].

Family CARABODIDAE Willmann, 1931

Dorsum sculptured or punctate. Propodosoma with leaflike protuberances. *Oligo-Rec.*

Carabodes C.L. KOCH, 1836 [**C. coriaceus*]. *Oligo.-Rec.* [*C. gerberi* SELLNICK, Oligo., Balt.].

Otocepheus BERLESE, 1904 [**O. longior*]. *Oligo.-Rec.* [*O. niger* SELLNICK, Oligo., Balt.].

Tectocepheus BERLESE, 1913 [**Tegeocranus velatus* MICHAEL, 1888]. *Oligo.-Rec.* [*T. similis* SELLNICK, Oligo., Balt.].

Cepheus C.L. KOCH, 1836 [**C. latus*]. *Oligo.-Rec.* [*C. implicatus* (SELLNICK), Oligo., Balt.].

Xenillus ROBINEAU-DESVOIDY, 1839 [**Notaspis tegeocranus* HERMANN, 1804]. *Oligo.-Rec.* [*X. tegeocraniformis* (SELLNICK), Oligo., Balt.].

Carabodites PAMPALONI, 1902 [**C. pavesii*]. *M.Mio.*, Sicily.

Plateocranus SELLNICK, 1919 [**Nothrus sulcatus* KARSCH]. *Oligo.* [*P. sulcatus* (KARSCH), Oligo., Balt.].

Scutoribates SELLNICK, 1919 [**S. perornatus*]. *Oligo.*, Balt.

Family CERATOZETIDAE Jacot, 1925

With pteromorphs attached to hysterosoma and curving ventrally. *Oligo.-Rec.*

Chamobates HULL, 1916 [**Oribata cuspidata* MICHAEL, 1884]. *Oligo.-Rec.* [*C. difficilis* SELLNICK, Oligo., Balt.].

Melanozetes HULL, 1916 [**Oribates mollicomus* C.L. KOCH, 1840]. *Oligo.-Rec.* [*M. foderatus* SELLNICK, Oligo., Balt.].

Sphaerozetes BERLESE, 1885 [**Oribates orbicularis* C.L. KOCH, 1835]. *Oligo.-Rec.*—FIG. 63,2. *S. convexulus* (KOCH & BERENDT), Oligo., Balt., ×20 (71).

Family CYMBAEREMEIDAE Willmann, 1931

Pteromorphs wanting. Dorsal shield turned under ventrally. *Oligo.-Rec.*

Cymbaeremaeus BERLESE, 1896 [**Eremaeus cymba* NICOLET, 1855]. *Oligo.-Rec.* [*C. acuminatus* (SELLNICK), Oligo., Balt.].

Tectocymba SELLNICK, 1919 [**T. rara*]. *Oligo.*, Balt.

Family ORIBATIDAE Kramer, 1877

[Type: *Oribata* LATREILLE, 1802] [=Eremaeidae WILLMAN, 1931 (type: *Eremaeus* C.L. KOCH, 1836)]

Dorsal shield not turned under. Hysterosoma without pteromorphs. *Oligo.-Rec.*

Caleremaeus BERLESE, 1910 [**Notaspis monilipes* MICHAEL, 1882]. *Oligo.-Rec.* [*C. gleso* SELLNICK, Oligo., Balt.].

Ceratoppia BERLESE, 1908 [**Notaspis bipilis* HERMANN, 1804]. *Oligo.-Rec.* [*C. bipilis fossilis* SELLNICK, Oligo., Balt.].

Lucoppia BERLESE, 1908 [**Zetes lucorum* C.L. KOCH, 1840]. *Oligo.-Rec.* [*L. simplex* SELLNICK, Oligo., Balt.].

Oppia C.L. KOCH, 1836 [**O. nitens*]. *Oligo.-Rec.* [*O. curvicornum* (SELLNICK), Oligo., Balt.].

Eremaeus C.L. KOCH, 1836 [**E. hepaticus*]. *Oligo.-Rec.* [*E. oblongus fossilis* SELLNICK, Oligo., Balt.].

Licneremaeus PAOLI, 1908 [**Notaspis lichnophorus* MICHAEL, 1882]. *Oligo.-Rec.* [*L. fritschi* SELLNICK, Oligo., Balt.].

Suctobelba PAOLI, 1908 [**Notaspis trigona* MICHAEL, 1888]. *Oligo.-Rec.* [*S. subtrigonus fossilis* SELLNICK, Oligo., Balt.].

Gradidorsum SELLNICK, 1919 [**G. asper*]. *Oligo.*, Balt.

Strieremaeus SELLNICK, 1919 [**S. illibatus*]. *Oligo.*, Balt.

Oppites PAMPALONI, 1902 [**O. melilli*]. *M.Mio.*, Sicily.

Family GALUMNIDAE Grandjean, 1936

Hysterosoma with movably hinged pteromorphs, their ends projecting far beyond line of attachment to body. *Oligo.-Rec.*

Galumna VAN HEYDEN, 1826 [**Notaspis alata* HERMANN, 1804]. *Oligo.-Rec.* [*G. clavata* SELLNICK, Oligo., Balt.].

Family HAPLOZETIDAE Grandjean, 1936

With movable pteromorphs. Tectopedium IV with leaflike extension. *Oligo.-Rec.*

Protoribates BERLESE, 1908 [**Oribata monodactyla* HALLER, 1884]. *Oligo.-Rec.* [*P. longipilis* SELLNICK, Oligo., Balt.].

Family HYPOCHTHONIIDAE Berlese, 1910

Genital and anal openings on common, narrow plate. Hysterosoma flat below, weakly arched above. *Oligo.-Rec.*

Trhypochthonius BERLESE, 1904 [**Hypochthonius tectorum* BERLESE, 1896]. *Oligo.-Rec.* [*T. badiformis* SELLNICK, Oligo., Balt.].

Family HERMANNIELLIDAE Grandjean, 1934

Pair of lateral tubes on hysterosoma, with openings of oil glands. *Oligo.-Rec.*

Hermanniella BERLESE, 1908 [**Hermannia granulata* NICOLET, 1855]. *Oligo.-Rec.* [*H. concamerata* SELLNICK, Oligo., Balt.].

Family LIACARIDAE Willmann, 1931

Hysterosoma without pteromorphs. Dorsal shield not bent under. Coxae 3 and 4 ventral, far from sides. *Oligo.-Rec.*

Cultroribula BERLESE, 1908 [**Notaspis juncta* MICHAEL, 1885]. *Oligo.-Rec.* [*C. lauta* SELLNICK, Oligo., Balt.].

Family NEOLIODIDAE Willmann, 1931

Dorsal shield not bent under. Adults carry exuviae of preceding instars on back. *Oligo.-Rec.*

Neoliodes BERLESE, 1888 [**Notaspis theleproctus* HERMANN, 1804]. *Oligo.-Rec.* [*N. quadriscutatus* SELLNICK, Oligo., Balt.].

Platyliodes BERLESE, 1916 [**Nothrus doderleinii* BERLESE, 1883]. *Oligo.-Rec.* [*P. ensigerus* SELLNICK, Oligo., Balt.].

Embolocarus SELLNICK, 1919 [**E. pergratus*]. *Oligo.*, Balt.

Family ORIBATELLIDAE Jacot, 1925

Pteromorphs not extending anteriorly beyond opisthosoma. Lamellae cover most of propodosoma. *Oligo.-Rec.*

Oribatella BANKS, 1895 [**O. quadridentata*]. *Oligo.-Rec.* [*O. mirabilis* SELLNICK, Oligo., Balt.].

Tectoribates BERLESE, 1910 [**Oribata tecta* MICHAEL, 1883]. *Oligo.-Rec.* [*T. parvus* SELLNICK, Oligo., Balt.].

Family ORIBATULIDAE Jacot, 1929

Pteromorphs do not bend under. Four pairs of genital setae. *Oligo.-Rec.*

Eporibatula SELLNICK, 1928 [**Eremaeus rauschensis* SELLNICK, 1928]. *Oligo.-Rec.* [*E. pellucida* SELLNICK, Oligo., Balt.].

Liebstadia OUDEMANS, 1906 [**Notaspis similis* MICHAEL, 1888]. *Oligo.-Rec.* [*L. similiformis* SELLNICK, Oligo., Balt.].

Scheloribates BERLESE, 1908 [**Zetes latipes* C.L. KOCH, 1844]. *Oligo.-Rec.* [*S. areatus* SELLNICK, Oligo., Balt.].

Family ORIPODIDAE Jacot, 1925

Pteromorphs coalesced along anterior edge of opisthosoma, forming bridge over propodosoma. *Oligo.-Rec.*

Oripoda BANKS, 1904 [**O. elongata*]. *Oligo.-Rec.* [*O. baltica* SELLNICK, Oligo., Balt.].

Family PARAKALUMMIDAE Grandjean, 1936

Pteromorphs large, hinged. A suture between propodo- and hysterosoma. *Oligo.-Rec.*

Neoribates BERLESE, 1914 [**Oribates roubali* BERLESE, 1900]. *Oligo.-Rec.* [*N. borussicus* SELLNICK, Oligo., Balt.].

Family PHENOPELOPIDAE Petrunkevitch, nov.

[*nom. subst.* PETR., herein (*pro* Pelopidae EWING, 1917, *ex Pelops* C.L. KOCH, 1836, *non Pelops* GISTL, 1834)]

Pteromorphs large, hinged, posteriorly not extending beyond line of attachment. Chelicerae long, with minute shears. *Oligo.-Rec.*

Phenopelops PETRUNKEVITCH, 1955 [*nom. subst.* PETR., herein (*pro Pelops* C.L. KOCH, 1836, *non* GISTL, 1834)] [**Notaspis hirsutus* C.L. KOCH, 1836]. *Oligo.-Rec.* [*P. punctulatus* (SELLNICK), Oligo., Balt.].

Family PHTHIRACARIDAE Perty, 1841

Cephalothorax withdrawable into notogaster; opening of latter closed by cephalic shield (aspis). *Oligo.-Rec.*

Hoploderma MICHAEL, 1898 [**Hoplophora laevigata* C.L. KOCH, 1844]. *Oligo.-Rec.* [*H. multipunctata* SELLNICK, Oligo., Balt.].

Oribotritia JACOT, 1924 [**Hoplophora decumana* C.L. KOCH, 1836]. *Oligo.-Rec.* [*O. translucida* SELLNICK, Oligo., Balt.].

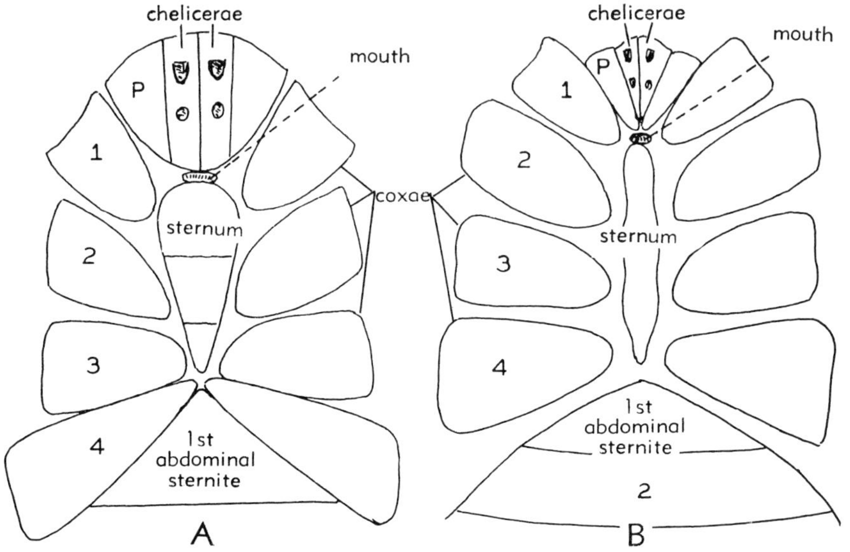

FIG. 64. Coxosternal region of Stethostomata; *A, Plesiosiro madeleyi,* U.Carb., Eng., ×20; *B, Cryptomartus priesti,* U. Carb., Eng., ×14 (76).

Suborder TETRAPODILINA Bremi, 1872

[*nom. correct.* PETRUNKEVITCH, herein (*pro* Tetrapodili BREMI, 1872)] [=Eriophyiformes REUTER, 1909] [Type: *Eriophyes* VON SIEBOLD, 1850]

Body annulate, with only 2 pairs of legs. No respiratory system. Chelicerae stylet-like. Plant-eaters. No fossil representatives of this suborder are known. *Rec.*

Subclass STETHOSTOMATA Petrunkevitch, 1949

Arachnida with broad juncture between cephalothorax and abdomen as in Latigastra, but with configuration of coxosternal region differing from that of other subclasses in having basal joints of chelicerae wedged in between pedipalpal coxae and placed on one level with all other coxae (Fig. 64). Mouth between 1st pedal coxae, immediately in front of sternum. Carapace entire. Abdomen segmented, with fixed number of segments, 11 in one order, 10 in the other. Respiration by means of book lungs. *Carb.*

Order HAPTOPODIDA Pocock, 1911

[*nom. correct.* PETRUNKEVITCH, herein (*pro* Haptopoda POCOCK, 1911)] [Type: *Plesiosiro* POCOCK, 1911]

Carapace with a pair of eyes (Fig. 65, *2,4*). Abdomen composed of 11 segments. Tergites entire, 1st tergite abbreviated. Sternites 3 to 10 bisected by a median longitudinal line; 2nd sternite by far the largest, with strongly convex posterior edge and a pair of round genital openings. Two pairs of book lungs, 1st pair on second sternite, 2nd pair on 3rd sternite, both pairs close to lateral edges (Fig. 65,*1,3*). Lateral pleura unsegmented, soft. Chelicerae 3-jointed, chelate (Fig. 65,*7*). All tarsi subsegmented (Fig. 65,*5,6*). *Carb.*

Family PLESIOSIRONIDAE Pocock, 1911

First tarsus 6-jointed, 2nd to 4th tarsi 4-jointed. Sternum wedge-shaped, pointed behind. *Carb.*

***Plesiosiro** POCOCK, 1911 [**P. madeleyi*]. Carapace longer than wide, with sides converging to a point, straight posterior edge, median longitudinal ridge and pair of lateral ridges. Pair of round eyes

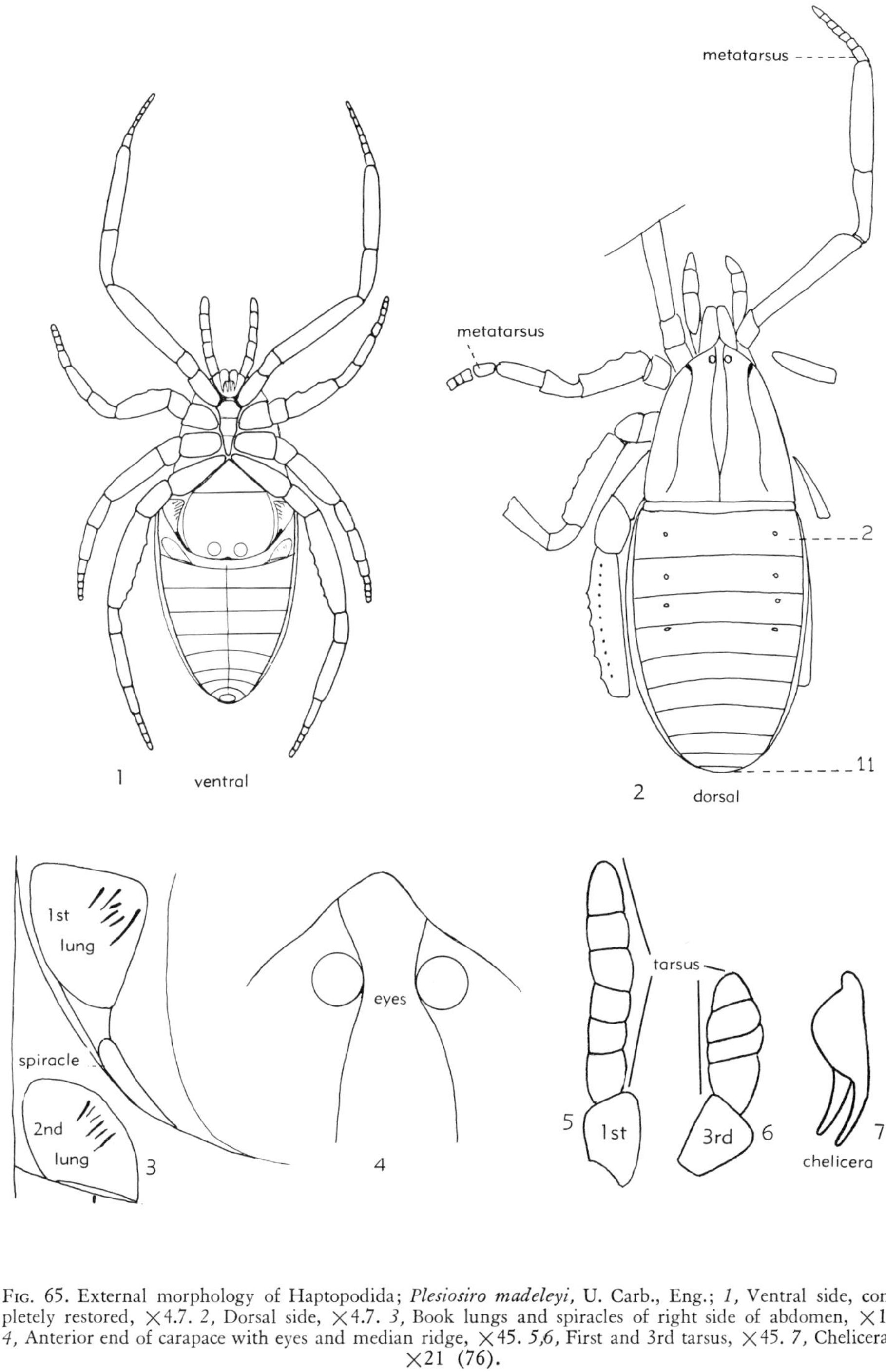

FIG. 65. External morphology of Haptopodida; *Plesiosiro madeleyi*, U. Carb., Eng.; *1*, Ventral side, completely restored, ×4.7. *2*, Dorsal side, ×4.7. *3*, Book lungs and spiracles of right side of abdomen, ×17. *4*, Anterior end of carapace with eyes and median ridge, ×45. *5,6*, First and 3rd tarsus, ×45. *7*, Chelicerae, ×21 (76).

close to anterior edge (Fig. 65,*2,4*). *Carb.*, Eng.—FIG. 64*A*, 65,*1,2*. **P. madeleyi*; 64*A*, coxosternal region, ×20; 65,*1*, ventral side, ×4.7; 65,*2*, dorsal side, ×4.7 (76).

Order ANTHRACOMARTIDA Karsch, 1882

[*nom. correct.* PEARSE, 1936 (*pro* Anthracomarti KARSCH, 1882), emend. PETRUNKEVITCH, 1949] [Type: *Anthracomartus* KARSCH, 1882]

Carapace entire, without eyes. Abdomen composed of 10 segments; 2nd and 3rd tergites fused without trace of division line, forming a common tergite for both segments; 1st and 10th tergites entire, 2nd to 9th divided by 2 pairs of longitudinal lines into a median, a pair of submarginal and a pair of marginal plates (Fig. 66,*2a*); 10th tergite bent under ventrally, its ventral portion wedged in between marginal plates of 9th sternite (Fig. 66,*2b*). All 10 sternites present, clearly separated by intersegmental lines; 1st sternite triangular, entire; 3rd to 9th sternites divided by a pair of longitudinal lines into a median and a pair of lateral (marginal) plates (Fig. 66,*2b*); 10th sternite completely enclosed between strongly curved median 9th plate and ventral portion of 10th tergite. Genital openings on 2nd sternite (Fig. 66,*1c,3c*), a single opening in male, a genital opening followed by a pair of openings leading into seminal receptacles in female. Anus on 10th sternite, disc-shaped, closed by an operculum consisting of a large anterior and a small posterior valve (Fig. 66,*2b,3b*). Chelicerae 3-jointed, their 3rd joint not known, but its existence indicated by articulations at end of 2nd joint (Fig. 66,*1d*). Basal joint wedged in between pedipalpal coxae and these between 1st pedal coxae (Figs. 64*B*, 66,*1a*). Pedipalpi 6-jointed, considerably shorter and more slender than legs (Fig. 68,*5*). Legs 7-jointed, stout and shorter than total length of body. Tarsi with a pair of claws on onychium (Fig. 66,*3a*). Three pairs of book lungs with openings on 2nd, 3rd and 4th sternites (Fig. 66,*1b*), with apodemes for attachment of pulmonary muscles. Sexes separate, recognizable by secondary sexual characters of the 2nd, 3rd and 4th sternites (Fig. 66, *2c,2d,3c*) and by proportions of carapace and legs. Males with narrow depression *"A"* on 2nd sternite and forked depression *"B"* on 3rd sternite. Females with broadly triangular depression *"A"* and semilunar depression *"B."*

The classification of Anthracomartida is based on the shape of the carapace and of the abdomen, on the structure of the 9th tergite and the proportions of the central area formed by the median plates of tergites 4 to 8. The scalloping of the abdominal edge, once stressed by POCOCK as an important character, is of very minor value. *Carb.*

DISCUSSION

For some unaccountable reason Anthracomartida were often considered to be primitive Arachnida. This is decidedly wrong and the reverse is true. The configuration of their coxosternal region, similar to that in Haptopodida and comparable only with that of Xiphosura, represents an extreme stage of the trend consisting in a backward displacement of the mouth. The dorsoventrally flattened body, combined with a sclerotization of the body wall, must have required for proper functioning of the processes of nourishment, respiration and reproduction, the retention of abdominal segmentation, and subdivision of hard tergites and sternites into transverse rows of smaller, hinged plates held together by soft membranes, thus permitting expansion of the body cavity. Even so, hampered by the clumsy shape of their body (Fig. 69) and the awkward position of the mouth, requiring some peculiar way of eating and presumably special type of food; confronted with changing environment and new exigencies of meeting them, yet unable to do so because of the law of irreversibility of evolution which in their case reached its end, these contemporaries of other Carboniferous animals succeeded only in producing minor, generic changes insufficient to prevent early extinction. This becomes strikingly clear when one compares Anthracomartida with scorpions which, having exhausted all macroevolutionary possibilities during the same geological period, managed to survive and to produce new families from one of the intermediate macroevolutionary stages of a trend which in its further stages also proved to be lethal, but which did not involve a conflict between progressive adaptation and irreversibility of evolution.

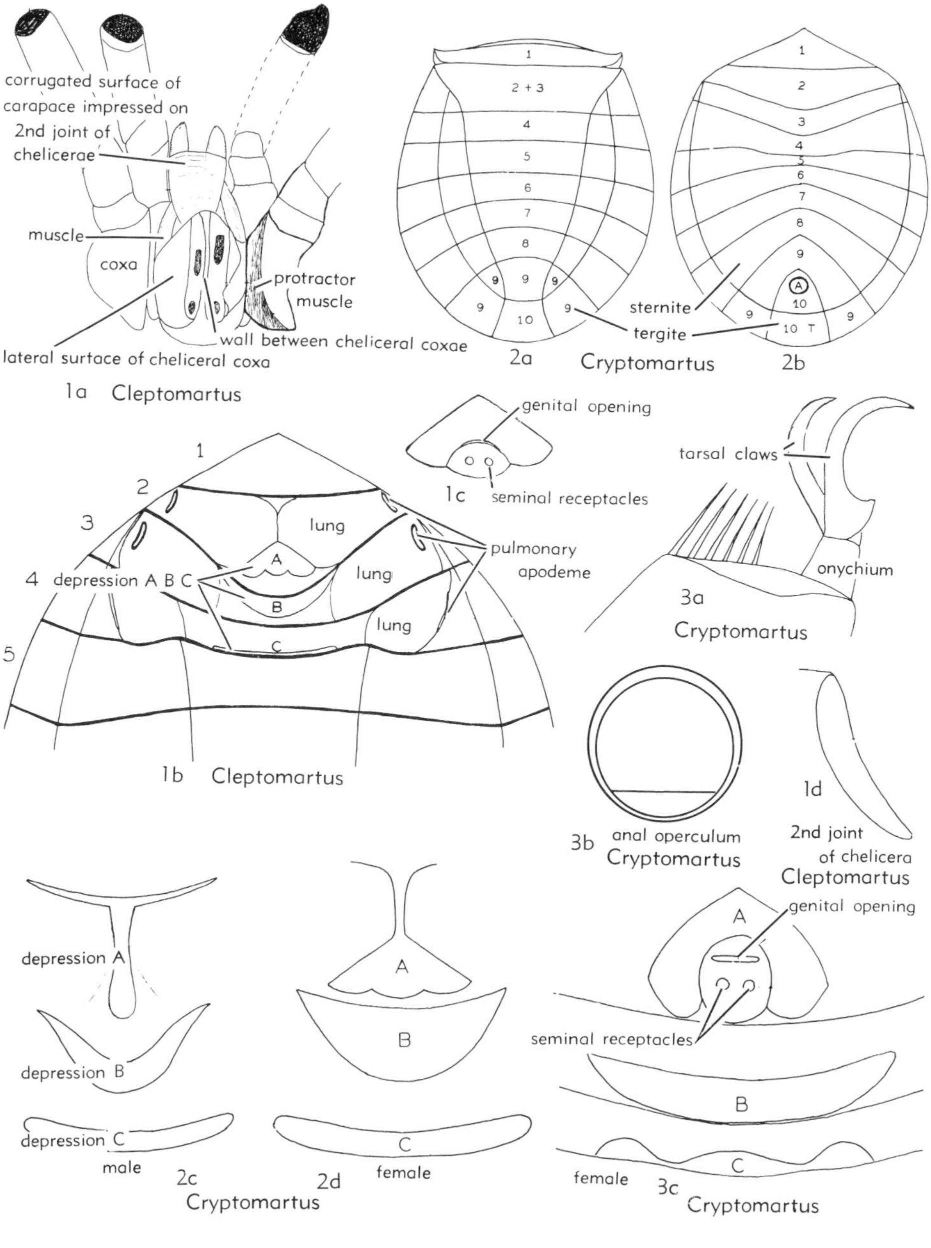

FIG. 66. External morphology of Anthracomartida. *1, Cleptomartus plautus,* U.Carb., Eng.; *1a,* anterior end of coxosternal region, ×8.5; *1b,* genital region, ×12; *1c,* genital opening and seminal receptacles of female, enlarged; *1d,* 2nd joint of chelicerae, enlarged. *2, Cryptomartus priesti,* U.Carb., Eng.; *2a,* segmentation of dorsal side of abdomen; *2b,* ventral side; *2c,d,* ventral abdominal depressions of male and female, ×15.5. *3, Cryptomartus hindi,* U.Carb., Eng.; *3a,* claws; *3b,* anal operculum; *3c,* genital region with ventral depressions of female, ×13 (76) (p. P105).

Family ANTHRACOMARTIDAE Haase, 1890

[=Anthracomartidae + Brachypygidae Pocock, 1911 + Pleomartidae + Coryphomartidae Petrunkevitch, 1945; emend. Petr., 1953]

Characters of the order. *Carb.*

Brachypyge Woodward, 1878 [**B. carbonis*]. Median plate of 9th tergite hexagonal, formed by complete fusion with submarginal plates. Carapace subtriangular, with rounded anterior end. Abdomen distinctly longer than wide, with edge escalloped in posterior half (45). *Carb.*—Fig. 67, *6*. **B. carbonis,* Belg.; dorsal side, ×1.5 (76).

Coryphomartus Petrunkevitch, 1945 [**Anthracomartus triangularis* Petr., 1913]. Ninth tergite as in *Brachypyge*. Carapace triangular with concave sides. Abdomen with smooth edge (38, 45). *U. Carb.,* N.Am.—Fig. 67,*7*. **C. triangularis* (Petr.), N.Scot.; dorsal side, ×3.6 (76).

Maiocercus Pocock, 1911 [**Eophrynus carbonis* Howard & Thomas, 1896]. Carapace subtriangular, sculptured, 9th tergite with submarginal plates. Abdomen orbicular, wider than long, with escalloped edge (45). *Carb.*—Fig. 68,*6*. **M. carbonis* (Howard & Thomas), Eng.; dorsal side, ×1.6 (76).

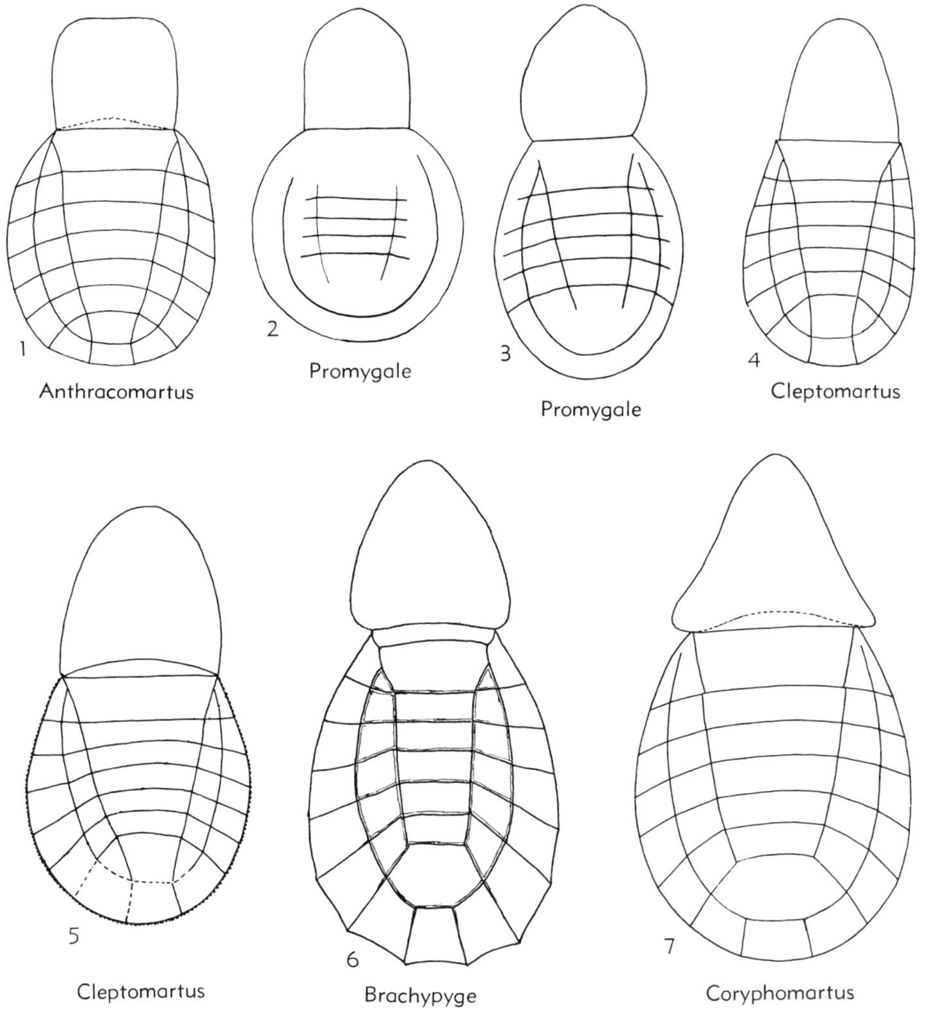

Fig. 67. Anthracomartidae. *1, Anthracomartus völkelianus,* U.Carb., Ger., ×2.5; *2, Promygale bohemica,* U.Carb., Czech., ×3.2; *3, Promygale elegans,* U.Carb., Czech., ×4.4; *4, Cleptomartus planus,* U.Carb., Eng., ×4.7; *5, Cleptomartus denuiti,* Carb., Belg., ×3.4; *6, Brachypyge carbonis,* U.Carb., Belg., ×1.5; *7, Coryphomartus triangularis,* U.Carb., Nova Scotia, ×3.6 (76) (p. P104-P107).

Cryptomartus PETRUNKEVITCH, 1949 [**Anthracomartus hindi* POCOCK, 1911]. Carapace with almost vertical sides converging anteriorly, with median crest on cephalic portion (Fig. 68,*3*). Ninth tergite as in *Maiocercus* (45). *Carb.*—FIGS. 66,*3*; 68,*3*. **C. hindi* (POCOCK), Eng.; 66,*3c*, genital openings and depressions of female, ×13 (76); 68,*3*, dorsal side of female, ×2.7 (76).—FIGS. 64*B*; 66,*2c*,*2d*; 68,*5*. *C. priesti* (POCOCK), Eng.; 64*B*, coxosternal region, ×14; 66,*2c*,*d*, median depressions on 2nd to 4th sternites of male and female, ×15.5; 68,*5*, ventral side, ×2.9 (76).

Cleptomartus PETRUNKEVITCH, 1949 [**C. plautus*]. Carapace flat, as wide as long, evenly rounded in front. Ninth tergite as in *Maiocercus* (43, 45). *Carb.*—FIGS. 66,*1*; 68,*2*; 69. **C. plautus*, Eng.;

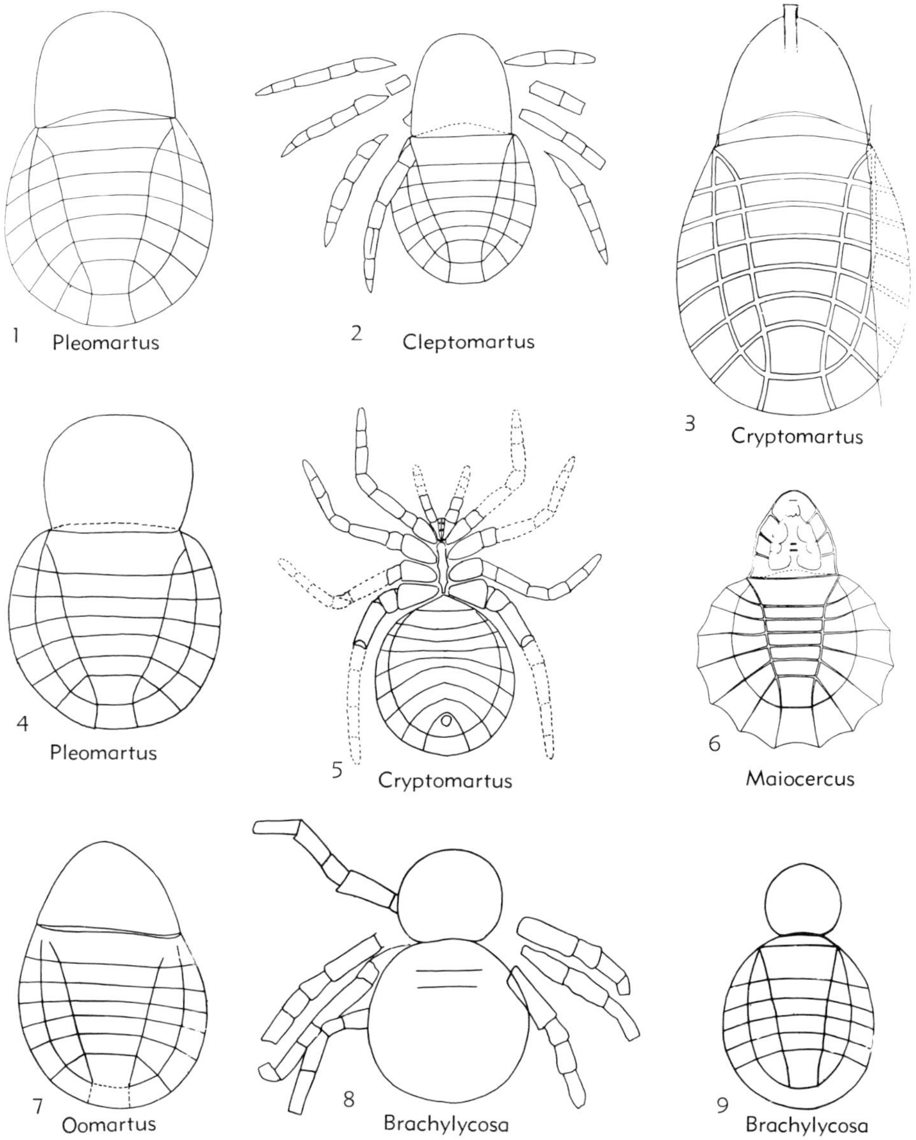

FIG. 68. Anthracomartidae. *1*, *Pleomartus trilobitus*, Penn., Ark., ×3.1; *2*, *Cleptomartus plautus*, U.Carb., Eng., female, ×3; *3*, *Cryptomartus hindi*, U.Carb., Eng., female, ×2.7; *4*, *Pleomartus palatinus*, U.Carb., Ger., ×2.3; *5*, *Cryptomartus priesti*, U.Carb., Eng., ventral side, ×2.9; *6*, *Maiocercus carbonis*, U.Carb., Eng., ×1.6; *7*, *Oomartus nyranensis*, U.Carb., Czech., ×1; *8*, *Brachylycosa carcinoides*, U.Carb., Czech., ×1.3; *9*, *Brachylycosa kustae*, U.Carb., Czech., ×1.3 (76) (p. P104-P107).

FIG. 69. Anthracomartidae. *Cleptomartus plautus,* U.Carb., Eng.; *A,B,* Photographs of dorsal and ventral sides, ×7.5 (76) (p. P105).

66,*1a*, coxae of chelicerae wedged in between pedipalpal coxae, ×8.5; 66,*1b*, external morphology of anterior 5 sternites of female, ×12 (76); 68,*2*, female, ×3; 69, dorsal and ventral sides of holotype, ×7.5 (76).—FIG. 67,*4*. *C. planus* PETR., Eng.; ×4.7 (76).—FIG. 67,*5*. *C. denuiti* (PRUVOST), Belg.; ×3.4 (76).

Anthracomartus KARSCH, 1882 [**A. völkelianus*]. Carapace as wide as long, with parallel sides and almost truncated front. Ninth tergite as in *Maiocercus* (34, 38, 45). *Carb.*, Eu.—FIG. 67,*1*. **A. völkelianus*, Ger.; ×2.5 (76).

Pleomartus PETRUNKEVITCH, 1945 [**Anthracomartus trilobitus* SCUDDER, 1884]. Carapace transversely rectangular, much wider than long. Ninth tergite as in *Maiocercus* (38, 45). *U.Carb.*, N.Am.-Eu.—FIG. 68,*1*. **P. trilobitus* (SCUDDER), Penn. (Pottsville), Ark.; ×3.1 (76).—FIG. 68,*4*. *P. palatinus* (AMMON), Ger.; ×2.3 (76).

Promygale FRITSCH, 1899 [**Kreischeria bohemica* FRITSCH, 1899]. Carapace considerably longer than wide, narrowed anteriorly. Ninth tergite as in *Maiocercus* (9, 45). *Carb.*, Eu.—FIG. 67,*2*. **P. bohemica* (FRITSCH), Czech.; ×3.2;—FIG. 67,*3*. *P. elegans* (FRITSCH), Czech.; ×4.4 (76).

Brachylycosa FRITSCH, 1904 [**Arthrolycosa carcinoides* FRITSCH, 1899]. Carapace more or less disclike. Ninth tergite as in *Maiocercus* (9, 45). *Carb.*, Eu.—FIG. 68,*8*. **B. carcinoides* (FRITSCH), Czech.; ×1.3 (76).—FIG. 68,*9*. *B. kustae* PETRUNKEVITCH, Czech.; ×1.3 (76).

Oomartus PETRUNKEVITCH, 1953 [**O. nyranensis*]. Carapace parabolic, body oval. Ninth tergite as in *Maiocercus* (45). *Carb.*, Eu.—FIG. 68,*7*. **O. nyranensis*, Czech.; ×1 (76).

Anthracophrynus ANDRÉE, 1913 [**A. tuberculatus*]. Abdomen with pair of tubercles on all median plates. Ninth tergite as in *Maiocercus*. *Carb.*, Ger.

Subclass SOLUTA Petrunkevitch, 1949

Arachnida with 2 types of juncture between cephalothorax and abdomen: broad juncture in some and juncture restricted to middle third in others while width of abdomen remains same as that of carapace. Abdomen segmented, composed of 8 to 11 segments. Tergites divided by 2 longitudinal lines into median and a pair of marginal plates. Sternites entire. Coxosternal region of same type as in Recent spiders, i.e., consisting of sternum with lower lip and 5 pairs of coxae in same ventral plane. Chelicerae 2-jointed, retrovert, 2nd joint developed as fang. Pedipalpi 6-jointed, their coxa without maxillary lobe. Legs 7-jointed, with a pair of claws. Respiration by means of book lungs (45). *Dev.-Rec.*

DISCUSSION

All Soluta have the same configuration of the coxosternal region, almost identical with that of Araneida. The question may be raised, *why* are they not placed as an order in the same class with the Araneida? There are several different configurations of the coxosternal region in Caulogastra, just as there are several in Latigastra. The reason for the segregation of Soluta in a subclass of their own is the still labile status of other important characters by which they differ from all Caulogastra and which are not present in any other fossil or living Arachnida. It is the combination of their characters, rather than any particular character, that makes the Soluta different from the other 3 subclasses. As long as no fossil Soluta are found with a different configuration of the coxosternal region, the class may be considered as being represented by a single order, the Trigonotarbida. In accordance with our present knowledge we may assume that the subclass Soluta became extinct in the Late Carboniferous, but left sufficient evidence to show the main trends of its evolution since its appearance in the Devonian; these are the loss of posterior segments in a progressive series until only 8 segments were left and an arrested loss of the 1st anterior segment.

When imperfectly preserved, Soluta may be confused with spiders and (in case of only 4 genera) with Anthracomartida. They can be always distinguished from the latter by the absence of marginal plates in abdominal sternites, and from spiders by the presence of marginal plates in abdominal tergites.

Order TRIGONOTARBIDA Petrunkevitch, 1949

[*nom. correct.* PETR., herein (*pro* Trigonotarbi PETR., 1949)] [Type: *Trigonotarbus* POCOCK, 1911]

Configuration of coxosternal region as in Araneida. *Dev.-Carb.*

Family PALAEOCHARINIDAE Hirst, 1923

Abdomen composed of 11 segments, the last 2 of which are reduced, forming a pygidium (Fig. 70, *1a,2c*). Carapace with pair of median eyes and pair of lateral organs (interpreted by HIRST as compound eyes, 17) (Fig. 70,*2a*). *Dev.*

Palaeocharinoides HIRST, 1923 [**P. hornei*]. Sternum pointed in front, angular behind (17, 45). *Dev.*, Eu.—FIG. 70,*1*. **P. hornei*, Old Red Sandstone, Scot.; *1a*, ventral side with dorsal segmentation shown by dotted lines, ×30; *1b*, carapace in side view, showing crest, ×40; *1c*, sternum, ×60; *1d*, chelicera, ×50 (70).

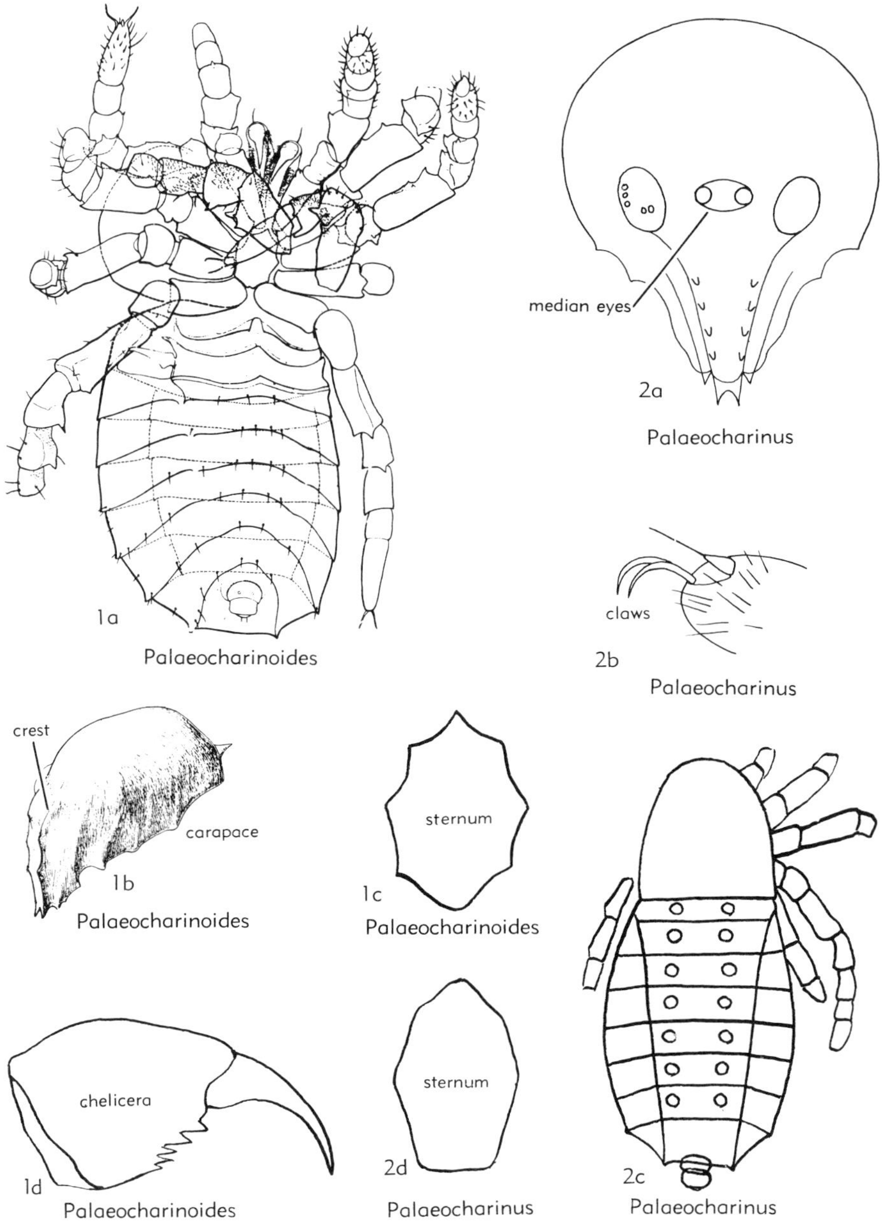

FIG. 70. Order Trigonotarbida; Paleocharinidae. *1, Paleocharinoides hornei*, Dev., Old Red Sandstone, Scot.; *1a*, ventral side of a complete specimen, ×30; *1b*, side view of carapace, ×40; *1c*, sternum, ×60; *1d*, chelicera, ×50. *2, Palaeocharinus* sp., Dev., Old Red Sandstone, Scot.; *2a*, front view of carapace, ×25; *2b*, claws, ×50; *2c*, dorsal side of specimen, ×18; *2d*, sternum, ×60 (70, 76) (p. P108-P109).

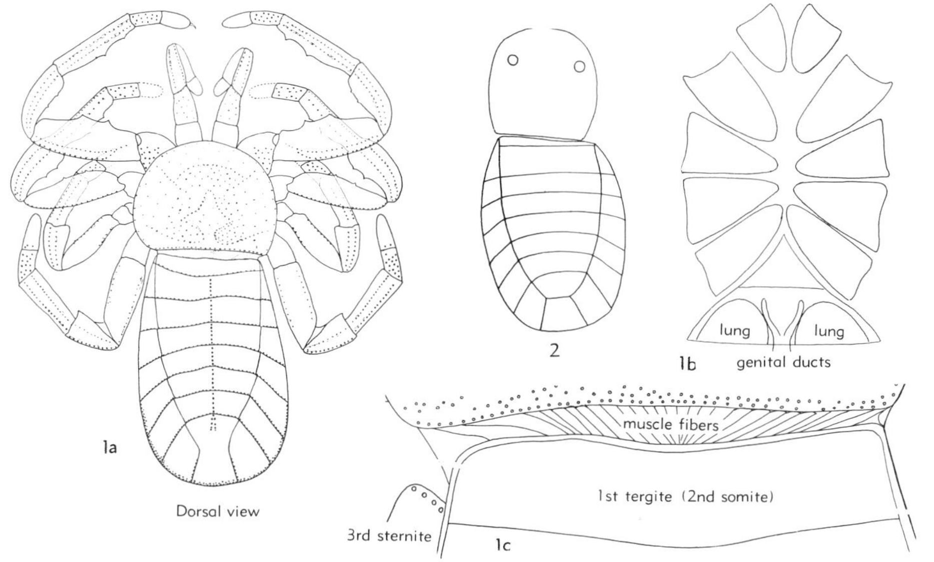

FIG. 71. Anthracosironidae. *1, Anthracosiro woodwardi,* U.Carb., Eng.; *1a,* female, ×2.7; *1b,* coxosternal region, ×6; *1c,* junction between cephalothorax and abdomen showing muscles, ×12. *2, Anthracosiro fritschi,* U.Carb., Eng.; dorsal side, showing eyes, ×6 (76) (p. P109).

Palaeocharinus HIRST, 1923 [**P. rhyniensis*]. Sternum bluntly pointed in front, truncated behind (17, 45) (Fig. 70,*2d*). *Dev.,* Eu.—FIG. 70,*2d.* **P. rhyniensis,* Old Red Sandstone, Scot.; sternum, ×60.—FIG. 70,*2a-c. P.* sp.; *2a,* front view of carapace showing crest, ocular tubercle and pair of special organs (HIRST's "compound eyes"), ×25; *2b,* claws, ×50; *2c,* dorsal side, ×18 (70).

Family ANTHRACOSIRONIDAE Pocock, 1903

Abdomen broadly joined to cephalothorax; 10 abdominal segments, but tergite of 1st segment reduced to a muscular membrane, sclerotized 1st visible tergite in reality belongs to 2nd segment. Single pair of book lungs on 2nd sternite. All 10 sternites present. Anus with round operculum on 10th sternite (45). *Carb.*

Anthracosiro POCOCK, 1903 [**A. woodwardi*]. Carapace flat, rounded in front, with straight posterior edge. Legs stout, granular (45). *U.Carb.,* Eu.—FIGS. 71, 72. **A. woodwardi,* Eng.; 71,*1a,* dorsal side, ×2.7 (legs partly reconstructed); *1b,* sternocoxal region, ×6; *1c,* juncture of abdomen with carapace, ×12; 72, photo of dorsal side, ×5.5.—FIG. 71,*2. A. fritschi* POCOCK, Eng.; dorsal side, showing eyes characteristic of the species, ×6 (76).

Family EOPHRYNIDAE Karsch, 1882

[emend. PETRUNKEVITCH, 1949]

Abdomen composed of 9 segments; anterior tergite as wide as carapace; but juncture reduced to middle third. Ninth tergite entire, enclosed between median and lateral plates of 8th tergite and often fused with median plate of the latter, forming with it a single, hourglass-shaped plate. Carapace usually sculptured and with an anterior, pointed, often spikelike process. Abdomen usually tubercular, often with a pair of posterior thorns. Four pairs of book lungs with openings on 2nd to 5th sternites (45). *Carb.*

Subfamily AREOMARTINAE Petrunkevitch, nov.

Abdominal tergites smooth or granular, but not with regularly arranged rows of large tubercles. *Carb.*

Areomartus PETRUNKEVITCH, 1913 [**A. ovatus*]. Tergites smooth. Carapace triangular, with shallow, hexagonal depressions. No eyes (34, 45). *Penn.,* N.Am.—FIG. 73,*2.* **A. ovatus,* W.Va.; ×5 (76).

Vratislavia FRITSCH, 1904 [**Architarbus silesiacus* RÖMER, 1876]. Carapace transversely rectangular, presumably without eyes. Four slender spurs at posterior end of abdomen. *Carb.,* Eu.—FIG. 74,*3.* **V. silesiaca* (RÖMER), Ger.; ventral side, ×4 (66a).

Anzinia PETRUNKEVITCH, 1953 [**Kreischeria theveninì* PRUVOST, 1919]. Carapace triangular, finely granular in front, ornamented at base, presumably without eyes. Abdomen with 4 short, pointed projections at posterior end (45). *Carb.*, Eu.—FIG. 73,*3*. **A. thevenini* (PRUVOST), Fr.; X4.7 (76).

Hemiphrynus FRITSCH, 1899 [**H. longipes*]. Carapace with anterior, spikelike projection. Abdomen with 4 short, pointed thorns at posterior end (9, 45). *Carb.*, Eu.—FIGS. 73,*6*; 77,*4*. **H. longipes*, Czech.; 73,*6*, ventral side, X1.35; 77,*4*, photo of holotype, X0.9 (76).—FIG. 74,*1*. *H. hofmanni* FRITSCH, Czech.; ventral side, X1.35 (76).

Pseudokreischeria PETRUNKEVITCH, 1953 [**Eophrynus pococki* GILL, 1924]. Carapace ornamented, with short anterior spike. Abdomen with 4 posterior projections (45). *U.Carb.*, Eu.—FIG. 75,*1*. **P. pococki* (GILL), Eng.; carapace and front part of abdomen, X2.5—FIG. 75,*2*. *P. varia* (PETR.), Eng.; dorsal side of posterior portion of abdomen, X2.5 (76).

Subfamily EOPHRYNINAE Karsch, 1882

[*nom. transl.* PETRUNKEVITCH, herein (*ex* Eophrynidae KARSCH, 1882)]

Abdominal tergites with rows of conspicuous, large tubercles. *Carb.*

Cyclotrogulus FRITSCH, 1904 [**Eophrynus sturii* HAASE, 1890]. Carapace triangular, without eyes. Abdomen rounded behind, with 6 rows of dorsal abdominal tubercles and a pair of spurious divi-

FIG. 72. *Anthracosiro woodwardi*, U.Carb., Eng.; photograph of dorsal side, X5.5 (76) (p. P109).

sion lines (9, 14, 45). *Carb.*, Eu.—FIG. 75,*3*. **C. sturii* (HAASE), Ger.; ×2.4 (76).

Poconomia PETRUNKEVITCH, 1953 [**Trigonomartus whitei* EWING, 1930]. Carapace rounded in front, ornamented, with pair of eyes. Abdomen rounded behind, with 6 rows of tubercles and pair of spurious division lines. Median 2 rows of tubercles largest. *Penn.*, N.Am.—FIG. 76,*1*. **P. whitei* (EWING), ×6.6 (76).

Kreischeria GEINITZ, 1882 [**K. wiedei*]. Carapace ornamented, subtriangular, presumably with rounded anterior end, longer than wide, with pair of eyes. Abdomen with 4 rows of tubercles and 4 posterior spurs (45). *Carb.*, Eu.—FIG. 76,*3*. **K. wiedei*, Ger.; ×1 (76).

Petrovicia FRITSCH, 1904 [**P. proditoria*]. Carapace with pointed anterior spike. Abdomen with 4 rows of tubercles and 4 posterior spurs which form a

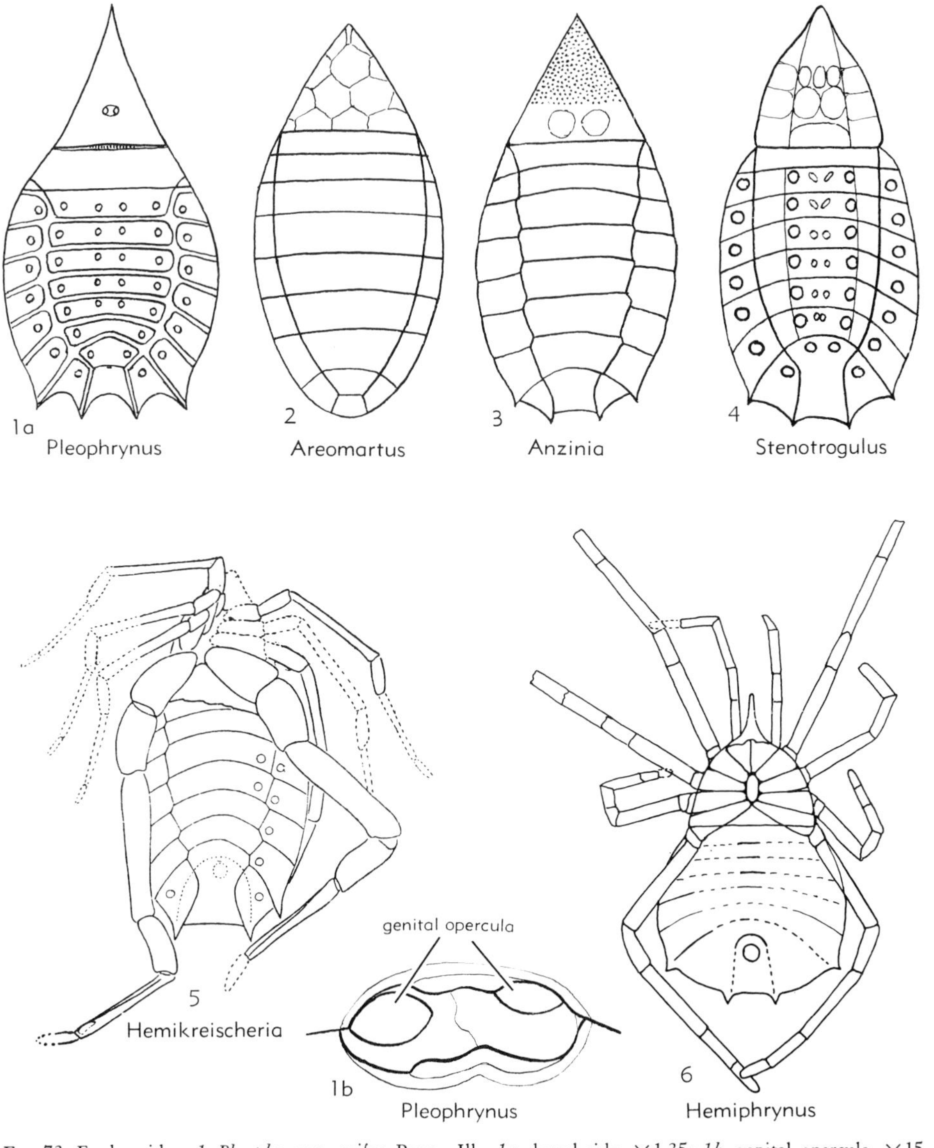

FIG. 73. Eophrynidae. *1, Pleophrynus ensifer*, Penn., Ill.; *1a*, dorsal side, ×1.35; *1b*, genital opercula, ×15. *2, Areomartus ovatus*, Penn., Ill., ×5. *3, Anzinia thevenini*, U.Carb., Fr., ×4.7. *4. Stenotrogulus salmii*, U. Carb., Czech., ×1.9. *5, Hemikreischeria geinitzi*, U.Carb., Fr., ×1.8. *6, Hemiphrynus longipes*, U. Carb., Czech., ×1.35 (76) (p. P109-P112).

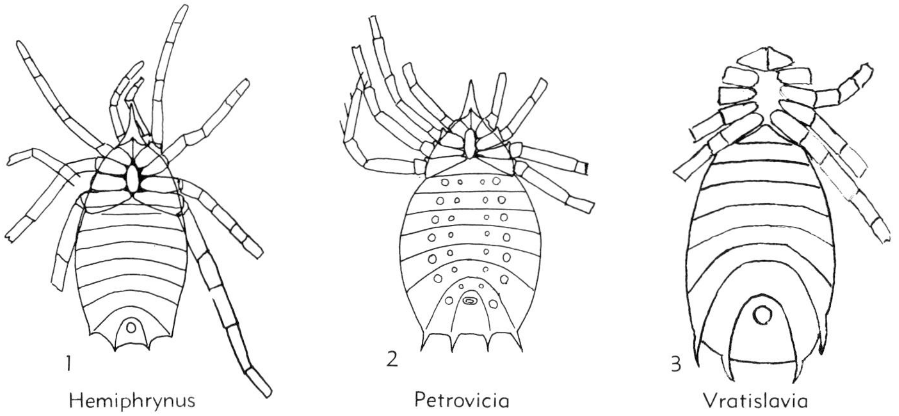

FIG. 74. Eophrynidae. *1, Hemiphrynus hofmanni,* U.Carb., Czech., ×1.35. *2, Petrovicia proditoria,* U.Carb., Czech., ×1.35. *3, Vratislavia silesiaca,* U.Carb., Pol. (Silesia), ×4 (76) (p. P109-P111).

transverse row. Tibiae with ventral spines (9, 45). *Carb.,* Eu.—FIGS. 74,*2;* 77,*1.* **P. proditoria,* Czech.; 74,*2,* ventral side, ×1.35; 77,*1,* photo of holotype, ×1.3 (76).

Hemikreischeria FRITSCH, 1904 [**Kreischeria geinitzi* THEVENIN, 1902]. Shape of carapace not known. Abdomen with 4 rows of tubercles and 4 posterior spurs. Fourth legs much stouter and longer than the others. *Carb.,* Eu.—FIG. 73,*5.* **H. geinitzi* (THEVENIN), Fr.; ×1.8 (79).

Stenotrogulus FRITSCH, 1904 [**Eophrynus salmii* STUR, 1877]. Carapace subtriangular, ornamented, longer than wide, without eyes. Six rows of tubercles, medial row composed of smaller and elongated, not round, tubercles. Abdomen with a pair of spurious division lines and 4 posterior spurs (9, 45). *Carb.,* Eu.—FIG. 73,*4.* **S. salmii* (STUR), Ger.; ×1.9 (76).

Acrokreischeria PETRUNKEVITCH, 1953 [**Kreischeria verrucosa* POCOCK, 1911]. Carapace acutely drawn out in front, with concave sides and rounded corners, ornamented, without eyes. Abdomen with 6 rows of tubercles, a pair of spurious division lines and 4 short projections behind (45). *U.Carb.,* Eu.—FIG. 76,*2.* **A. verrucosa* (POCOCK), Wales, ×1.8 (76).

Pleophrynus PETRUNKEVITCH, 1945 [**P. ensifer*]. Carapace triangular, longer than wide, acutely pointed in front, ornamented, with pair of eyes on tubercle situated behind middle. Abdomen with 6 rows of tubercles and 4 posterior spurs (38, 45). *Penn.,* N.Am.—FIG. 73,*1.* **P. ensifer,* Ill.; *1a,* dorsal side, ×1.35; *1b,* genital atrium with a pair of opercula, ×15 (76).

Eophrynus WOODWARD, 1871 [**Curculioides prestvici* BUCKLAND, 1837]. Carapace subtriangular, with pointed spike in front, ornamented, as long as wide, with pair of eyes on tubercle slightly behind middle. Abdomen with 6 rows of tubercles and 4 spurs at end (12, 45). *U.Carb.,* Eu.—FIGS. 32; 78. **E. prestvici* (BUCKLAND), Eng.; 32; photo, dorsal and ventral sides of lectotype, ×3; 78, drawing of same, showing details of structure, ×1.5 (76).

Family TRIGONOTARBIDAE Petrunkevitch, 1949

Abdomen composed of 8 segments and broadly joined to cephalothorax. Eighth tergite entire, enclosed between median and marginal plates of strongly curved 7th tergite and turned under ventrally, where it occupies the space between the ends of the strongly curved 7th sternite, behind the 8th sternite (Fig. 79,*1*). Posterior edge of abdomen evenly rounded. Carapace subtriangular, with a pair of eyes. Two pairs of book lungs with openings on 2nd and 3rd sternites (43, 45). *Carb.*

Trigonotarbus POCOCK, 1911 [**T. johnsoni*]. Carapace as long as wide, with convex sides and pair of eyes on tubercle in middle (45). *U.Carb.,* Eu.—FIGS. 77,*2;* 79. **T. johnsoni,* Eng.; 77,*2,* photo of holotype, ×6; 79,*1,* ventral side, ×6.8; 79,*2,* dorsal side, ×9 (76).

Family TRIGONOMARTIDAE Petrunkevitch, 1949

Abdomen composed of 8 segments, and broadly joined to cephalothorax. Eighth tergite with marginal plates and limited to dorsal surface, lacking ventrally folded part. *Carb.*

Trigonomartus PETRUNKEVITCH, 1913 [**Anthracomartus pustulatus* SCUDDER, 1884]. Carapace triangular, high, with median crest and pustulose

FIG. 75. Eophrynidae. *1, Pseudokreischeria pococki,* U.Carb., Eng.; anterior part of dorsal side, ×2.5. *2, Pseudokreischeria varia,* U.Carb., Eng.; posterior dorsal side of abdomen, ×2.5. *3, Cyclotrogulus sturii,* U.Carb., Czech., ×2.4 (76) (p. P110-P111).

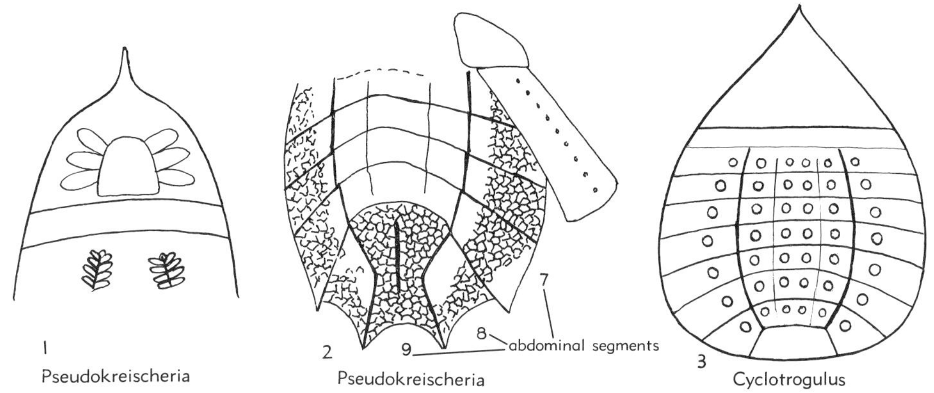

surface, without eyes. Abdomen with pustulose surface (34, 45). *U.Carb.,* Eu.-N.Am.—FIGS. 80,*1;* 81,*1.* **T. pustulatus* (SCUDDER), Ill.-Ger.; 80,*1,* dorsal side, ×2.8; 81,*1,* ventral side, ×2.9 (76).

Aphantomartus POCOCK, 1911 [**A. areolatus*]. Carapace ornamented, subtriangular, with pair of eyes ⅓ its length from anterior end. Dorsal lines dividing abdomen strongly converging, straight. Surface of abdomen with flattened tubercles (45). *Carb.,* Eu.—FIG. 80,*3.* **A. areolatus,* Wales; ×3 (76).

Lissomartus PETRUNKEVITCH, 1949 [**Trigonomartus schucherti* PETR., 1913]. Carapace and abdomen smooth, not ornamented (43, 45). *Penn.,* N. Am.—FIG. 80,*2.* **L. schucherti* (PETR.), Ill.; *2a,* dorsal side, ×2.4; *2b,* ventral side, ×2 (76).

Phrynomartus PETRUNKEVITCH, 1945 [**Eophrynus waechteri* GUTHÖRL, 1938]. Carapace ornamented, abdomen with 4 rows of tubercles and 4 posterior spurs. *Carb.,* Eu.—FIG. 80,*4.* **P. waechteri* (GUTHÖRL), Saar; ×3.3 (76).

Elaverimartus PETRUNKEVITCH, 1953 [**E. pococki*]. Carapace semicircular, without eyes, granular. Second and 3rd coxae rectangular (cylindrical), 4th coxae triangular (45). *U.Carb.,* Eu.—FIGS. 80,*6;* 81,*2.* **E. pococki,* Scot.; 80,*6,* dorsal side, ×7.5; 81,*2,* ventral side, ×6 (76).

Planomartus PETRUNKEVITCH, 1953 [**Anthraco-*

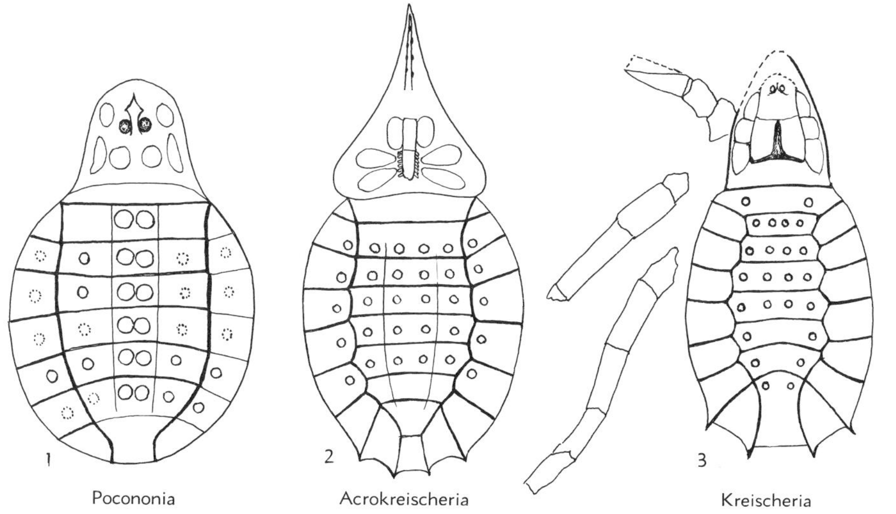

FIG. 76. Eophrynidae. *1, Pocononia whitei,* Penn., Va., ×6.6. *2, Acrokreischeria verrucosa,* U.Carb., Wales, ×1.8. *3, Kreischeria wiedei,* U.Carb., Ger., ×1 (76) (p. P111-P112).

martus krejcii Kušta, 1883]. Carapace subtriangular, rounded in front. Eighth tergite entire, more or less hexagonal (45). *Carb.*, Eu.—Fig. 80,*5*. **P. krejcii* (Kušta), Czech.; dorsal side (dotted lines indicate original derivation of 8th tergite from a median and a pair of marginal plates fused into a single plate), ×3 (76).

Subclass CAULOGASTRA Pocock, 1893

[Type: *Araneus* Clerck, 1757]

Arachnida with a constriction between cephalothorax and abdomen, due to a modification of the 1st abdominal segment (or

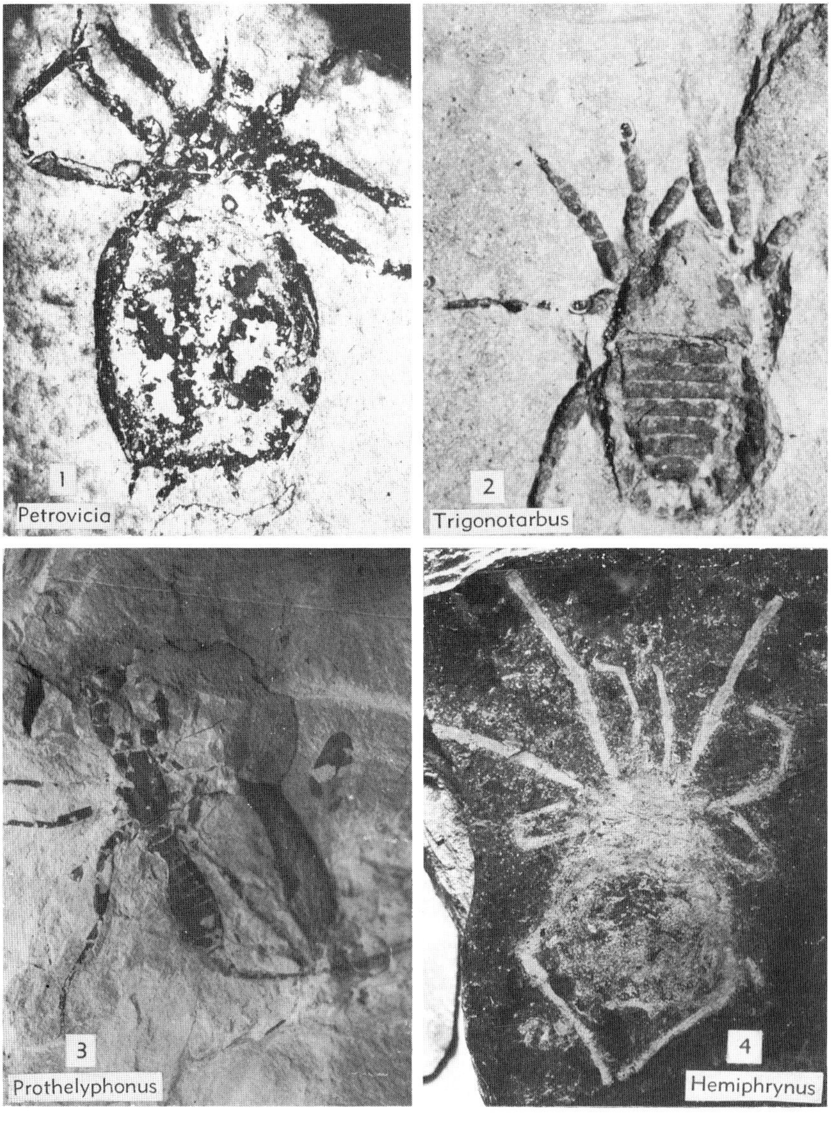

Fig. 77. Trigonotarbida *(1,2,4)*; Thelyphonida *(3)*. *1, Petrovicia proditoria,* U.Carb., Czech.; photograph of holotype, ×1.3. *2, Trigonotarbus johnsoni,* U.Carb., Eng.; photograph of holotype, ×6. *3, Prothelyphonus bohemicus,* U.Carb., Czech.; photograph of holotype, ×0.65. *4, Hemiphrynus longipes,* U.Carb., Czech.; photograph of holotype, ×0.9 (76) (p. P110-P120).

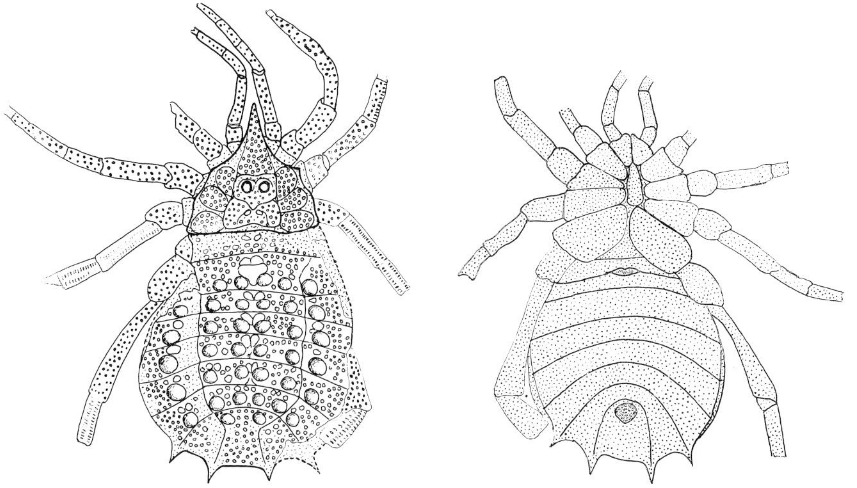

FIG. 78. *Eophrynus prestvici,* U.Carb., Eng.; dorsal and ventral sides of lectotype, ×1.5 (76) (p. P112).

2nd, when the 1st is lost) into a pedicel permitting greater independent motion of the abdomen as a unit. Existence of a pedicel not always apparent when the latter is concealed under the posterior edge of the carapace or under the projecting anterior edge of the tergite of the abdominal segment following upon the pedicel (45). *?Dev., Carb.-Rec.*

Superorder LATISTERNA Petrunkevitch, 1949

[Type: *Eukoenenia* BÖRNER, 1901]

Caulogastra with pedipalpal coxae widely separated by anterior sternal plate and devoid of maxillary lobes (Fig. 82,*2b*). *Jur.-Rec.*

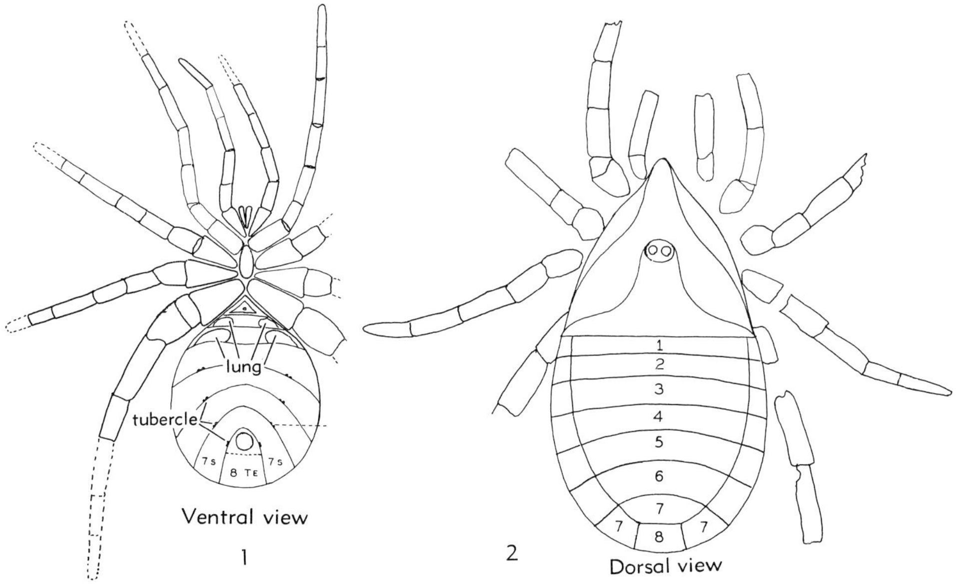

FIG. 79. *Trigonotarbus johnsoni,* U.Carb., Eng.; *1,* ventral side, ×6.8; *2,* dorsal side showing details of external morphology, ×9 (76) (p. P112).

Order PALPIGRADIDA Thorell, 1888

[*nom. correct.* PEARSE, 1936 (*pro* Palpigradi THORELL, 1888)] [=Microthelyphonida[1] GRASSI & CALANDRUCCIO, 1885] [Type: *Eukoenenia* BÖRNER, 1901 (=*Koenenia* GRASSI-C., 1885, *non* BEUSHAUSEN, 1884)]

Carapace segmented, composed of 4 sclerites: large propeltidium extending just beyond 2nd pair of legs; pair of very small, triangular mesopeltidia belonging to the segment bearing 3rd pair of legs; and metapeltidium belonging to last thoracic segment, bearing 4th pair of legs (Fig. 82, *2a,b*). Abdomen composed of 11 segments; 12th segment lost, 9th to 11th segments reduced to a pygidium bearing at its end a dorsal, multijointed whip or flagellum which gives the creature the appearance of a minute whip scorpion (Fig. 82,*1a*). Mouth anteroventral, provided with anterior and posterior lip, the latter hinged to 1st sternal plate which is in reality sternite 2+3 separating widely the pedipalpal and 1st pedal coxae. This is followed by 2 independent mesosternites and a metasternite (Fig. 82,*2b*). All coxae movable, none with maxillary lobes. Chelicerae 3-jointed, chelate (Fig. 82,*2c*),

[1] The name Microthelyphonida has priority over Palpigradi, but after the publication of THORELL's paper, in which the latter name was proposed, has not been used nor is generally known at present. Since there is no International Rule prohibiting the use of a later name in the case of an order, it is not reasonable to replace a generally accepted name by an obsolete one.

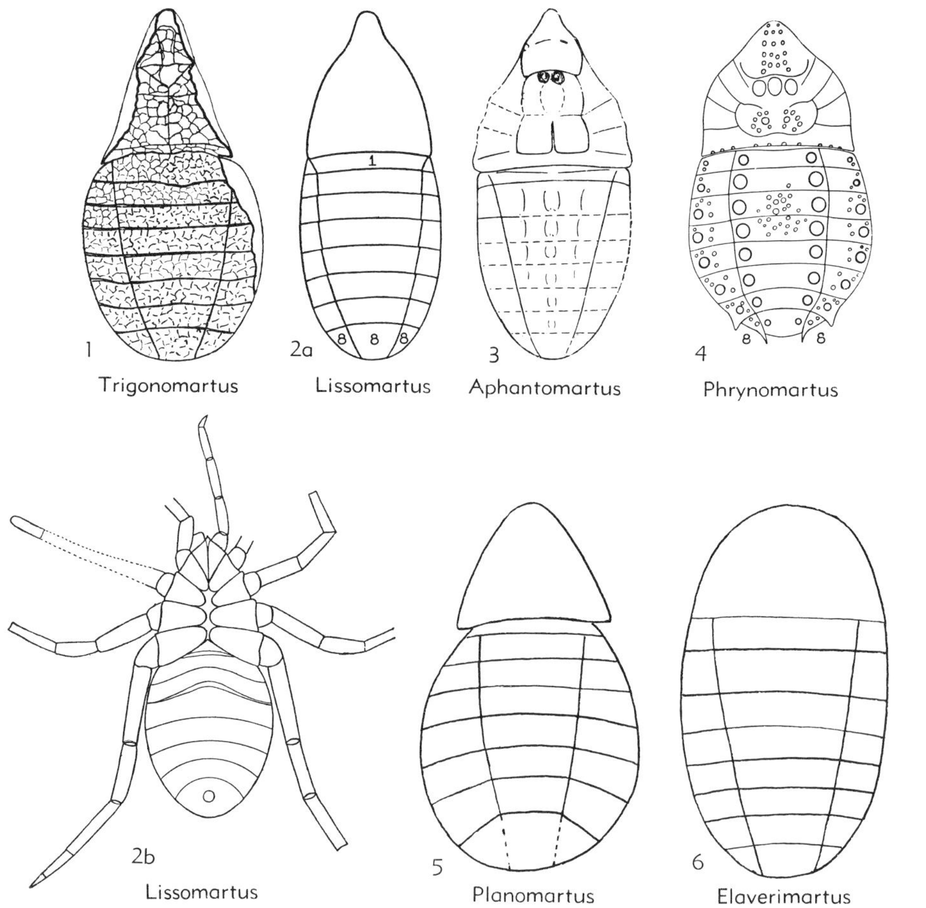

FIG. 80. Trigonomartidae. *1, Trigonomartus pustulatus,* Penn., Ill., ×2.8; *2, Lissomartus schucherti,* Penn., Ill.; *2a,b,* dorsal and ventral sides, ×2.4; *3, Apantomartus araeolatus,* U.Carb., Eng., ×3; *4, Phrynomartus waechteri,* U.Carb., Saar, ×3.3; *5, Planomartus krejcii,* U.Carb., Czech., ×3; *6, Elaverimartus pococki,* U.Carb., Eng.; dorsal side of body, ×7.5 (76) (p. P113).

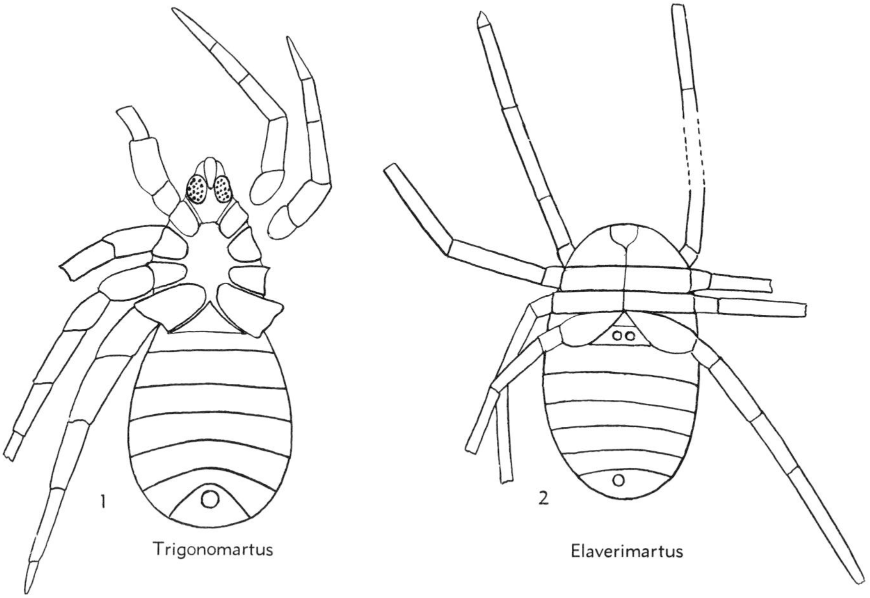

FIG. 81. Trigonomartidae. *1, Trigonomartus pustulatus,* Penn., Ill., ventral side, ×2.9; *2, Elaverimartus pococki,* U.Carb., Eng., ventral side, ×6 (76) (p. P113).

their movable finger lateral in position. Pedipalpi pediform, used for walking (hence the name Palpigradi), 6-jointed, but without patella, with metatarsus subdivided into 2 segments and tarsus into 3 segments and with a pair of claws on a pseudonychium (Fig. 82,*2d*). First legs tactile, longer and stouter than pedipalpi, 7-jointed, with a patella, but metatarsus subdivided into 4 segments and tarsus into 3; 2nd and 3rd legs 7-jointed, without subsegmentation; 4th legs also 7-jointed, but with tarsus subdivided into 2 segments. All pedal tarsi end in 2 claws on a pseudonychium.

Fore-gut without a gizzard, esophagus opening directly into mid-gut, while the pharynx serves as a filter. Thoracenteron with single pair of pouchlike diverticles; chylenteron in the shape of a lobed sac. Stercoral pouch wanting; hind-gut short; anus terminal, without operculum. Heart restricted to abdomen, with 4 pairs of ostia. Book lungs and tracheal tubes wanting, respiration taken care of by 3 pairs of ventral evertible sacs situated on 4th to 6th segments. Nervous system concentrated in the cephalothorax, but with 3 neuromeres in 2nd abdominal segment. Sexes separate. Oviparous. *Jur.-Rec.*

DISCUSSION

The Palpigradida are soft-bodied arachnids of very small size, varying from 0.65 to 2.8 mm., and represented by a single Recent family with 4 genera and 21 species. Most of these species are found in southern Europe, southern United States of America and south to Paraguay and Chile, one in Asia and 2 in Africa. A single fossil species was found in the Solnhofen lithographic shale of Germany (Jur.), mistaken for an insect by its discoverer, later correctly referred by HAASE to the order Palpigradida and considered by him to be sufficiently different from its Recent relative *Eukoenenia* to require the erection of a special family. The difference between the Recent family Eukoeneniidae and the fossil family Sternarthronidae is in the presence of separate sternites in the latter, corresponding to the single, large sternite 2+3 of *Eukoenenia.*

Family EUKOENENIIDAE Petrunkevitch, 1955, nov.

[*nom. nov.* PETR., *pro* Koeneniadae GRASSI & CALANDRUCCIO, 1885] [Type: *Eukoenenia* BÖRNER, 1901, proposed as subgenus of *Koenenia* GRASSI-C. without knowledge that *Koenenia* was preoccupied]

First sternite large, resulting from complete fusion of sternites 2 and 3. *Rec.*

Eukoenenia Börner, 1901 [**Koenenia mirabilis* Grassi & Calandruccio, 1885] [=*Koenenia* Grassi-C., 1885 (*non* Beushausen, 1884)]. *Rec.*—Fig. 82,*1*. **E. mirabilis* (Grassi-C.); *1a*, dorsal side, ×17 (68); *1b*, side view of cephalothorax, ×17 (64).

Family STERNARTHRONIDAE Haase, 1890

Thoracic sternites 2 and 3 separated from each other by intersegmental membrane. *Jur.*, Eu.

Sternarthron Haase, 1890 [**S. zitteli*]. Ger.,—Fig. 83. **S. zitteli*; *1*, dorsal side, ×1; *2*, same, ventral side showing sternites, ×3 (67).

Superorder CAMAROSTOMATA Petrunkevitch, 1949

[Type: *Thelyphonus* Latreille, 1802]

Caulogastra with pedipalpal coxae permanently fused along line of contact, their common dorsal wall forming a camarostome in the shape of an elongated depression semicircular in cross section into which the upper lip fits snugly, leaving a slitlike cavity (Fig. 84,*2*). The periphery of this cavity opens to the outside and serves for admission of food, while the mouth opens into it at its posterior end. The camarostome serves not only for admission of food to the mouth, but also as a filter, for which pur-

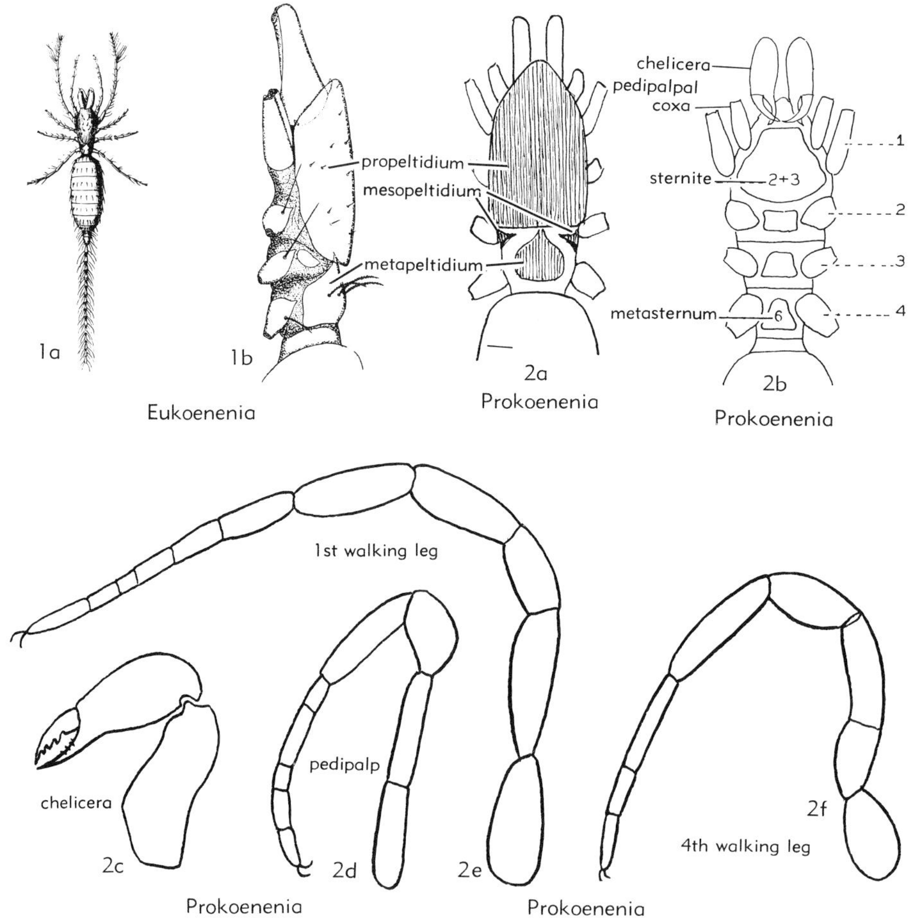

Fig. 82. External morphology of Palpigradida. *1, Eukoenenia mirabilis*, Rec., Italy; *1a*, ×17 (68); *1b*, side view of its cephalothorax, enlarged (64). *2, Prokoenenia wheeleri*, Rec., Calif.; *2a*, dorsal side of cephalothorax, showing segmented carapace, enlarged; *2b*, ventral side showing coxosternal region; *2c*, chelicera; *2d*, pedipalp; *2e*, 1st leg; *2f*, 4th leg (76) (p. P116-P118).

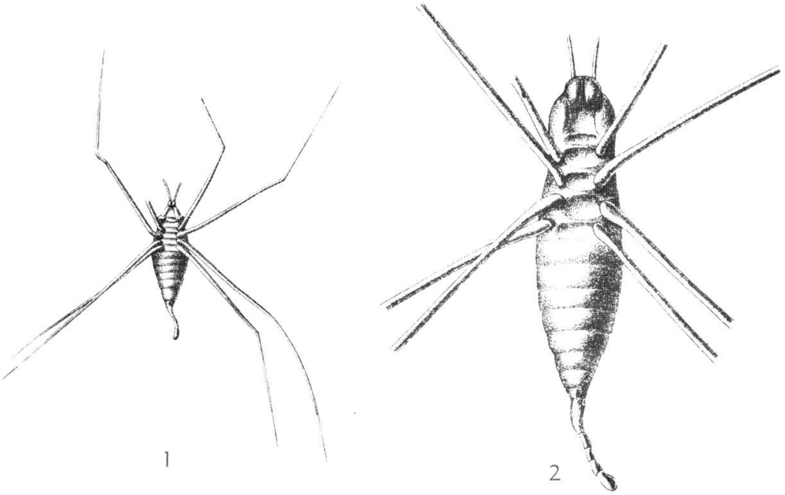

FIG. 83. *Sternarthron zitteli,* only known fossil palpigrade, U.Jur., Ger.; *1,* complete specimen, ×1; *2,* ventral side of body, ×3 (67) (p. P118).

pose its wall, as well as the wall of the upper lip, is covered with fine bristles. Neither pedipalpal nor pedal coxae have any maxillary lobes or maxillary glands, but in some males the pedipalpal coxae are produced into long processes with only ornamental use (Fig. 84,*1*).

Camarostomata breathe by means of book lungs, wherein they differ from Ricinuleida which also possess a camarostome, but breathe by means of tracheal tubes. *Carb.-Rec.*

Order THELYPHONIDA Latreille, 1804

[*nom. transl.* CAMBRIDGE, 1872 (*ex et correct. pro* Thelyphoni LATREILLE, 1804)] [=Uropygi THORELL, 1882 (*partim*); Holopeltidia BÖRNER, 1904] [Type: *Thelyphonus* LATREILLE, 1802]

Carapace entire, with doublure in front and slight infolding on sides, usually longer than wide, with or without eyes. Abdomen composed of 12 segments, 1st segment greatly shortened, 10th to 12th segments reduced to a pygidium ending in a dorsal, multijointed whip or flagellum (Fig. 33). Chelicerae 2-jointed, retrovert. Pedipalpi 6-jointed, stout, raptorial, with terminal joint pointed for piercing and holding by flexion against preceding joint which sometimes has a short projection at its anteroventral end, serving to hold an object stronger and making the palp subchelate. Pedipalpal coxae fused permanently along line of contact, which remains visible, while both contact walls disappear and leave a common cavity within the fused coxae. First pedal coxae slender and small, far apart and like all other coxae devoid of maxillary lobes. First legs tactile, without patella, with tarsus subdivided into 9 segments; 2nd to 4th legs, with patella, their tarsi always subdivided into 3 segments and bearing 2 claws. Sternum composed of 3 or 4 successive sternites; 1st large, triangular, with apex directed backward, 2nd and 3rd minute; 4th (called metasternite) situated between 4th coxae and easily mistaken for 1st abdominal sternite which is usually concealed in a fold of intersegmental membrane (Fig. 84,*1*). Lower lip lacking; mouth opens directly into camarostome. Fore-gut with pharynx, esophagus and gizzard, which is poorly developed, so that pharynx is main pumping organ. Thoracenteron with 4 pairs of diverticles; chylenteron with 7 or 8 pairs, its median tube enlarged posteriorly and serving in place of a stercoral pouch which is lacking. Hind-gut short. Anus terminal, without operculum. Heart with 9 pairs of ostia, 1st and 2nd pairs in cephalothorax. Two pairs of book lungs in 2nd and 3rd abdominal segments. Seven abdominal neuromeres incorporated in thoracic ganglionic mass, last 5 (8th to 12th) neuromeres forming an abdominal ganglionic mass in

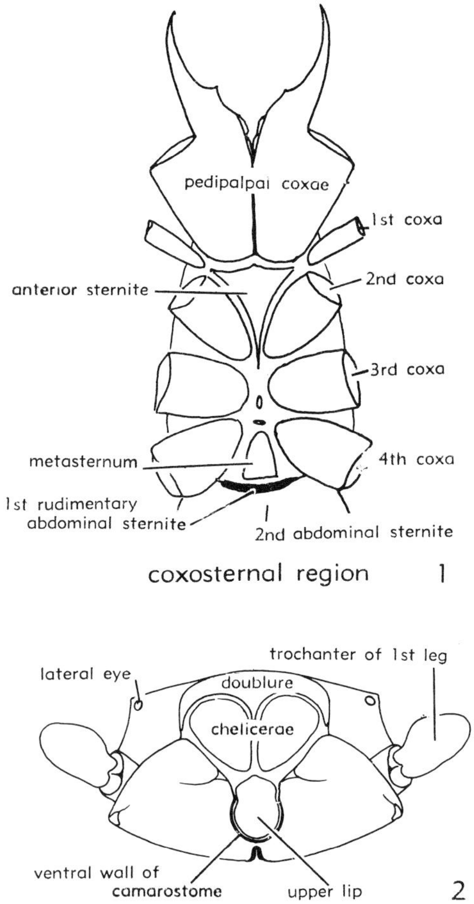

FIG. 84. Morphology of Thelyphonida. *1*, Coxosternal region of male *Labochirus proboscideus*, a whip scorpion from Ceylon, enlarged. *2*, Transverse section through the head of *Mastigoproctus giganteus*, a whip scorpion from Arizona, showing camarostome formed by a dorsal depression of the fused pedipalpal coxae and upper lip fitting the depression (76).

8th and 9th segments. Pair of median and 3 pairs of lateral eyes on carapace usually present in Recent species and some fossil ones. Pair of excretory coxal glands in the cephalothorax with openings at base of 1st coxae. Poison glands lacking, but so-called anal glands, situated in abdomen, with a pair of openings each closed by an operculum on 12th segment, produce a volatile and highly irritating fluid that turns immediately into gas used for protection. Sexes separate, recognizable by secondary sexual characters. Males with accessory sexual glands of fenestrated type, situated dorsally to pair of tubular testes. Females with pair of tubular ovaries. Oviparous. *Carb.-Rec.*

DISCUSSION

Thelyphonida, commonly called whip scorpions, are tropical and subtropical arachnids varying in size from 18 to 65 mm. They represent one of the most stable orders among all arthropods, having retained their fundamental structure from the Carboniferous to the present, the Carboniferous genera belonging to the same family with the Recent ones. Nevertheless, the number of genera increased from 2 in the Carboniferous to 10 in the present fauna with 70 living species, distributed over Asia, Oceania and America; not found in Europe, Africa and Australia.—FIG. 33,*2*. *Mastigoproctus giganteus* (LUCAS), Rec., Mex.; ×1.5.

Family THELYPHONIDAE Lucas, 1835

Characters of order. *Carb.-Rec.*

Geralinura SCUDDER, 1884 [**G. carbonaria*]. Carapace oval, without eyes. *Penn.*, N.Am.—FIG. 85,*1*. **G. carbonaria*, Ill.; *1a*, dorsal side, ×4; *1b*, sternum, ×8.2 (76).—FIG. 85,*3*. *G. similis* PETRUNKEVITCH, Ill.; *3a*, ventral side showing coxosternal region, appendages and flagellum, ×2.6; *3b*, dorsal side of same, ×2.6 (76).

Prothelyphonus FRITSCH, 1904 [**Thelyphonus bohemicus* KUŠTA, 1884]. Carapace an elongated hexagon with eyes. *Carb.*, Eu.-N.Am.—FIGS. 77,*3*; 85,*4*. **P. bohemicus* (KUŠTA), Czech.; 77,*3*, holotype, photo, ×0.67; 85,*4*, dorsal side with missing portions of appendages restored, ×1.9 (76).—FIG. 85,*5*. *P. britannicus* (POCOCK), Eng.; ventral side, ×5.5 (76).—FIG. 85,*2*. *P. giganteus* (PETR.), Ill.; *2a*, dorsal side, ×3; *2b*, ventral side with appendages, ×3 (76).

Thelyphonus LATREILLE, 1802 [**Phalangium caudatum* LINNÉ, 1758]. Carapace with eyes. Pair of anal gland openings (formerly called "ommatidia") on last pygidial segment. *Mio.-Rec.* [*T. hadleyi* PIERCE, Mio., Calif.].

FIG. 85. Thelyphonida. *1*, *Geralinura carbonaria*, Penn., Ill.; *1a*, ×4; *1b*, sternum, ×8.2. *2*, *Prothelyphonus giganteus*, Penn., Ill.; *2a,b*, dorsal and ventral, ×3. *3*, *Geralinura similis*; *3a,b*, Penn., Ill., ventral and dorsal, ×2.6. *4*, *Prothelyphonus bohemicus*, U.Carb., Czech., ×1.9. *5*, *Prothelyphonus britannicus*, U.Carb., Eng., ×5.5 (76) (p. P120).

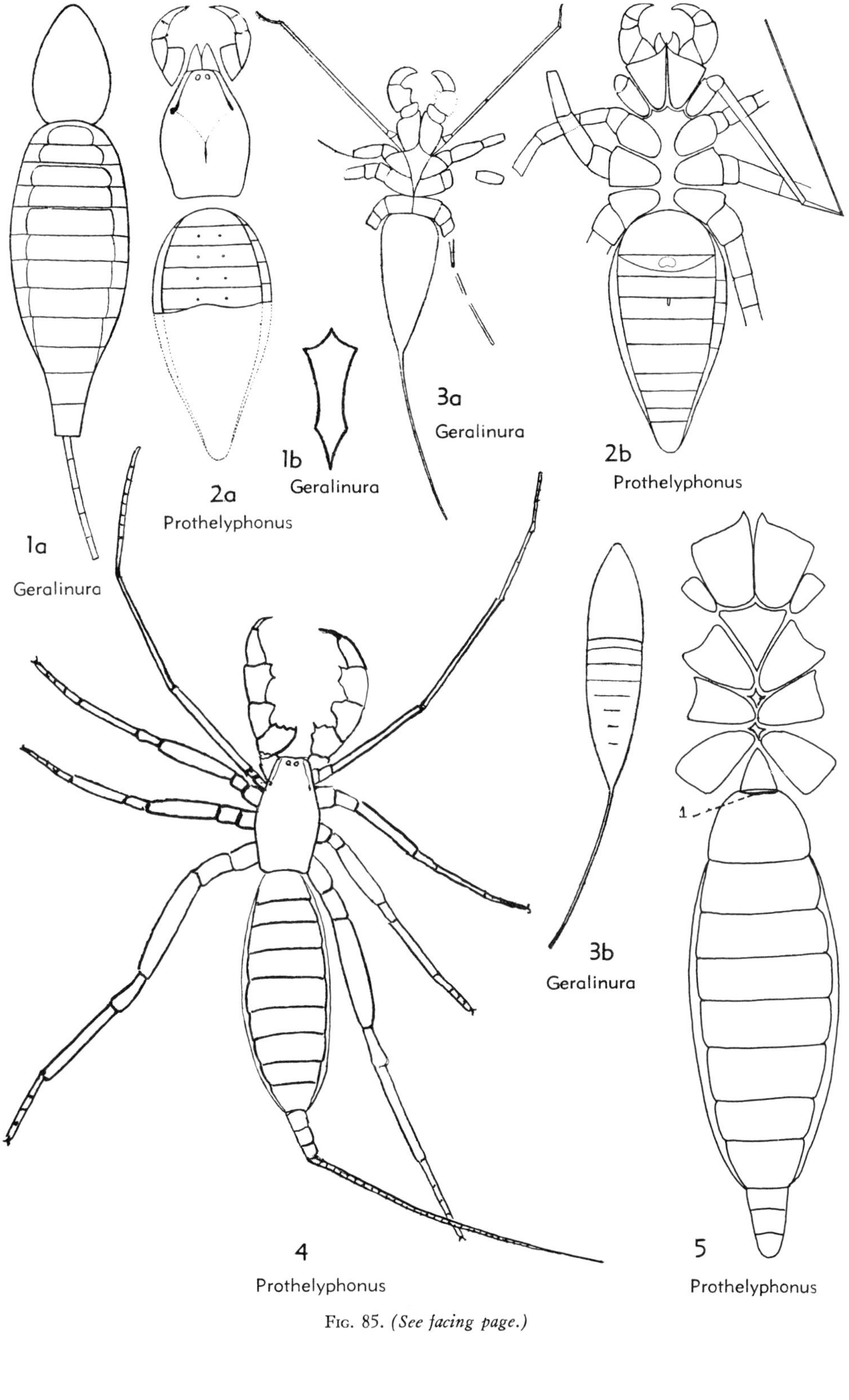

FIG. 85. *(See facing page.)*

Order SCHIZOMIDA Petrunkevitch, 1945

[=Tartarides CAMBRIDGE, 1872; Colopyga COOK, 1899; Schizopeltidia BÖRNER, 1904] [Type: *Schizomus* COOK, 1899 (*nom. subst. pro Schizonotus* THORELL, 1897, *nom. subst. pro Nyctalops* CAMBRIDGE, 1872 (*non Nyctalops* WAGLER, 1832)]

Carapace segmented (Fig. 86,*1*), without doublure and without eyes. Abdomen composed of 12 segments, the 1st considerably abbreviated, the 9th to 12 reduced to a pygidium. Telson short, with 1 to 7 segments. Pedipalpal coxae fused and forming a camarostome as in Thelyphonida. First pedal coxae wide apart, 2nd coxae separated by triangular sternite whose apex is directed backwards, 3rd coxae contiguous in middle line, 4th coxae separated by metasternite (Fig. 86,*2*). Chelicerae 2-jointed, chelate. Pedipalpi raptorial, their terminal joint pointed. First pair of legs tactile, without patella, tarsus subdivided into 3 to 8 segments. Second to 4th legs with patella, tarsus 2- to 5-jointed, ending with onychium bearing 3 claws. Pair of book lungs in 2nd abdominal segment. Heart restricted to abdomen with 5 pairs of ostia. Anterior 4 abdominal neuromeres incorporated in the thoracic ganglionic mass, 5th to 12th neuromeres forming an abdominal ganglionic mass in the 2nd abdominal segment. Thor-

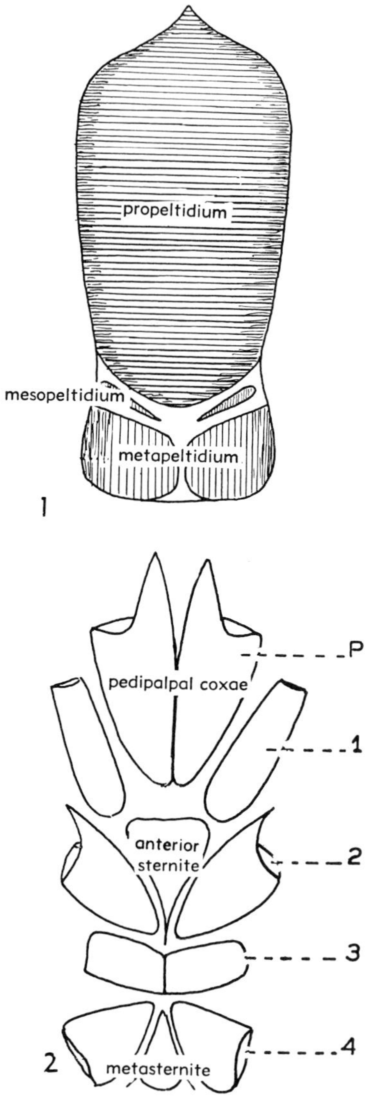

FIG. 86. External morphology of Schizomida. *1*, Segmented carapace of *Schizomus crassicaudatus*, from Ceylon. *2*, Coxosternal region of *Trithyreus pentapeltis*, from California (76).

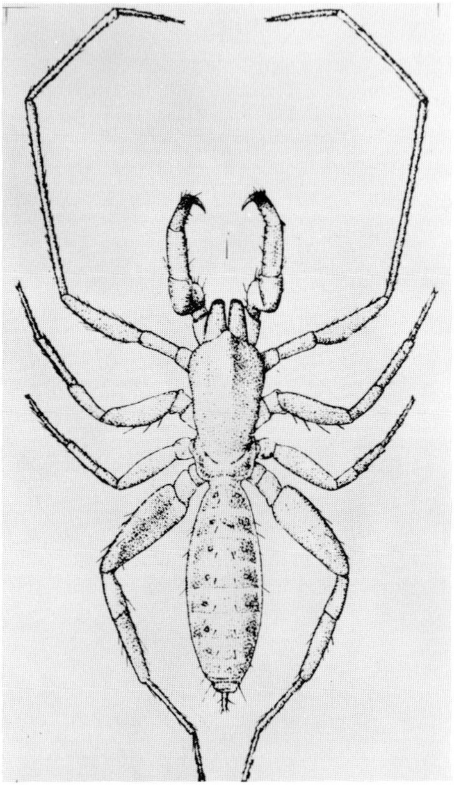

FIG. 87. *Schizomus peradeniyensis*, Rec., from Ceylon; female, ×17 (by permission from *Traité de Zoologie*, tome 6, sous la direction du PR. GRASSÉ, Masson & Cie., Editeurs, Paris) (p. P123).

acenteron with 1 pair of diverticles, chylenteron with 7 pairs. Anal glands lacking. Sexes separate, outwardly distinguished by structure of telson (flagellum). *U.Tert. (?Plio.)-Rec.*

Schizomida are soft-bodied, small arachnids from 2 to 7 mm. in length. They are found in the tropical and subtropical regions of the world. Recent Schizomida represent a single, very uniform family with only 28 species and 3 closely related genera. Fossil Schizomida are known only from the Upper Tertiary.

Family SCHIZOMIDAE Chamberlin, 1922

[=Schizonotidae THORELL, 1888; Hubbardiidae[1] COOK, 1899]

First tarsus 8-jointed, others 3-jointed. Telson 1- to 4-jointed (39). *?Plio., Rec.*

Calcoschizomus PIERCE, 1950 [**C. latisternum*]. Telson rodlike, unsegmented (46). *?Plio.,* N.Am. —FIG. 90,*2*. **C. latisternum,* Ariz.; ×4 (77).

Schizomus COOK, 1899 [*pro Schizonotus* THORELL, 1888 (*non* RATZEBURG, 1852); *etiam Nyctalops* CAMBRIDGE, 1872 (*non* WAGLER, 1832)] [**Nyctalops crassicaudatus* CAMBRIDGE, 1872]. *Rec.*—FIG. 86,*1*. *S. crassicaudatus* (CAMBRIDGE), Ceylon; carapace, ×30 (76).—FIG. 87. *S. peradeniyensis* GRAVELY, Ceylon; female, ×17 (75).

Family CALCITRONIDAE Petrunkevitch, 1945

First tarsus 7-jointed, 2nd 5-jointed, 3rd and 4th 4-jointed. Telson 3- to 7-jointed (39). *?Plio.*

[1] The name Hubbardiidae was proposed by COOK for his genus *Hubbardia,* which was later proved to be a synonym of *Nyctalops* CAMBRIDGE, a preoccupied name changed by THORELL to *Schizonotus*. This name also proved to be preoccupied and was changed to *Schizomus*. Still later, CHAMBERLIN quite correctly changed the family name Schizonotidae to Schizomidae.

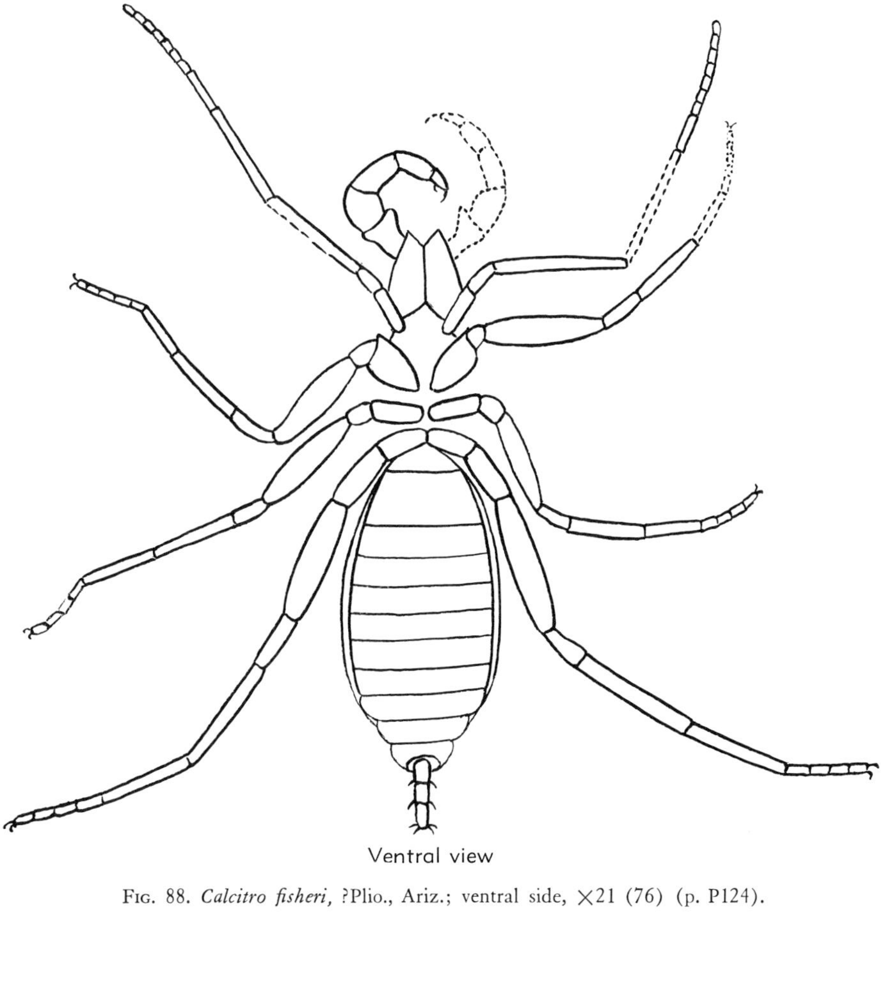

FIG. 88. *Calcitro fisheri,* ?Plio., Ariz.; ventral side, ×21 (76) (p. P124).

Calcitro PETRUNKEVITCH, 1945 [**C. fisheri*]. Telson 3-jointed. Order of legs 1432. *?Plio.*, N.Am.—FIGS. 88, 89. **C. fisheri*, Ariz.; 88, ventral side, ×15; 89, photo of holotype, ×21 (76).

Onychothelyphonus PIERCE, 1950 [**O. bonneri*]. Telson 7-jointed. Order of legs 4132 (46, 47). *?Plio.*, N.Am.—FIG. 90. **O. bonneri*, Ariz.; *1a*, ventral side, ×10; *1b*, telson, ×50 (77).

Order KUSTARACHNIDA Petrunkevitch, 1913

[*nom. correct.* PETR., herein (*pro* Kustarachnae PETR., 1913)] [Type: *Kustarachne* SCUDDER, 1890]

Presumptive Camarostomata with pedipalpal coxae completely fused along the line of contact, leaving no trace of their paired origin. All coxae triangular, contiguous, radiating fanlike from a minute sternum. Abdomen segmented, ending in a small pygidium, without telson. Carapace entire. Pedipalpi chelate. Trochanters of legs 2-jointed, distal joint conical. Legs slender and long. Chelicerae not known (34, 45). *Penn.*

Family KUSTARACHNIDAE Petrunkevitch, 1913

Characters of order. *Penn.*, N.Am.

Kustarachne SCUDDER, 1890 [**K. tenuipes*]. Carapace evenly rounded, with pair of eyes on transversely ellipsoidal tubercle. *Penn.*, N.Am.—FIG. 91,*1*. **K. tenuipes*, Ill.; *1a*, ventral side, ×4; *1b*, carapace, ×4 (76).—FIG. 91,*2*. *K. conica* PETRUNKEVITCH, Ill.; ventral side, ×4 (76).—FIG. 91,*3*. *K. extincta* (MELANDER), Ill.; *3a*, ventral side, ×4; *3b*, carapace, ×4 (76).

Superorder LABELLATA Petrunkevitch, 1949

[Type: *Araneus* CLERCK, 1757]

Caulogastra with mouth anteroventral between 2 lips, with pedipalpal coxae contiguous, but free (43). *?Dev., Carb.-Rec.*

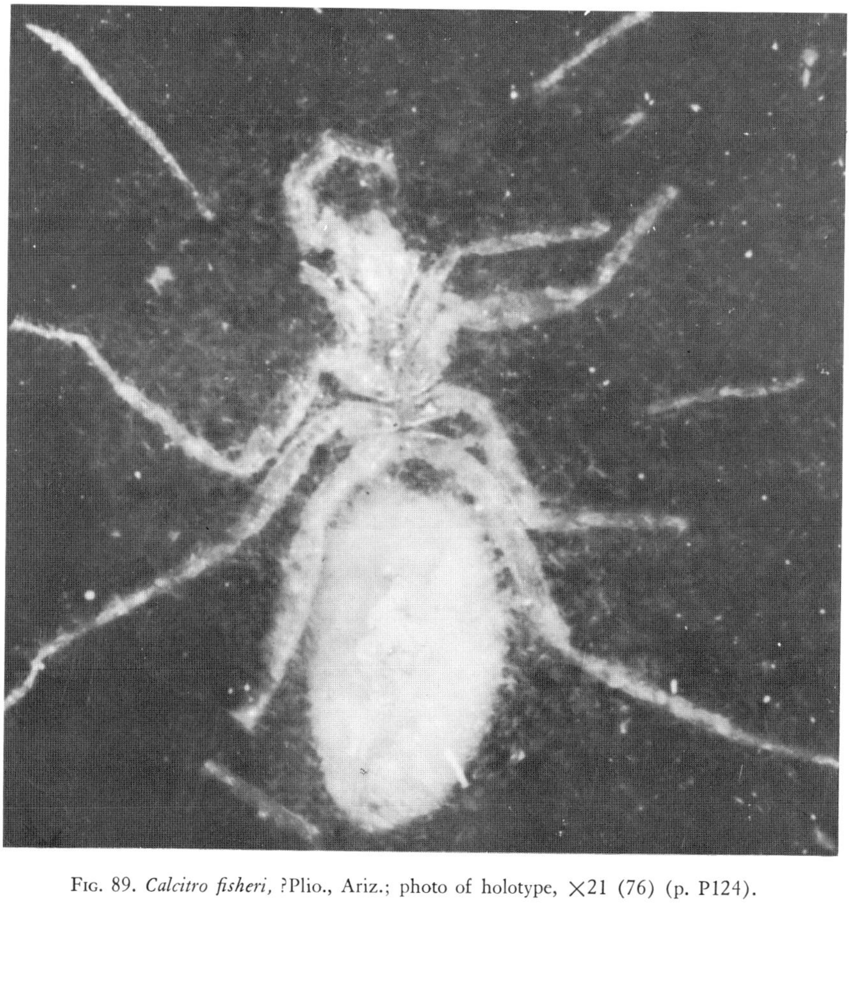

FIG. 89. *Calcitro fisheri*, ?Plio., Ariz.; photo of holotype, ×21 (76) (p. P124).

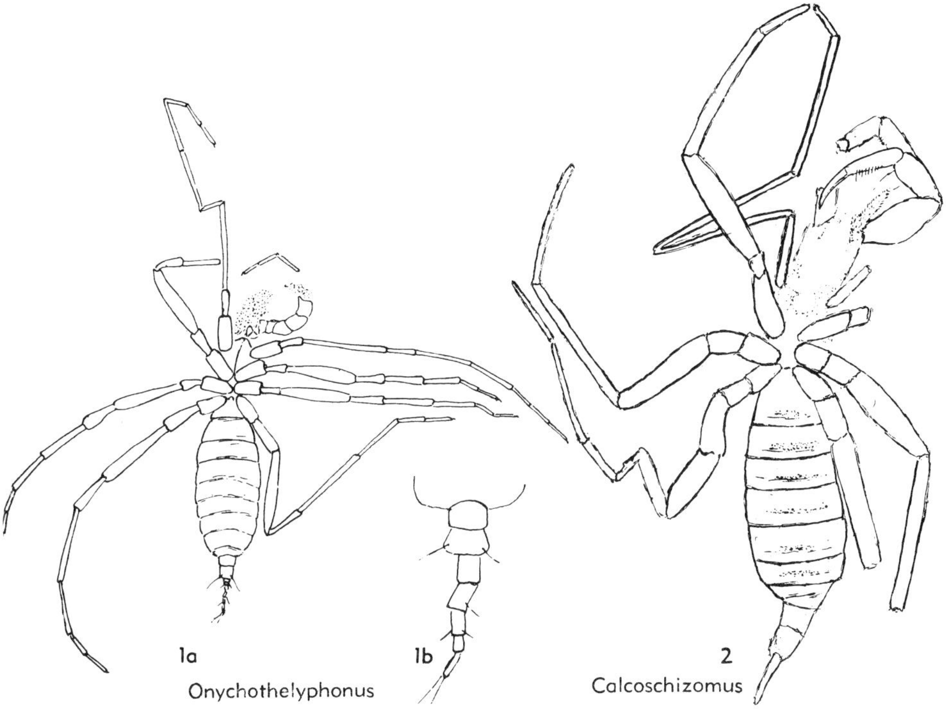

FIG. 90. Schizomida. *1, Onychothelyphonus bonneri,* ?Plio., Ariz., *1a,* ×10; *1b,* flagellum, ×50. *2, Calcoschizomus latisternum,* ?Plio., Ariz., ×4 (p. P123-P124).

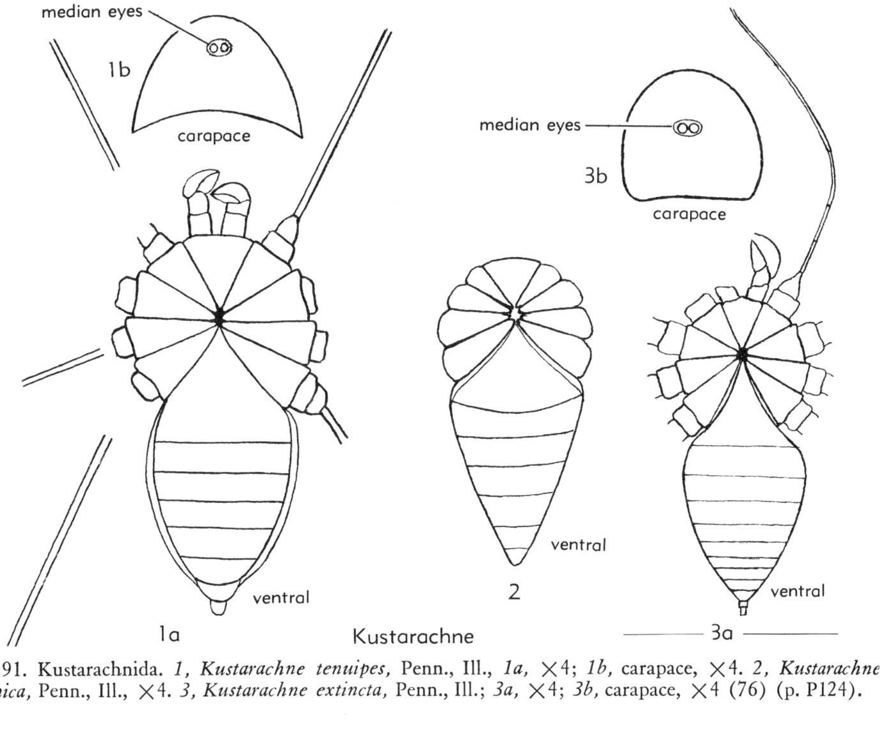

FIG. 91. Kustarachnida. *1, Kustarachne tenuipes,* Penn., Ill., *1a,* ×4; *1b,* carapace, ×4. *2, Kustarachne conica,* Penn., Ill., ×4. *3, Kustarachne extincta,* Penn., Ill.; *3a,* ×4; *3b,* carapace, ×4 (76) (p. P124).

Order PHRYNICHIDA Petrunkevitch, 1945

[=Phryneides[1] GERVAIS, 1844; Amblypygi THORELL, 1883] [Type: *Phrynichus* KARSCH, 1879 (=*Phrynus* LAMARCK, 1801; =*Phrynus* OLIVIER, 1793, *secundum* GERVAIS, 1844, p. 2, footnote; source not mentioned and not known)]

Labellata with raptorial pedipalpi, modified 1st legs and segmented abdomen ending in a pygidium.

Carapace entire, flat, usually wider than long, with pair of median eyes and 3 pairs of lateral eyes. Chelicerae 2-jointed, retrovert. Pedipalpi raptorial, powerful, long, 6-jointed, with terminal joint pointed. Pedipalpal coxae free, movable, without maxillary lobes or glands. All pedal coxae movable, 1st pair small, widely separated and wedged in between pedipalpal and 2nd coxae (Fig. 92). Sternal plate single, with 3 or 4 separate sclerites arising from it in a median row and representing trito-, tetra-, penta-, and when present, hexasternite, usually called metasternite. Tritosternite long and narrow, extending forward between pedipalpal coxae. First legs slender, with very long, multijointed, flexible, antenna-like tarsi, held away from the ground and used as organs of touch; 2nd, 3rd and 4th tarsi 4- or 5-jointed, ending in a pair of claws which are in some genera accompanied by a pulvillus. Abdomen composed of 12 segments, the 12th reduced to a pygidium, without a telson.

Fore-gut with pumping pharynx, short esophagus and pumping gizzard. No pharyngeal filter, but a rough preoral brush serving as filter. Thoracenteron with 4 pairs of diverticles, chylenteron highly branched. Anus terminal, without operculum. Two pairs of book lungs with openings on 2nd and 3rd sternites. Pair of ventral, evertible sacs present in some Recent species. Heart limited to abdomen, with 6 pairs of ostia. Sexes separate. Oviparous. Female carries eggs attached to venter of abdomen. *Carb.-Rec.*

Phrynichida are harmless creatures possessing neither offensive nor defensive poison glands of any kind. Denizens of tropical countries, varying in size of body from 8 to 37 mm.; 60 living and 4 fossil species. Two families: Tarantulidae, strictly Recent, and Phrynichidae.

[1] GERVAIS' name Phryneides, used by him as the name of an order, was based on the generic name *Phrynus* LAMARCK, 1801 (=*Phrynus* OLIVIER, 1793). It had for its type *Phalangium reniforme* LINNÉ and it has been shown later that *Phalangium lunatum* PALLAS, 1772, placed by FABRICIUS in his genus *Tarantula*, 1793, is a synonym of *P. reniforme*, but that the specimen of *P. reniforme* used by FABRICIUS as type of his genus *Tarantula* was in reality a *Phalangium palmatum* HERBST. So *palmatum* is now the type of *Tarantula* FABRICIUS and *reniforme* the type of *Phrynichus, nom. nov.* proposed by KARSCH because the 2 species, *reniforme* and *lunatum*, are not co-generic. Phryneides of GERVAIS has not only priority over Amblypygi THORELL, but is derived from the name of the genus in which LINNÉ's species was correctly placed. The name Phrynichida is simply a substitute for Phryneides in accordance with rule 54(b) of the Copenhagen Decisions.

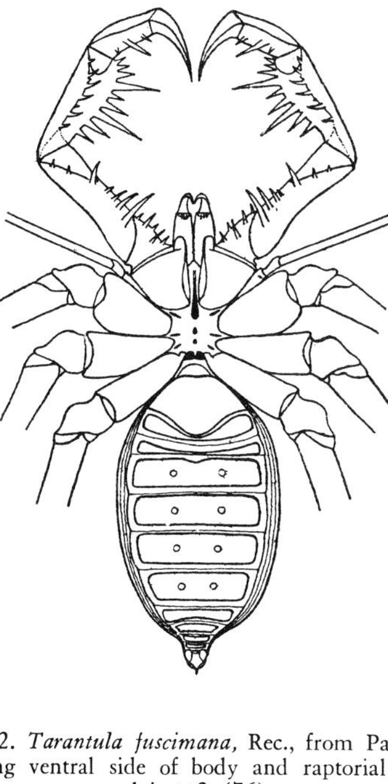

FIG. 92. *Tarantula fuscimana*, Rec., from Panama, showing ventral side of body and raptorial pedipalpi, ×2 (76).

Family PHRYNICHIDAE Pocock, 1900

[=Phrynidae SUNDEVALL, 1833]

Tarsi without pulvillus. *Carb.-Rec.*

Phrynichus KARSCH, 1879 [**Phalangium reniforme* LINNÉ, 1758]. Fourth tibia not subsegmented. *Rec.*—FIG. 36. *P. reniformis* (LINNÉ), Asia; dorsal view (Cuvier).

Graeophonus SCUDDER, 1890 [**G. carbonarius*]. Carapace with rounded median projection bearing tubercle with pair of eyes and with longitudinal groove in posterior half (45). *Carb.*, Eu.-N.Am.—FIG. 93,*1*. **G. carbonarius*, Ill.; *1a,c*, dorsal and ventral sides, ×3; *1b*, dorsal side of abdomen of another specimen, showing muscular attachments, ×3 (76).—FIG. 93,*2*. *G. anglicus* POCOCK, Eng.; *2a*, carapace (eyes not preserved) and appendages, ×4 (76); *2b,c*, holotype, dorsal side showing eyes and ventral side showing all but 1st legs, ×2.5 (78).

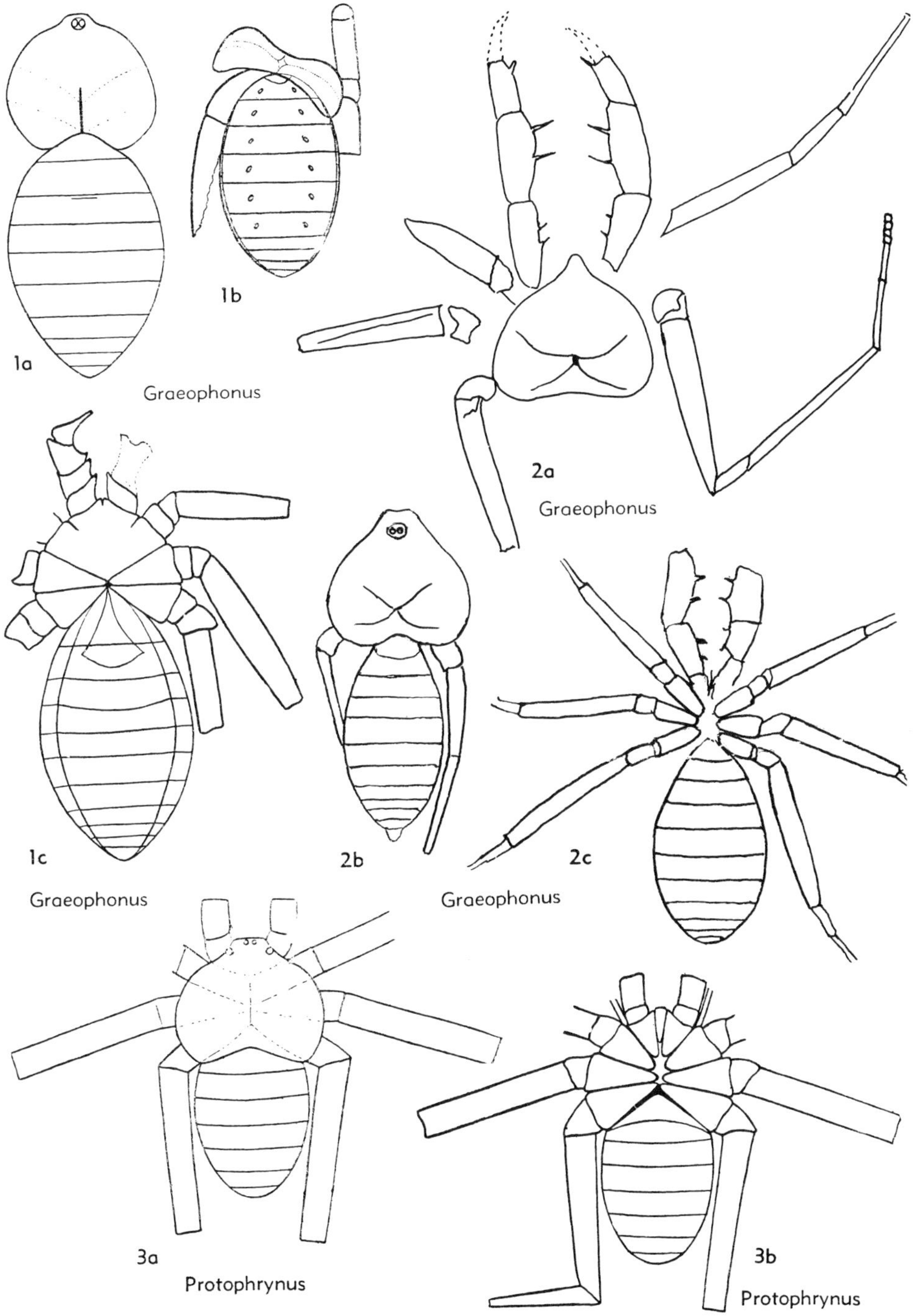

FIG. 93. Phrynichida. *1, Graeophonus carbonarius,* Penn., Ill.; *1a,c,* dorsal and ventral sides of holotype, ×3; *1b,* dorsal side of another specimen showing paired attachments of dorsoventral muscles, ×3. *2, Graeophonus anglicus,* U.Carb., Eng.; *2a,* carapace and appendages, ×4; *2b,* dorsal side showing eyes, ×2.5; *2c,* ventral side of holotype, ×2.5. *3, Protophrynus carbonarius,* Penn., Ill.; *3a,b,* dorsal and ventral, ×3 (76) (p. P126-P128).

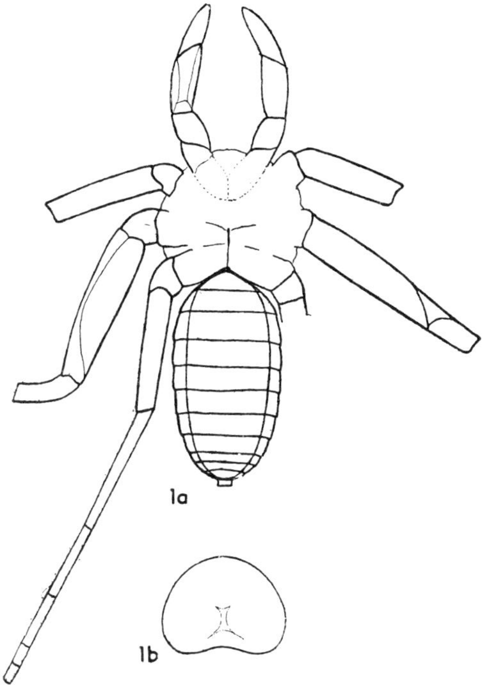

FIG. 94. *Thelyphrynus elongatus,* Penn., Ill.; *1a,* ×3; *1b,* carapace, ×3 (76) (p. P128).

Protophrynus PETRUNKEVITCH, 1913 [**P. carbonarius*]. Anterior projection of carapace truncated, bearing pair of sessile eyes; 3 pairs of lateral eyes (34). *Penn.,* N.Am.—FIG. 93,*3*. **P. carbonarius,* Ill.; *3a,b,* dorsal and ventral sides, ×3 (76).

Thelyphrynus PETRUNKEVITCH, 1913 [**T. elongatus* (misspelled *Telyphrynus* on p. 60, correctly spelled on p. 66)]. Carapace evenly rounded in front, without eyes. *Penn.,* N.Am.—FIG. 94. **T. elongatus,* Ill.; *1a,b,* ventral side and carapace, ×3 (76).

Order ARANEIDA Clerck, 1757

[*nom. correct.* DALLAS, 1864 (*pro* Araneidea BLACKWALL, 1861; *pro* Araneides LATREILLE, 1801; *pro* Aranei CLERCK, 1757)] [=Araneae *auctt.* (CLERCK's name validated by action of Intern. Zool. Congress, Paris, 1948)] [Type: *Araneus* CLERCK, 1757]

Labellata with maxillary lobes and glands, 1st pair of legs similar to others and abdomen with spinnerets, without pygidium.

Carapace entire, with up to 4 pairs of eyes. Abdomen in primitive families segmented, composed of 12 distinct segments; in all other families reduced to 5 or 6 segments completely fused in the adult, without trace of external segmentation. First abdominal segment always reduced to slender pedicel usually with only dorsal tergite (*lorum pediculi*), but in some genera with ventral sternite (*plagula sternalis*). Anus at end of abdomen, without operculum. Coxosternal region (Fig. 97*B*) similar to that in Trigonotarbida, with lip and sternum. All coxae movable. Pedipalpal coxae with maxillary glands and either rudimentary or more commonly well-developed maxillary lobes. Chelicerae 2-jointed, retrovert (Fig. 97*F*), 2nd joint developed as fang, with opening of poison gland below tip. Pedipalpi 6-jointed, terminal joint, with or without claw in females and immature specimens; with more or less complicated copulatory organ in mature males (Fig. 106,*1b*). Legs 7-jointed, either with 3 claws or with 2 claws and claw tufts serving as organ of adhesion. Abdominal appendages modified as spinnerets on 4th and 5th segments (Fig. 97*B,C*) with spinning tubes of various kinds for passage of silk from silk glands situated in abdomen. Maximum number of spinnerets 4 pairs, in segmented spiders, 3 pairs in most spiders, 2 pairs in majority of Theraphosina, 1 pair in few genera scattered over several families. In so-called cribellated spiders a perforated plate called cribellum and corresponding to 1st pair of spinnerets of segmented spiders, is situated in front of the 3 pairs of normal spinnerets and serves for passage of special silk. Presence of cribellum (Fig. 103,*5b*) is always correlated with presence of a so-called calamistrum or brush of specialized hairs on 4th metatarsi (Fig. 103, *1b,2c*). The calamistrum is used to brush the silk out of the cribellum (Figs. 95, 96).

Mouth anteroventral, between rostrum (upper lip) and labium. Fore-gut with pharynx, esophagus and gizzard. Pharynx composed of 2 plates and serves as pumping organ and filter. Gizzard has 3 plates and acts as main pumping organ. Its dorsal dilator muscles are attached to apodeme visible on outside of carapace and called thoracic groove (Fig. 97*A*). Mid-gut with several diverticles in cephalothorax, forming thoracenteron, and several greatly ramified diverticles in abdomen, forming chylenteron, often called liver. Mid-gut ends in a stercoral pouch in which excrements are stored. Hind-gut short, opening at true end of abdomen, which often is far in front of actual end when abdomen is dorsally prolongated far beyond anus. Pair of poison glands in cephalothorax, with ducts opening just below tip of fangs. Malpighian tubes in abdomen, opening into digestive tube at end of mid-gut. Coxal glands in cephalothorax, with pair of openings at base of 1st coxae and another pair at base of 3rd coxae. Respiration either by means of 2 pairs of book lungs situated in 2nd and 3rd abdominal segments, or 1 pair of book lungs and 1 pair of branched tracheal tubes in place of 2nd pair of book lungs, or 2 pairs of branched tracheal tubes in place of book lungs.

Circulatory system highly complex, of open type, but with strictly morphologically predetermined veinous channels for return of blood to the heart. Heart limited to abdomen, with 2 to 5 pairs of ostia, complicated cephalothoracic arterial system and 3 posterior, abdominal arteries. Blood pumped forward and backward. Nervous system highly concentrated, with all abdominal neuromeres incorporated in thoracic ganglionic mass, from which nerves pass through petiolus into abdomen. Known sense organs

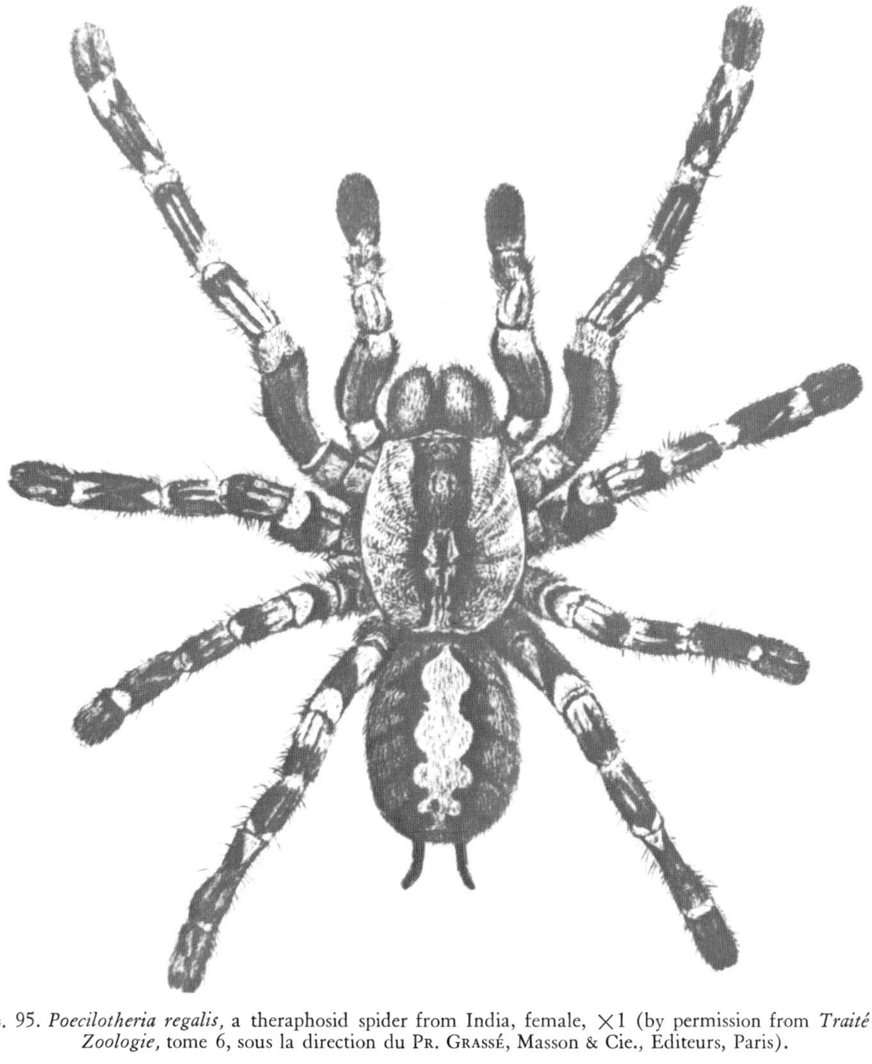

FIG. 95. *Poecilotheria regalis,* a theraphosid spider from India, female, ×1 (by permission from *Traité de Zoologie,* tome 6, sous la direction du PR. GRASSÉ, Masson & Cie., Editeurs, Paris).

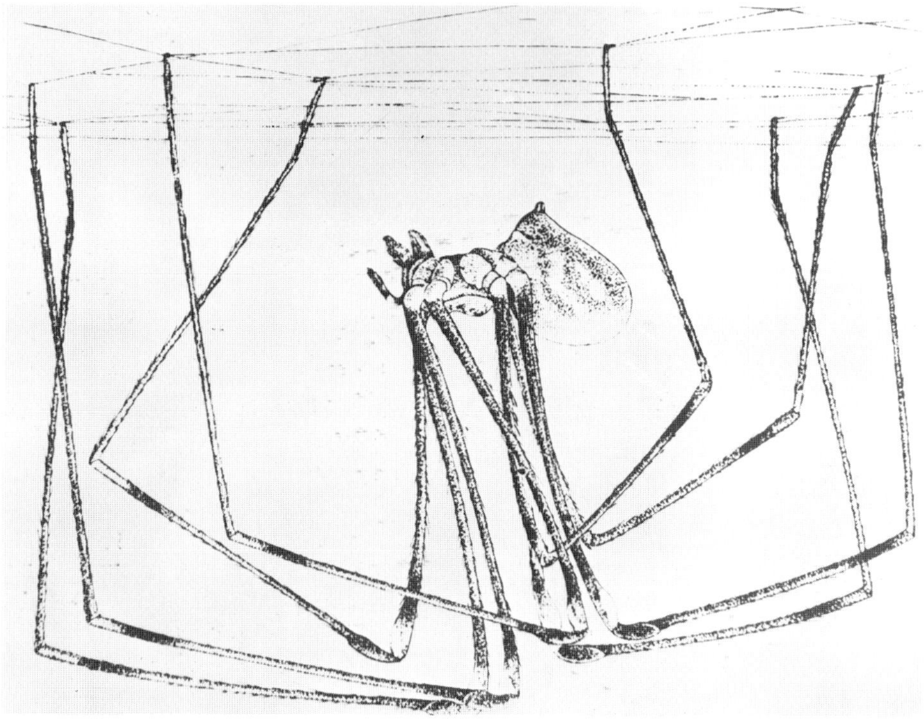

FIG. 96. *Artema mauricia,* a pholcid spider from tropical Africa, ×2 (by permission from *Traité de Zoologie,* tome 6, sous la direction du PR. GRASSÉ, Masson & Cie., Editeurs, Paris).

comprise eyes, organs of touch and lyriform organs. Eyes are of 2 types: pair of median eyes with direct retina, rods facing refractive apparatus, and 3 pairs of lateral eyes with inverted retina, rods facing away from lens. Configuration of eye group and their number of great systematic value. Organs of touch consist of various hairs; type called trichobothria, sensitive to gentle touch and especially to currents of air, have different distribution in different families. Sound-producing organs of stridulating type are common, but auditory organs remain unknown. Lyriform or slit organs and tarsal organs called by me "drums" are presumably olfactory in function.

Reproductive organs with openings on 2nd abdominal segment. In females the genital opening, through which eggs are laid, is usually accompanied by a pair of openings leading into seminal receptacles, and by an external chitinous organ, epigynum. Sperm introduced by the male into the receptacles remains in them for some time and is added to eggs as they are being laid. Males have their organs of copulation on the terminal joint of the pedipalpi. In their simplest form these organs consist of a soft sac (hematodocha) which becomes distended by blood in the process of copulation, and a chitinous bulb attached to the hematodocha, and ending in a fine tube, embolus, with an opening at its end. The cavity of the bulb is the receptacle which has to be filled with sperm before copulation can take place. This receptacle has no connection whatsoever with the rest of the reproductive system. Consequently the act of filling requires the construction of a special web, into which sperm is ejected from the genital opening and allowed to pass into the bulb through the embolus. In copulation the embolus is introduced into the seminal receptacle of the female, the sperm is injected and the process repeated with the other palp.

All spiders are oviparous. Embryonic development leads, with very few exceptions, to the formation of a spiderling having all essential characteristics of a spider, but differing in many respects from the adult, especially by the fact that their mid-gut is still a closed sac, isolated by its wall from both fore-gut and hind-gut and filled with embryonic yolk. During the 1st instar, spiderlings live either in a common web made by their mother before her death, or are carried on her back, as in Lycosidae, or

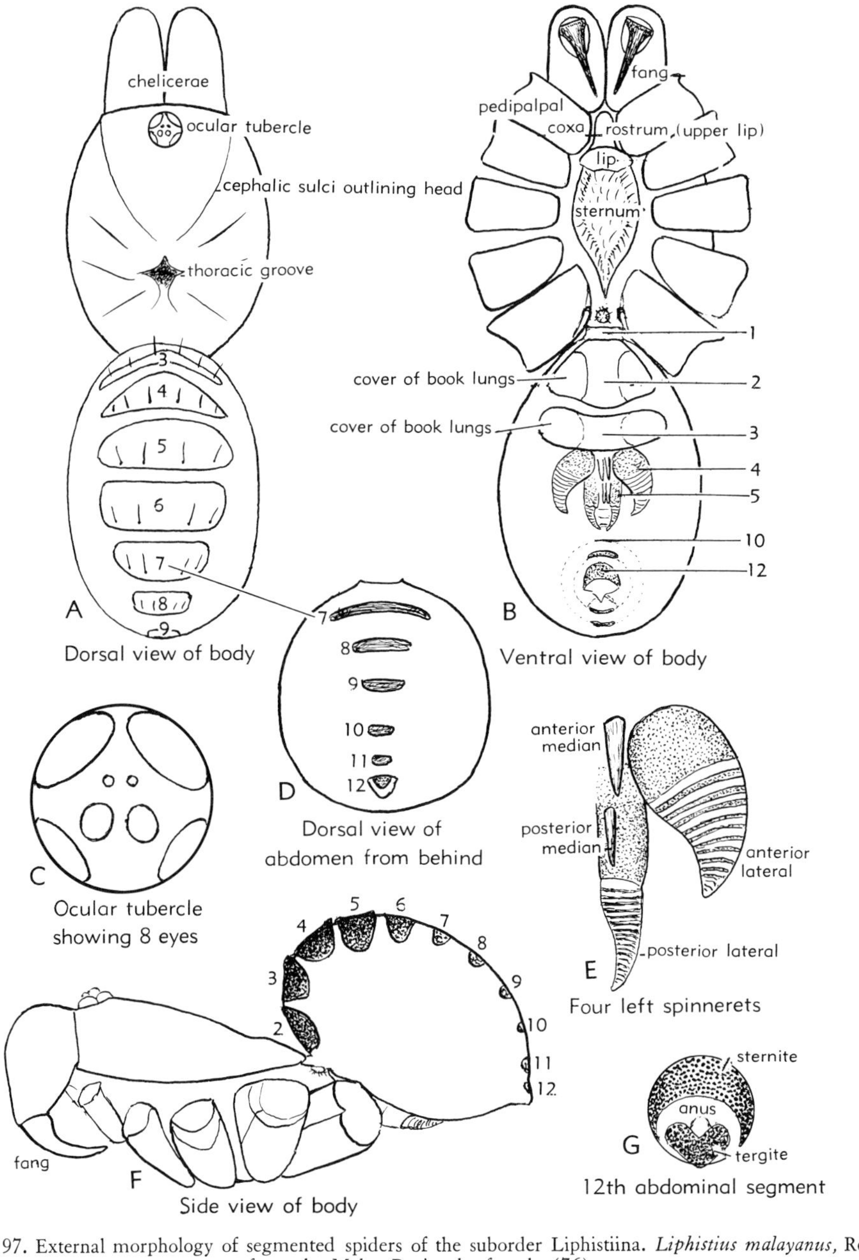

FIG. 97. External morphology of segmented spiders of the suborder Liphistiina. *Liphistius malayanus*, Rec., from the Malay Peninsula, female (76).

simply stay together when the egg cocoon is attached to the underside of a rock or bark and left to itself. The proportions of body and legs of 1st-instar spiderlings differ greatly from those of the adult, and the embryonic yolk is their only food. The advent of the 2nd instar induces hunger, independent feeding and dispersal as a natural preventive of extermination of the species through cannibalism. Maturity is normally attained after about 7 molts, but this number varies even within the same species. *?Dev., Carb.-Rec.*

DISCUSSION

The natural classification of spiders, proposed by me in 1933, is based on correlation of external characters with internal anatomy, supplemented by evaluation of evolutionary trends common to all Arachnida. Five suborders are recognized, 2 of which have no fossil representatives. The classification of fossil spiders is far from simple. Characters, easily visible in living spiders and imperative for correct identification even of suborders, are often missing. This applies especially to Paleozoic and such Tertiary spiders as are preserved in sedimentary rocks and not in amber. The structure of their internal organs is quite unknown and consequently their placement even in a suborder is not always certain. The total number of living genera is 2,735 and of fossil genera 84, or a total of 2,819 genera. Practically all Tertiary spiders belong to the suborder Dipneumonina (91 genera), the suborder Theraphosina being represented by only 3 genera. The Paleozoic spiders naturally belong to extinct genera. Two of the Paleozoic families are placed here in the suborder Liphistiina on account of their clearly segmented abdomen and general appearance. Their spinnerets are not preserved. The number of abdominal segments cannot be ascertained even in living specimens of *Liphistius* without rotating them, because only some of the segments can be seen in any one position (Fig. 97*A,D,F*). Two other Paleozoic families are tentatively placed here in the suborder Dipneumonina. Their representatives show abdominal segmentation, but their general appearance is that of dipneumone spiders, but not all important characters are preserved.

At present 81 families of spiders are recognized, 10 of them extinct. Of the 5 suborders, the Dipneumonina is by far the richest in genera and species (59 families, 2,401 genera). Theraphosina come second with 8 families and 370 genera. Apneumonina come third with 7 families and 27 genera. Liphistiina come fourth with 4 families and 14 genera. Hypochilina come last with 1 family and 2 genera. *Incertae sedis genera* are not included in this count.

Above figures are interesting inasmuch as they shed additional light on the evolution of spiders. Loss of abdominal segmentation, reduction in the number of cardiac ostia through loss of posterior cardiomeres, replacement of book lungs by tracheal tubes and the extension of the latter into the cephalothorax, and the loss of spinnerets are the important changes in the evolution of spiders. Liphistiina must be regarded as the most primitive spiders because they still have the maximum number of abdominal segments, of cardiac ostia, of book lungs and of spinnerets. At present they are represented by only 3 living genera with a few species. Their nearest relatives are undoubtedly the Theraphosina whose abdominal segments 6 to 12 are lost, segments 2 to 5 are fused without trace of external segmentation, cardiac ostia reduced to 4 or 3 pairs, spinnerets reduced to 3, or more commonly 2 pairs, while the book lungs are as in Liphistiina.

If the Paleozoic families referred tentatively to Dipneumonina have chelicerae of the diaxial type, which remains unknown because their chelicerae are not preserved, the change of chelicerae from their normal paraxial position, characteristic of Liphistiina and Theraphosina, to a diaxial one of all common spiders, must have taken place already in the Paleozoic. In Dipneumonina the 2nd pair of book lungs become replaced by tracheal tubes and the heart was reduced to 3 pairs of ostia in Trionychi and Dionychi, and even to 2 pairs in Quadrostiati, in the latter coupled with an extension of tracheae into the cephalothorax. In Apneumonina book lungs disappeared completely, tracheal tubes took their place and the heart became reduced to 2 pairs of ostia. It seems probable that Apneumonina evolved from Quadrostiati and are the youngest evolutionary group.

Suborder LIPHISTIINA Pocock, 1892

[*nom. correct.* PETRUNKEVITCH, herein (*pro* Liphistiomorphae POCOCK, 1892)] [Type: *Liphistius* SCHIÖDTE, 1849]

Chelicerae paraxial. Abdomen segmented. Four pairs of spinnerets. Two pairs of book lungs. Five or 4 pairs of cardiac ostia. *?Dev., Carb.-Rec.*

Family ARTHROLYCOSIDAE Fritsch, 1904

[*emend.* PETRUNKEVITCH, 1923]

Eyes on distinct tubercle. *Carb.*

Arthrolycosa HARGER, 1874 [**A. antiqua*]. Carapace with convex sides. Two pairs of eyes (45). *Penn.,* N.Am.—FIG. 98,*1*. **A. antiqua,* Ill.; ×1.3 (76).—FIG. 98,*3*. *A. danielsi* PETRUNKEVITCH, Ill.; *3a,b,* ventral and dorsal sides, ×2.3; *3c,* tubercle with 4 eyes, ×16 (76).

Eocteniza POCOCK, 1911 [**E. silvicola*]. Head clearly outlined, sides forming a concave line with anterior end of thorax. Sides of thorax convex. Single pair of eyes (45). *Carb.,* Eu.—FIG. 98,*2*. **E. silvicola,* Eng.; ×3.5 (76).

Family ARTHROMYGALIDAE Petrunkevitch, 1923

Eye tubercle wanting, presence of eyes doubtful. *?Dev., Carb.*

Protolycosa ROEMER, 1866 [**P. anthracophila*]. Abdominal tergites with transverse rows of tubercles (45). *Carb.,* Eu.—FIG. 99,*1*. **P. anthracophila,* Ger.; ×3 (76).

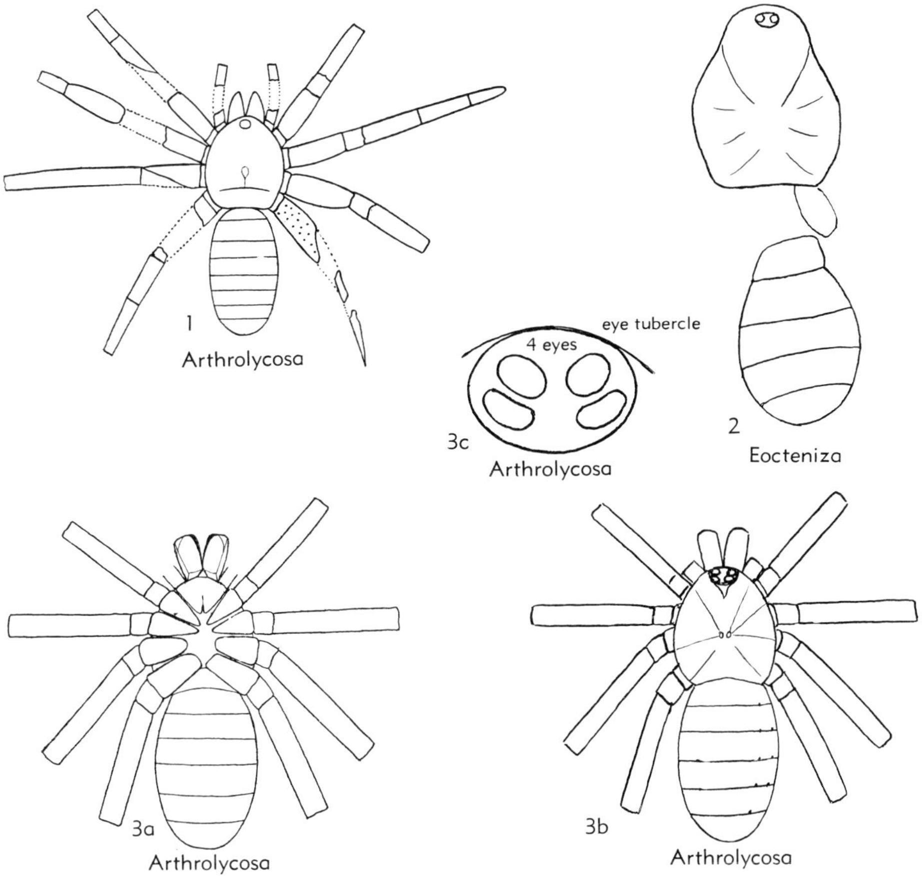

FIG. 98. Arthrolycosidae. *1, Arthrolycosa antiqua,* Penn., Ill., ×1.3; 2, *Eocteniza silvicola,* U.Carb., Eng., ×3.5; *3, Arthrolycosa danielsi,* Penn., Ill.; *3a,b,* ventral and dorsal, ×2.3; *3c,* eye tubercle, ×16 (76).

Arthromygale PETRUNKEVITCH, 1923 [**Arthrolycosa fortis* FRITSCH, 1904]. Carapace about as wide as long, with convex front and sides, without eyes. Legs short and stout (45). *Carb.*, Eu.—FIG. 99,*3*. **A. fortis* (FRITSCH), Czech.; ×1.7 (76).

Protocteniza PETRUNKEVITCH, 1949 [**P. britannica*].

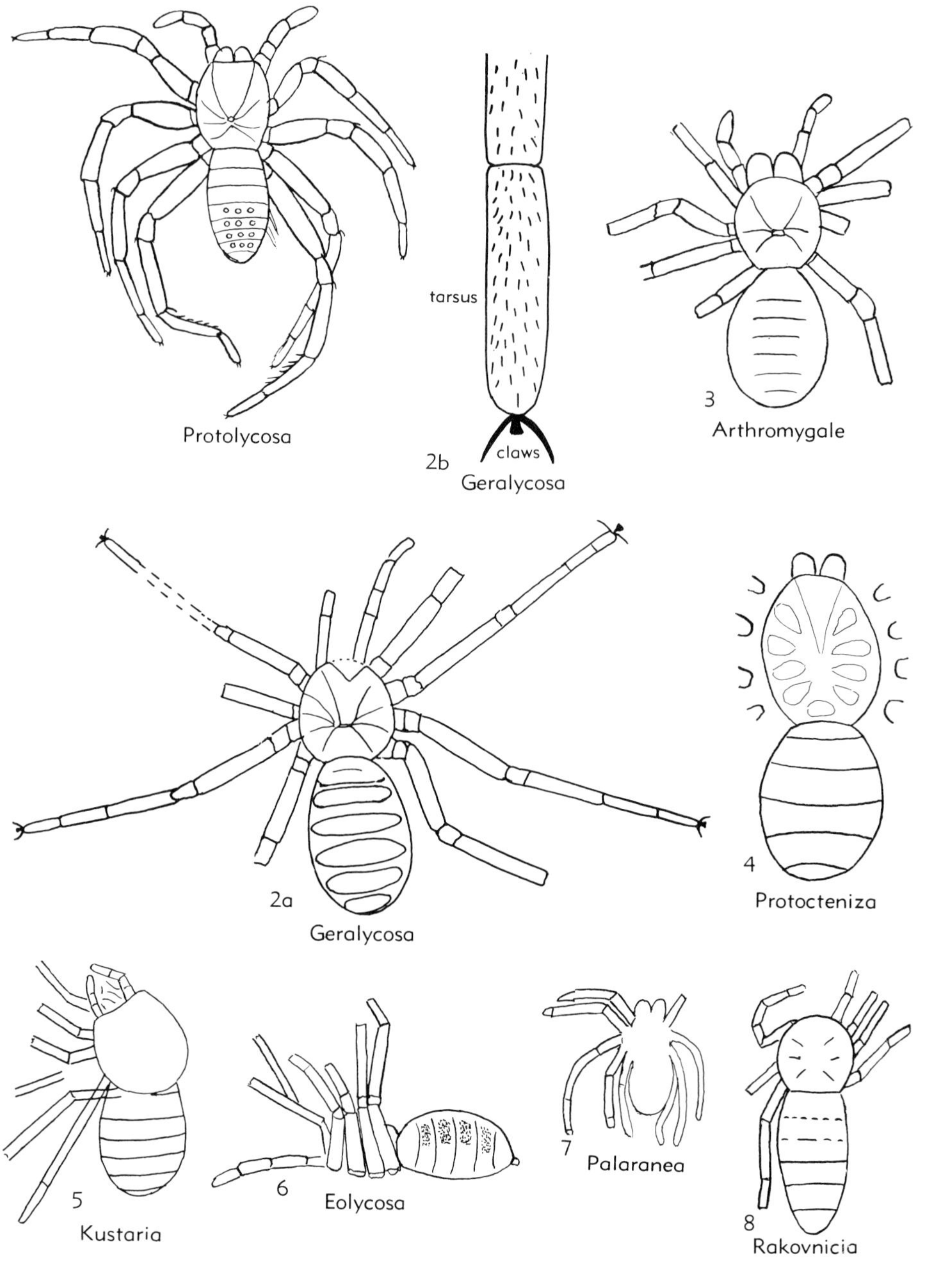

FIG. 99. Arthromygalidae. *1, Protolycosa anthracophila,* Carb., Ger., ×3 (66a). *2, Geralycosa fritschi,* U.Carb., Czech.; *2a,* ×2.3; *2b,* end of one of its legs, showing claws and arolium, ×13.5. *3, Arthromygale fortis,* U.Carb., Czech., ×1.7. *4, Protocteniza britannica,* U.Carb., Eng., ×7. *5, Kustaria carbonaria,* U. Carb., Czech, ×5.5. *6, Eolycosa lorenzi,* U.Carb., Czech., ×3.3. *7, Palaranea borassifoliae,* U.Carb., Czech., ×0.6. *8, Rakovnicia antiqua,* U.Carb., Czech., ×4.7 (76) (p. P133-P135).

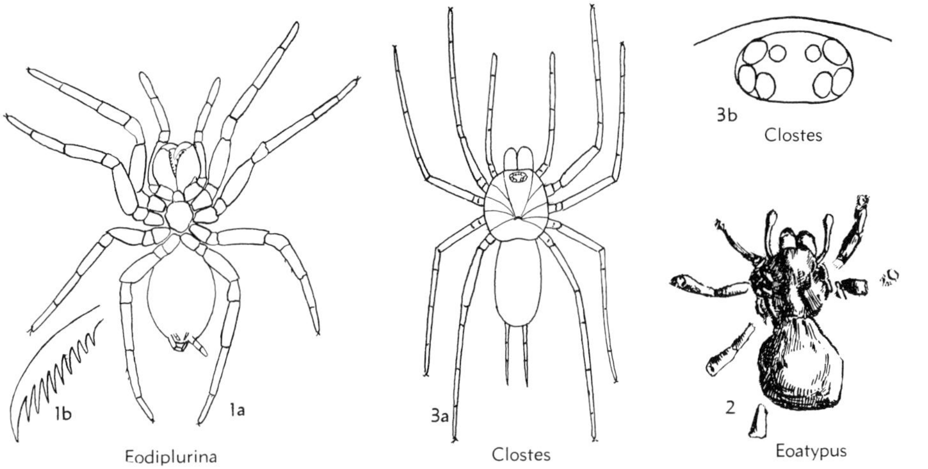

FIG. 100. Theraphosina. *1, Eodiplurina cockerelli,* Oligo., Colo. (Florissant); *1a,* ×6.5; *1b,* one of its claws, ×70 (76). *2, Eoatypus woodwardii,* Mio., Eng., ×3 (72). *3, Clostes priscus,* Oligo., Balt.; *3a,* ×5.5; *3b,* eye tubercle, ×20 (76) (p. P136).

Carapace considerably longer than wide (43). *Carb.,* Eu.—FIG. 99,*4.* **P. britannica,* Eng.; ×7 (76).

Palaranea FRITSCH, 1873 [**P. borassifoliae*]. Fourth leg as long as body. Order of legs 4321 (45). *Carb.,* Eu.—FIG. 99,*7.* **P. borassifoliae,* Czech.; outline of specimen as it appears impressed on leaf, ×0.6 (76).

Geralycosa KUŠTA, 1888 [**G. fritschi*]. Fourth leg longer than body by more than a fifth. Order of legs 4321. Tarsi with arolium (45). *Carb.,* Eu.—FIG. 99,*2.* **G. fritschi,* Czech.; *2a,* complete specimen, ×2.3; *2b,* tarsus with claws and arolium, ×13.5 (76).

Kustaria PETRUNKEVITCH, 1953 [**Scudderia carbonaria* KUŠTA, 1888]. Carapace with strongly convex posterior edge and slightly concave front (45). *Carb.,* Eu.—FIG. 99,*5.* **K. carbonaria* (KUŠTA), Czech.; ×5.5 (76).

Rakovnicia KUŠTA, 1884 [**R. antiqua*]. Carapace flat, with convex sides and front and truncated posterior edge. Abdomen twice as long as wide. Legs slender (45). *Carb.,* Eu.—FIG. 99,*8.* **R. antiqua,* Czech.; ×4.7 (76).

Eolycosa KUŠTA, 1885 [**E. lorenzi*]. Spinnerets terminal (45). *Carb.,* Eu.—FIG. 99,*6.* **E. lorenzi,* Czech.; ×3.3 (76).

Palaeocteniza HIRST, 1923 [**P. crassipes*]. Classification doubtful (17). *Dev.* (*Old Red Sandstone*), Scot.

Suborder THERAPHOSINA Sundevall, 1830

[*nom. correct.* PETRUNKEVITCH, herein (*pro* Theraphosae SUNDEVALL, 1830] [=Théraphoses WALCKENAER, 1805; Mygaliformes SIMON, 1864; Mygalomorphae POCOCK, 1893] [Type: *Theraphosa* WALCKENAER, 1805]

Chelicerae paraxial. Abdomen not segmented; 2 or 3 pairs of spinnerets. Two pairs of book lungs. Three or 4 pairs of cardiac ostia (36). *Oligo.-Rec.*

Family THERAPHOSIDAE Thorell, 1869

[=Aviculariidae SIMON, 1892][1]

Four pairs of cardiac ostia. Two pairs of spinnerets, posterior pair long, 3-jointed. Tarsi with 2 claws and claw tufts. *Oligo.-Rec.*

[1] The existing confusion of opinions concerning the status of the 2 generic names, *Mygale* and *Theraphosa,* is due to several causes, one of which is rooted in the historical change of the concept of the term *genus* and of its scope. LINNÉ did not use the term *family* in any way. He divided his orders directly into genera. DEGEER was the first to use the term *family,* but he placed it *under* the genus, dividing the genus into families, yet retaining binomial nomenclature for species, so that the family names appeared only in headings and the same genus in several families. WALCKENAER followed this principle. In 1802 he proposed the genus *Mygale.* In 1805 he proposed the genus *Theraphosa;* called "Théraphoses" all spiders with chelicerae of the type we call now paraxial, and subdivided them into 5 genera, the first of which is his *Mygale.* WALCKENAER did not mention the types of his genera, as that was not customary at his time. In 1850 C.L. KOCH restricted the genus *Mygale* to 2 species, *blondii* and *javanensis.* The name *Mygale* was found to be preoccupied by CUVIER for a mammal, so KOCH proposed for it the name *Scurria.* THORELL pointed out in 1869 that the name *Theraphosa* has priority, selected for its type the species *blondii* and erected the family Theraphosoidae which was later changed by POCOCK into Theraphosidae. SIMON retained in 1892 SUNDEVALL'S name Theraphosae for the suborder, selected for the name of the family the genus *Avicularia* LAMARCK, 1818. Before that, in 1885, SIMON thought that the type of *Mygale* was really *M. fasciata* and not *M. blondii* and gave the name *Poecilotheria* as substitute for *Mygale* and *Scurria* which also proved to be preoccupied. In 1923 I divided SIMON'S family Aviculariidae into 8 families, retaining the name Theraphosidae for one of them, following in this the example of POCOCK.

This family includes the largest spiders, known in America under the name of Tarantulae. A characteristic representative, *Poecilotheria regalis* POCOCK from Madras, India, is shown in Fig. 95.

Eodiplurina PETRUNKEVITCH, 1922 [**E. cockerelli*]. Legs spinose, in order 4123. Claws pectinated in a single row (35). *Oligo.*, N.Am.—FIG. 100,*1*. **E. cockerelli*, Colo.; *1a*, ventral side, ×6.5; *1b*, claw, ×70 (76).

Family DIPLURIDAE Pocock, 1894

Tarsi with 3 claws. Four pairs of cardiac ostia. Two or 3 pairs of spinnerets, posterior pair very long. *Oligo.-Rec.*

Clostes MENGE, 1869 [**C. priscus*]. Four spinnerets. Upper claws dissimilar, order of legs 4123. Eyes on tubercle, anterior median eyes smallest. No spines on legs (37). *Oligo.*—FIGS. 31,*3*, 100,*3*. **C. priscus*, Balt.; 31,*3*, dorsal side, ×8.4; 100,*3a*, dorsal side, ×5.5; 100,*3b*, tubercle with 8 eyes, ×20 (76).

THERAPHOSINA Incertae Sedis

Eoatypus McCOOK, 1888 [**E. woodwardi*]. *Eoc.*, Eu.—FIG. 100,*2*. **E. woodwardi*, Eng. (Wight); dorsal side, ×3 (72).

Suborder HYPOCHILINA Petrunkevitch, 1933

[*nom. correct.* PETR., herein (*pro* Hypochilomorphae PETR., 1933)] [Type: *Hypochilus* MARX, 1888]

Abdomen not segmented. Chelicerae di-

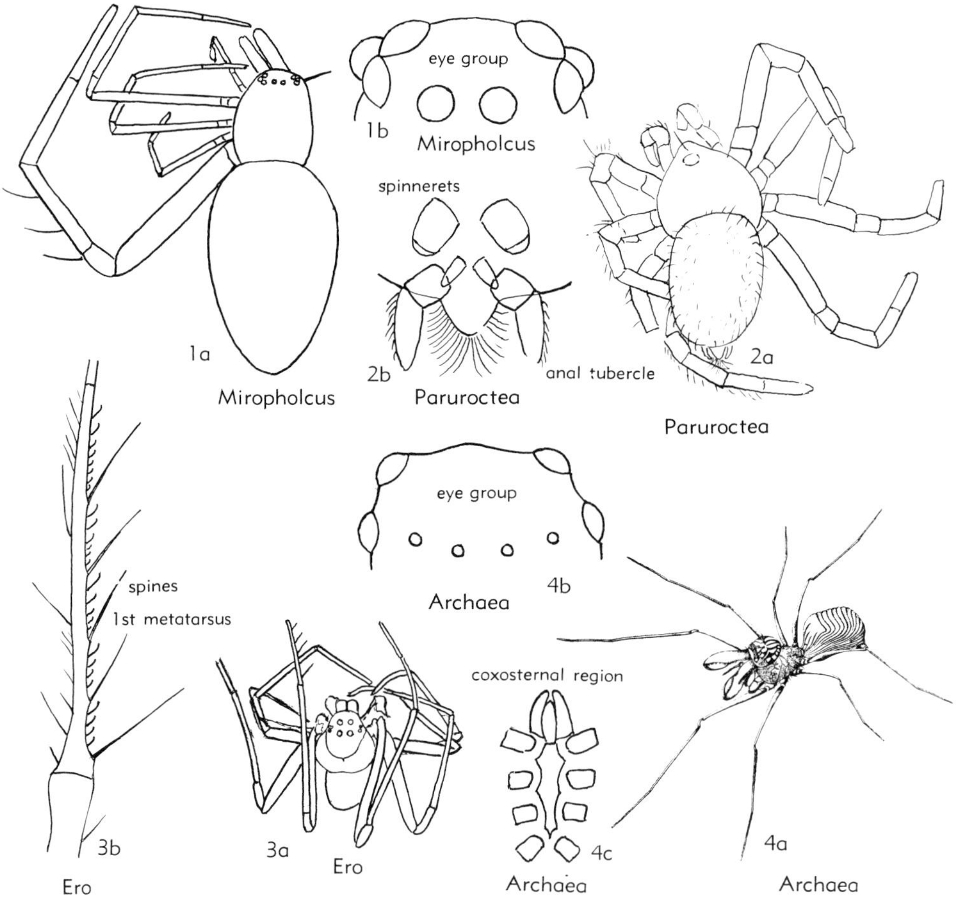

FIG. 101. Dipneumonina. *1, Miropholcus heteropus*, Oligo., Balt.; *1a*, ×20; *1b*, eye group, enlarged. *2, Paruroctea blauvelti*, Oligo., Balt.; *2a*, ×8.5; *2b*, spinnerets with anal tubercle, enlarged. *3, Ero permunda*, Oligo., Balt.; *3a*, ×5.2; *3b*, 1st metatarsus showing spines of 2 kinds, enlarged (76). *4, Archaea paradoxa*, Oligo., Balt.; *4a*, female, ×7.5 (71); *4b*, eye group, ×55; *4c*, coxosternal region, ×17 (76) (p. P138-P142).

axial. Two pairs of book lungs. Four pairs of cardiac ostia (36). *Rec.*

Suborder DIPNEUMONINA Latreille, 1817

[*nom. correct.* PETRUNKEVITCH, herein (*pro* Dipneumones LATREILLE, 1817] [=Dipneumonomorphae PETR., 1933] [Type: *Araneus* CLERCK, 1757]

Abdomen not segmented. Chelicerae diaxial. Single pair of book lungs and one pair of tracheal tubes. Three or 2 pairs of cardiac ostia. *?Carb., Oligo.-Rec.*

Division TRIONYCHI Petrunkevitch, 1933

Dipneumone spiders with 3 claws. Three pairs of cardiac ostia. *Oligo.-Rec.*

Superfamily HERSILIOIDEA Caporiacco, 1938

[*nom. correct.* PETRUNKEVITCH, herein (*pro* Hersiliaeformia CAPORIACCO, 1938)]

Posterior spinnerets wide apart, with long

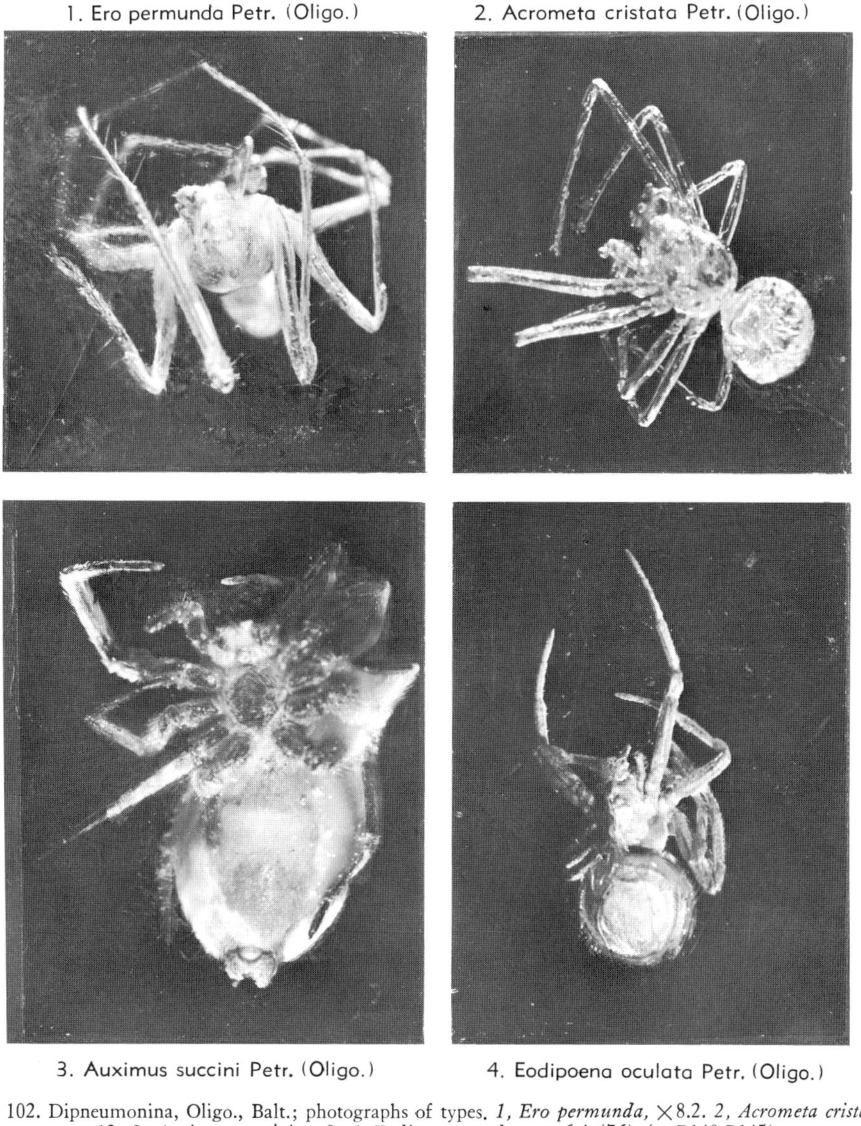

FIG. 102. Dipneumonina, Oligo., Balt.; photographs of types. *1, Ero permunda*, ×8.2. *2, Acrometa cristata*, ×12. *3, Auximus succini*, ×8. *4, Eodipoena oculata*, ×6.4 (76) (p. P140-P145).

2nd joint. Four pairs of heterogeneous eyes. *Oligo.-Rec.*

Family UROCTEIDAE Simon, 1875

Six spinnerets. Anal tubercle large, with fringe of long hair. *Oligo.-Rec.*

Paruroctea PETRUNKEVITCH, 1942 [**P. blauvelti*]. First row of eyes recurved. Fourth tarsi with ventral spines (37). *Oligo.*, Eu.—FIG. 101,*2*. **P. blauvelti*, Balt.; *2a*, dorsal side, ×8.5; *2b*, spinnerets and anal tubercle, much enlarged (76).

Family HERSILIIDAE Thorell, 1869

Head elevated. Posterior spinnerets very long, flexible. *Oligo.-Rec.*

Hersilia AUDOUIN, 1827 [**H. caudata*]. Clypeus very high. Metatarsi 2-jointed. *Oligo.-Rec.*—FIG. 105,*3*. **H. miranda* KOCH & BERENDT, Oligo., Balt.; ×2.1 (71).

Gerdia MENGE, 1869 [**G. myura*]. Anterior eyes half as large as posterior ones (74). *Oligo.*, Balt.

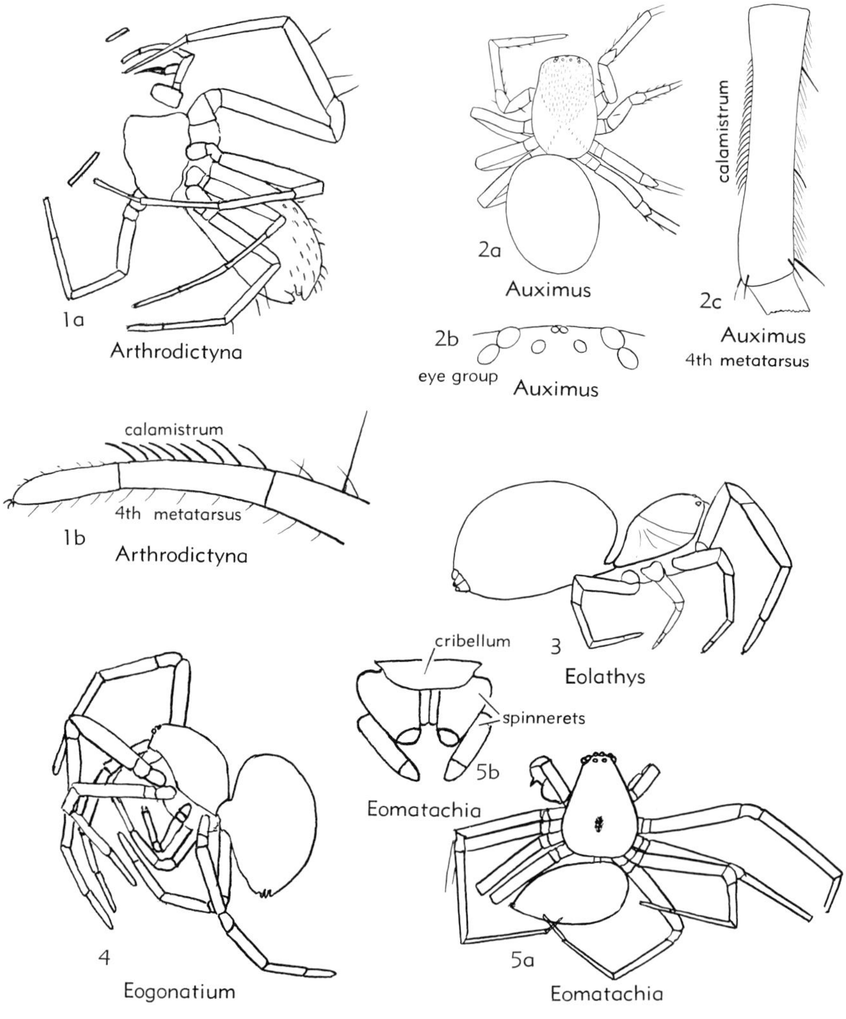

FIG. 103. Dipneumonina, Oligo., Balt. *1, Arthrodictyna segmentata; 1a*, ×25; *1b*, 4th metatarsus showing calamistrum, ×44. *2, Auximus fossilis; 2a*, ×4.6; *2b*, eye group, ×26; *2c*, 4th metatarsus showing calamistrum, ×26; *3, Eolathys succini*, ×11. *4, Eogonatium minutum*, ×24. *5, Eomatachia latifrons; 5a*, male, ×5; *5b*, its spinnerets, enlarged (76) (p. P140-P142).

Superfamily SCYTODOIDEA Caporiacco, 1938

[*nom. correct.* PETRUNKEVITCH, herein (*pro* Scytodiformia CAPORIACCO, 1938)]

Tracheal system rudimentary. Lip fused with sternum. *Oligo.-Rec.*

Family PHOLCIDAE Simon, 1874

Lateral eyes in triads. Legs long and slender. *Oligo.-Rec.*—FIG. 96. *Artema mauricia,* a Recent representative of the family, ×2 (75).

Miropholcus PETRUNKEVITCH, 1942 [**M. heteropus*]. Eye group full width of head. Fourth pair of legs

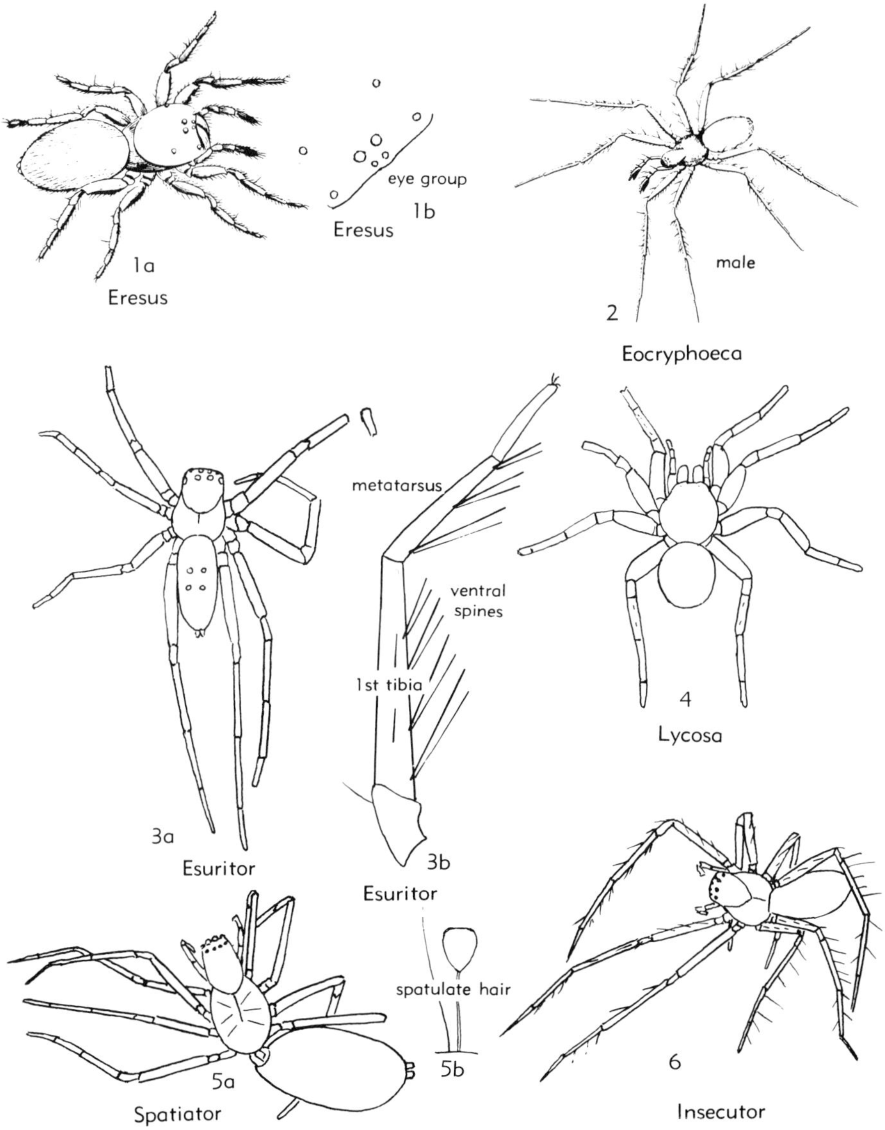

FIG. 104. Dipneumonina, Oligo. *1, Eresus monachus,* Balt.; *1a,* ×5; *1b,* eye group enlarged (71). *2, Eocryphoeca gracilipes,* Balt.; male, ×9 (71). *3, Esuritor longipes,* Balt.; *3a,* ×4.3; *3b,* 1st leg, showing ventral spines, ×14. *4, Lycosa florissanti,* Colo.; female, ×2.5. *5, Spatiator praeceps,* Balt.; *5a,* ×4.3; *5b,* spatulate hair, enlarged. *6, Insecutor aculeatus,* Balt.; ×6.5 (76) (p. P140-P141).

much stouter and longer than others (37). *Oligo.*, Eu.—FIG. 101,*1*. **M. heteropus*, Balt.; *1a*, dorsal side, ×26; *1b*, eye group, enlarged (76).

Superfamily ZODARIOIDEA Caporiacco, 1938

[*nom. correct.* PETRUNKEVITCH, herein (*pro* Zodariiformia CAPORIACCO, 1938)]

Anterior and median spinnerets rudimentary or lacking. Third claw lacking in some species. *Oligo.-Rec.*

Family ZODARIIDAE Simon, 1892

First pair of legs not stouter than the others. *Oligo.-Rec.*

Anniculus PETRUNKEVITCH, 1942 [**A. balticus*]. Only 2 spinnerets, on pedestal. Sternum almost circular. Integuments with plumose hair (37). *Oligo.*, Eu.—FIG. 105,*1*. **A. balticus*, Balt.; *1a*, dorsal side, ×4.3; *1b*, spinnerets on pedestal, enlarged (76).

Family SPATIATORIDAE Petrunkevitch, 1942

Head separated from thorax by semicircular groove. Eyes in 2 rows. Six spinnerets, anterior pair stoutest and longest (37). *Oligo.*

Spatiator PETRUNKEVITCH, 1942 [**S. praeceps*]. Legs without spines, 4th pair longest. Spatulate hair present (37). *Oligo.*, Eu.—FIG. 104,*5*. **S. praeceps*, Balt.; *5a*, dorsal side of male, ×4.3; *5b*, spatulate hair from leg, enlarged (76).

Adorator PETRUNKEVITCH, 1942 [**A. brevipes*]. Legs spinose, 1st pair longest (37). *Oligo.*, Balt.

Superfamily LYCOSOIDEA Sundevall, 1833

[*nom. correct.* PETRUNKEVITCH, herein (*pro* Lycosiformia CAPORIACCO, 1938, *nom. transl. ex* Lycosidae SUNDEVALL, 1833)]

Eyes in 2 or 3 rows. Tarsi without serrated bristles. *Oligo.-Rec.*

Family ERESIDAE C.L.Koch, 1837

Cribellum and calamistrum present. Eyes in 3 rows, homogeneous. *Oligo.-Rec.*

Eresus WALCKENAER, 1805 [**Aranea nigra* PETAGNA, 1787]. Posterior median eyes much larger than anterior median eyes. *Oligo.-Rec.*—FIG. 104,*1*. **E. monachus* KOCK & BERENDT, Oligo., Balt.; *1a*, female, ×5; *1b*, eye group, enlarged (71).

Family AMAUROBIIDAE C.L.Koch, 1868

Cribellum and calamistrum present. Eyes subequal, in 2 rows. *Oligo.-Rec.*

Amaurobius C.L. KOCH, 1837 [**Aranea fenestralis* STROEM, 1768]. Retromargin of chelicerae with 2 or 3 teeth (37). *Oligo.-Rec.* [*A. succini* PETRUNKEVITCH, Oligo., Balt.].

Auximus SIMON, 1892 [**A. dentichelis*]. Retromargin of chelicerae with 4 teeth. Anterior median eyes by far the smallest (37). *Oligo.-Rec.*—FIG. 102,*3*. *A. succini* PETRUNKEVITCH, Oligo., Balt.; photo of female, ventral side, ×8 (76).—FIG. 103,*2*. *A. fossilis* PETR., Oligo., Balt.; *2a*, holotype, ×4.6; *2b*, eye group, ×26; *2c*, 4th metatarsus with calamistrum, ×26 (76).

Family PSECHRIDAE Simon, 1892

Cribellum and calamistrum present. Three claws and claw tufts. Trichobothria few, on tarsi. *Oligo.-Rec.*

Eomatachia PETRUNKEVITCH, 1942 [**E. latifrons*]. Lateral eyes larger than median eyes. Cribellum entire (37). *Oligo.*, Eu.—FIG. 103,*5*. **E. latifrons*, Balt.; *5a*, dorsal side of male, ×5; *5b*, spinnerets with cribellum, enlarged (76).

Family AGELENIDAE C.L.Koch, 1837

[*nom. correct.* SIMON, 1875 (*pro* Agelenides C.L KOCH, 1837)] [=Agalenidae THORELL, 1869]

Six spinnerets, posterior pair longest, with 2nd joint cylindrical, long. Integument with plumose hair (37). *Oligo.-Rec.*

Agelena WALCKENAER, 1805 [**Araneus labyrinthicus* CLERCK, 1757]. Eyes homogeneous, in 2 strongly procurved rows (37). *Oligo.-Rec.* [*A. tabida* KOCH & BERENDT, Oligo., Balt.].

Eocryphoeca PETRUNKEVITCH, 1946 [**Tegenaria gracilipes* KOCH & BERENDT, 1954]. Anterior median eyes smallest. Second joint of posterior spinnerets conical, shorter than basal joint (40). *Oligo.*, Eu.—FIG. 104,*2*. **E. gracilipes* (KOCH & BERENDT), Balt.; male, ×9 (71).

Myro O. P. CAMBRIDGE, 1876 [**M. kerguelensis*]. Posterior spinnerets somewhat shorter than anterior ones (37). *Oligo.-Rec.* [*M. hirsutus* PETRUNKEVITCH, Oligo., Balt.].

Family INSECUTORIDAE Petrunkevitch, 1942

Spines numerous, long, erect. Sternum widely truncated in front. Eyes in 2 rows. *Oligo.*

Insecutor PETRUNKEVITCH, 1942 [**I. aculeatus*]. Anterior median eyes smallest. Quadrangle wider behind than in front (37). *Oligo.*, Eu.—FIG. 104,*6*. **I. aculeatus*, Balt.; immature female, ×6.5 (76).

Family PISAURIDAE Simon, 1897

Anterior eyes small; posterior row recurved and much wider than anterior row. Trochanters deeply notched. *Oligo.-Rec.*

Esuritor PETRUNKEVITCH, 1942 [**E. spinipes*]. Ventral spines on anterior tibiae and metatarsi very long (37). *Oligo.*, Eu.—FIG. 104,*3*. **E. spinipes*, Balt.; *3a*, immature female, ×4.3; *3b*, distal half of 1st leg, showing tibial and metatarsal spines, ×14 (76).

Family LYCOSIDAE Sundevall, 1933

Eyes in 3 rows, 1st row formed by 4 small eyes. Trochanters deeply notched. *Oligo.-Rec.*

Lycosa LATREILLE, 1804 [**Aranea tarentula* ROSSI, 1790]. Eyes of 2nd row separated by about their diameter. Legs stout, in order 4123 (35). *Oligo.-Rec.*—FIG. 104,*4*. *L. florissanti* PETRUNKEVITCH, *Oligo.*, Colo.; female, ×2.5 (76).

Superfamily ARANEOIDEA Leach, 1815

[*nom. transl.* PETRUNKEVITCH, herein (*ex* Araneides LEACH, 1815)] [=Argyopiformia CAPORIACCO, 1938]

Eyes in 2 rows. Tarsi with serrated bris-

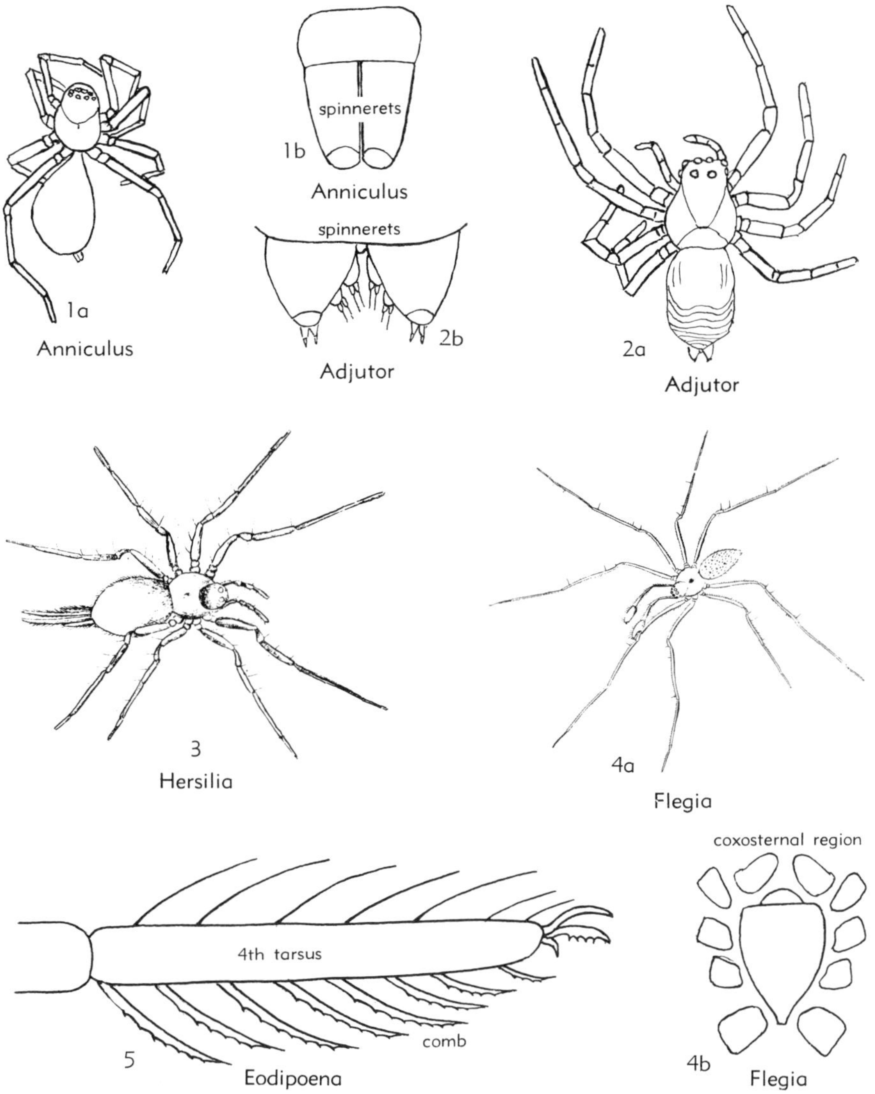

FIG. 105. Dipneumonina, Oligo., Balt. *1, Anniculus balticus; 1a*, ×4.3; *1b*, 2 spinnerets on pedestal, enlarged. *2, Adjutor mirabilis; 2a*, ×20; *2b*, spinnerets, enlarged (76). *3, Hersilia miranda*, ×2.1. *4, Flegia longimana; 4a*, male, ×3 (71); *4b*, its coxosternal region, ×15 (76). *5, Eodipoena oculata*, 4th metatarsus showing comb of serrated bristles, ×75 (76) (p. P138-P142).

tles. Pedipalpal coxae with maxillary serrula. *Oligo.-Rec.*

Family ARCHAEIDAE Koch & Berendt, 1854

Head elevated. Chelicerae inserted in an elliptic foramen far above mouth, strongly divergent. *Oligo.-Rec.*

Archaea Koch & Berendt, 1854 [**A. paradoxa*]. Sternum elongate. Maxillae converging over lip (37). *Oligo.-Rec.*—Fig. 101,*4*. **A. paradoxa,* Balt.; *4a,* female, ×7.5; *4b,* eye group, ×55; *4c,* coxosternal region, ×17 (76).

Family MIMETIDAE Simon, 1895

First and 2nd tibia and metatarsus with a row of long spines separated by much shorter, curved spines. *Oligo.-Rec.*

Ero C.L. Koch, 1837 [**Aranea tuberculata* DeGeer, 1778]. Posterior legs not much shorter than anterior ones (37). *Oligo.-Rec.*—Figs. 101,*3;* 102,*1. E. permunda* Petrunkevitch, Oligo., Balt.; 101, *3a,* male, ×5.2; 101,*3b,* 1st metatarsus showing row of spines of 2 kinds, ×20; 102,*1,* photo of holotype, ×8.2 (76).

Family ARTHRODICTYNIDAE Petrunkevitch, 1942

Cribellum and calamistrum present. Abdomen with imperfect segmentation (37). *Oligo.*

Arthrodictyna Petrunkevitch, 1942 [**A. segmentata*]. Cheliceral margins with scopula of 4 hairs. First coxae wide apart (37). *Oligo.,* Balt.—Fig. 103,*1*. **A. segmentata; 1a,* holotype, ×25; *1b,* end of 4th leg showing calamistrum on metatarsus, ×44 (76).

Family DICTYNIDAE Simon, 1874

Cribellum and calamistrum present. Abdomen not segmented. Tracheal system extends into cephalothorax. *Oligo.-Rec.*

Eolathys Petrunkevitch, 1950 [**E. succini*]. Quadrangle of median eyes much wider behind than in front and much wider than long (40). *Oligo.,* Eu.—Fig. 103,*3*. **E. succini,* Balt.; holotype, ×11 (76).

Family MICRYPHANTIDAE Bertkau, 1885

[=Erigonidae Simon, *nom. transl.* Petrunkevitch, 1939 (*ex* Erigoninae Simon, 1926)]

Tracheal system extends into cephalothorax. Tarsi with "drum." *Oligo.-Rec.*

Eogonatium Petrunkevitch, 1942 [**E. minutum*]. Legs without spines, with single trichobothrium on metatarsi, considerably beyond middle (37). *Oligo.,* Balt.—Fig. 103,*4*. **E. minutum;* holotype in side view, ×24 (76).

Family THERIDIIDAE Sundevall, 1833

Fourth tarsi with a "comb" of serrated bristles. Maxillae converging. Eyes heterogeneous, anterior median eyes diurnal. *Oligo.-Rec.*

Theridion Walckenaer, 1805 [**Araneus lineatus* Clerck, 1757] [=*Theridium auctt.*]. Abdomen ovoid with pointed posterior end. Fourth leg shorter than 1st (37). *Oligo.-Rec.* [*T. simplex* Koch & Berendt, Oligo., Balt.].

Flegia Koch & Berendt, 1854 [**F. longimana*]. Carapace flat. Quadrangle rectangular. First coxae wide apart. Femur of male palp as long as carapace. *Oligo.,* Eu.—Fig. 105,*4*. **F. longimana,* Balt.; *4a,* male, ×3; *4b,* coxosternal region, ×15 (76).

Eodipoena Petrunkevitch, 1942 [**E. oculata*]. Eyes equal. Eye group much narrower than head (37). *Oligo., Eu.*—Figs. 102,*4;* 105,*5*. **E. oculata,* Balt.; 102,*4,* photo of holotype, ×6.4; 105, *5,* 4th tarsus with "comb," ×75 (76).

Nactodipoena Petrunkevitch, 1942 [**N. dunbari*]. Eye group almost as wide as head. Posterior median eyes smaller than anterior median eyes (37). *Oligo.,* Balt.

Steatoda Sundevall, 1833 [**Aranea bipunctata* Linné, 1758]. Stridulatory organ between abdomen and carapace. Anterior median eyes larger than anterior lateral eyes (37). *Oligo.-Rec.* [*S. succini* Petrunkevitch, Balt.].

Municeps Petrunkevitch, 1942 [**M. pulcher*]. Eyes equal. Upper claws with single tooth (37). *Oligo.,* Balt.

Eomysmena Petrunkevitch, 1942 [**E. moritura*]. Posterior median eyes smallest. Upper claws smooth (37). *Oligo.,* Balt.

Family ADJUTORIDAE Petrunkevitch, 1942

Anterior spinnerets very stout. Legs with spines and very long trichobothria (37). *Oligo.*

Adjutor Petrunkevitch, 1942 [**A. mirabilis*]. Single trichobothrium on tibiae and metatarsi. Legs in order 1234. Third claw well developed (37). *Oligo.,* Eu.—Fig. 105,*2*. **A. mirabilis,* Balt.; *2a,* holotype, ×20; *2b,* spinnerets, enlarged (76).

Admissor Petrunkevitch, 1942 [**A. aculeatus*]. Anterior spinnerets not much stouter than posterior ones. Legs in order 2143 (37). *Oligo.,* Balt.

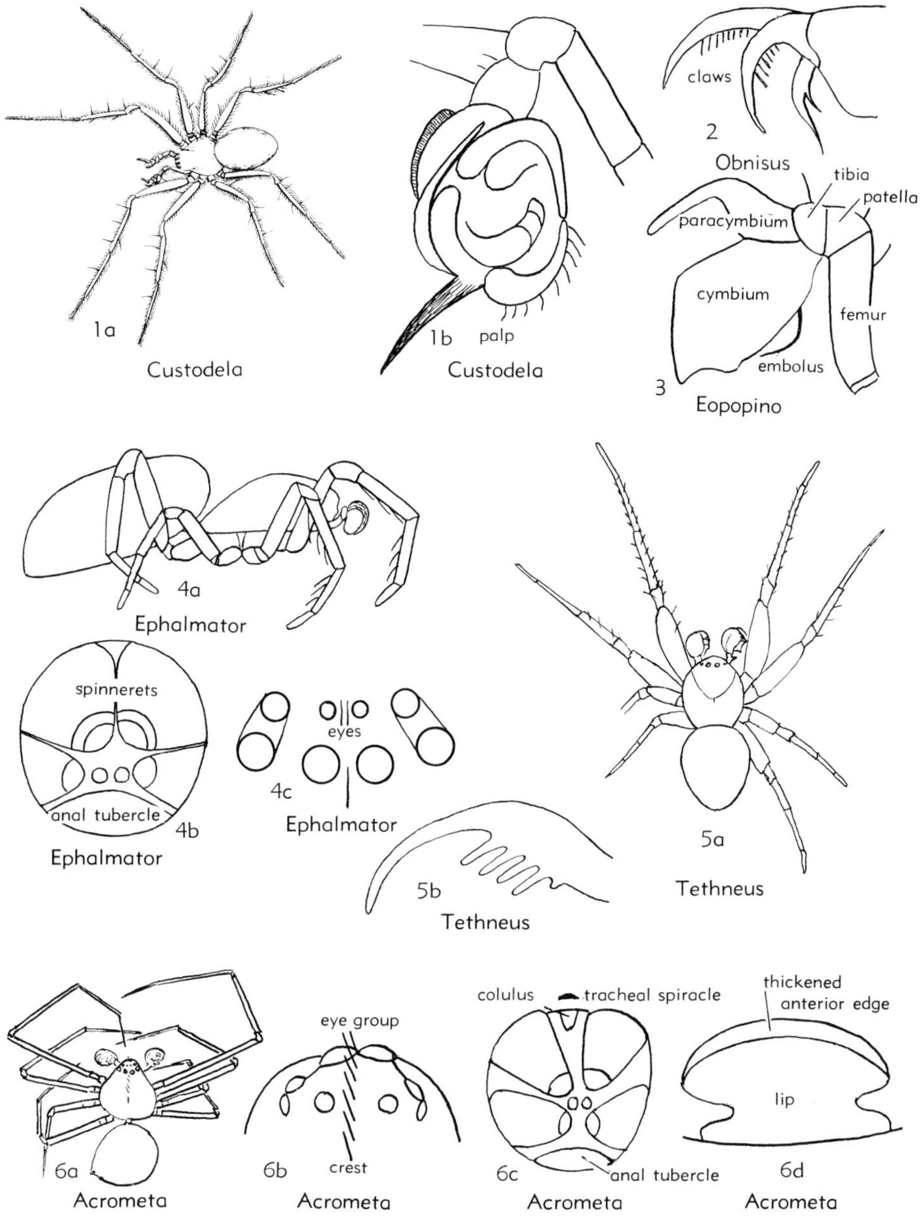

FIG. 106. Dipneumonina, Oligo. *1, Custodela cheiracantha,* Balt.; *1a,* male, ×3.3 (71); *1b,* its palp, enlarged (76). *2, Obnisus tenuipes,* Balt.; claws, enlarged. *3, Eopopino longipes,* Balt.; palp of male showing paracymbium and embolus, enlarged. *4, Ephalmator fossilis,* Balt.; *4a,* male, ×16; *4b,* its spinnerets, ×60; *4c,* its eye group, ×52. *5a, Tethneus robustus,* Colo.; male, ×3.3; *5b, Tethneus hentzi,* Colo.; claw, ×270. *6, Acrometa cristata,* Balt.; *6a,* male, ×8; *6b,* its eye group, ×25; *6c,* spinnerets, enlarged; *6d,* lip, showing thickened anterior edge, enlarged (76) (p. P144-P145).

Adjunctor PETRUNKEVITCH, 1942 [*A. *similis*]. Several trichobothria on tibiae and metatarsi. Legs in order 4123. Third claw reduced to a thorn (37). *Oligo.*, Balt.

Family EPHALMATORIDAE Petrunkevitch, 1950

Anterior spinnerets much stouter than posterior ones. Chelicerae with transverse margins. First and 2nd tibia with several pairs of very long ventral spines. *Oligo.*

Ephalmator PETRUNKEVITCH, 1950 [*E. *fossilis*]. Legs short, stout. Lateral eyes on common tubercle. Anterior median eyes smallest. Quadrangle much narrower in front (44). *Oligo.*, Eu.—FIG. 106,*4*. *E. *fossilis*, Balt.; *4a*, holotype in side view, ×16; *4b*, spinnerets, ×60; *4c*, eye group, ×52 (76).

Family LINYPHIIDAE Dahl, 1913

Eyes heterogeneous. Chelicerae with oblique margins. Maxillae parallel. Legs slender, with 1 or 2 bristles, without spines. Pedipalp of male with paracymbium, of female with terminal claw. *Oligo.-Rec.*

Linyphia LATREILLE, 1804 [*Araneus triangularis* CLERCK, 1757]. Posterior eyes well separated. Margins of chelicerae with at least 3 teeth. *Oligo.-Rec.* [*L. seclusa* (SCUDDER), Oligo., Colo.].

Custodela PETRUNKEVITCH, 1942 [*Linyphia cheiracantha* KOCH & BERENDT, 1854]. Head higher than thorax. Order of legs 1243. Male palp with powerful spike directed forward (37). *Oligo.*, Eu.—FIGS. 31,*6* 106,*1*. *C. *cheiracantha* (KOCH & BERENDT), Balt.; 31,*6*, photo of male, ×6; 106,*1a*, copy of figure of KOCH-B., ×3.3; *1b*, palp showing spike, enlarged (76).

Obnisus PETRUNKEVITCH, 1942 [*O. *tenuipes*]. Carapace slightly narrowed in front. Legs in order 1243. Claws slightly dissimilar. Male palp with apophysis shaped like a pick-axe (37). *Oligo.*, Eu.—FIG. 106,*2*. *O. tenuipes*, Balt.; claws of 2nd leg, enlarged (76).

Eopopino PETRUNKEVITCH, 1942 [*E. *longipes*]. Carapace almost circular. Lip much wider than long. Legs in order 1243 (37). *Oligo.*, Eu.—FIG. 106,*3*. *E. *longipes*, Balt.; male palp, enlarged (76).

Meditrina PETRUNKEVITCH, 1942 [*M. *circumvallata*]. Carapace with greatly thickened edges and pair of shoulder humps. Eyes on tubercles. Legs in order 1243 (37). *Oligo.*, Balt.

Malleator PETRUNKEVITCH, 1942 [*M. *niger*]. Carapace almost circular. Legs in order 4123. Femur of male palp longer than carapace, tibia with terminal brush of long bristles (37). *Oligo.*, Balt.

Impulsor PETRUNKEVITCH, 1942 [*I. *neglectus*]. Legs in order 4123. Lip pentagonal. Claws dissimilar (37). *Oligo.*, Balt.

Liticen PETRUNKEVITCH, 1942 [*L. *setosus*]. Legs in order 1243, setose. All eyes on common tubercle (37). *Oligo.*, Balt.

Mystagogus PETRUNKEVITCH, 1942 [*M. *glaber*] Legs in order 1243. Eyes on separate tubercles. Abdomen wider behind than in front (37). *Oligo.*, Balt.

Family ARANEIDAE Leach, 1819

[*nom. correct.* PETRUNKEVITCH, herein (*pro* Araneadae LEACH, 1819)] [=Epeireides SUNDEVALL, 1833; Epeiridae BLACKWELL, 1859; Argiopidae SIMON, 1892]

Six spinnerets sitting in a circle or square, with colulus in front and anal tubercle behind. Lip with thickened anterior edge. Tarsi with spurious claws in addition to regular ones. Cheliceral margins oblique, with teeth. Maxillae parallel, wide in front. Legs with spines. Orb weavers. *Oligo.-Rec.*

Subfamily TETRAGNATHINAE Simon, 1892

Femora with single or double row of trichobothria. Chelicerae long. Body elongate. Orb web horizontal. *Oligo.-Rec.*

Tetragnatha LATREILLE, 1804 [*Aranea extensa* LINNÉ, 1757]. Chelicerae divergent, with long fangs. Abdomen 3 or 4 times longer than wide. *Oligo.-Rec.* [*T. tertiaria* SCUDDER, Oligo., Colo.].

Palaeopachygnatha PETRUNKEVITCH, 1922 [*P. *scudderi*]. Abdomen oval. Carapace slightly longer than wide (37). *Oligo.*, Colo.

Palaeometa PETRUNKEVITCH, 1922 [*Theridium opertaneum* SCUDDER, 1890]. Carapace at least twice as long as wide. Eyes elevated (35). *Oligo.*, Colo.

Subfamily NEPHILINAE Simon, 1892

Metatarsus with tarsus longer than patella with tibia in all legs. Lip longer than wide. Body and legs long. Males many times smaller than females. Orb web vertical. *Oligo.-Rec.*

Nephila LEACH, 1815 [*Aranea maculata* FABRICIUS, 1793]. Carapace convex. Eye group almost as wide as head. Lateral eyes separated. *Oligo.-Rec.* [*N. pennatipes* SCUDDER, Oligo., Colo.].

Subfamily ARANEINAE Simon, 1892

Metatarsus with tarsus shorter than patella with tibia in all legs. Lip wider than long. Orb web vertical. *Oligo.-Rec.*

Epeira WALCKENAER, 1805 [*Aranea cornuta* LINNÉ, 1758]. First leg longer than 4th. Legs spinose; femora in females without ventral spines; tibiae and metatarsi in males with double row of ventral spines (37). *Oligo.-Rec.* [*E. vulcanalis* SCUDDER, Oligo., Colo.].

Tethneus SCUDDER, 1890 [*T. *guyoti*]. First and 2nd femora very stout. Legs spinose (35). *Oligo.*, N.Am.—FIG. 106,*5a*. *T. robustus* PETRUNKEVITCH,

Colo.; male, ×3.3 (76).—FIG. 106,*5b*. *T. hentzii* SCUDDER; claw of 1st left foot, ×270 (76).

Eustaloides PETRUNKEVITCH, 1942 [**E. setosus*]. Carapace narrowed in front, with shoulder humps. Lateral eyes on common tubercle. Sternum triangular. *Oligo.*, Balt.

Subfamily METINAE Simon, 1892

Chelicerae with rudimentary boss. Sternum pointed behind. Legs with spines. *Oligo.-Rec.*

Acrometa PETRUNKEVITCH, 1942 [**A. cristata*]. Carapace greatly narrowed in front, in middle twice as wide as eye group. Quadrangle of median eyes wider behind than in front. Male palp with spiral embolus (37). *Oligo.*, Eu.—FIGS. 31,*4;* 102,*2*; 106,*6*. **A. cristata,* Balt.; 31,*4,* photo of male, ×12; 102,*2,* photo of holotype, ×12; 106, *6a,* dorsal side of male, ×8; 106,*6b,* eye group, ×25; 106,*6c,* spinnerets, enlarged; 106,*6d,* lip showing thickened edge, enlarged (76).

Theridiometa PETRUNKEVITCH, 1942 [**T. edwardsi*].

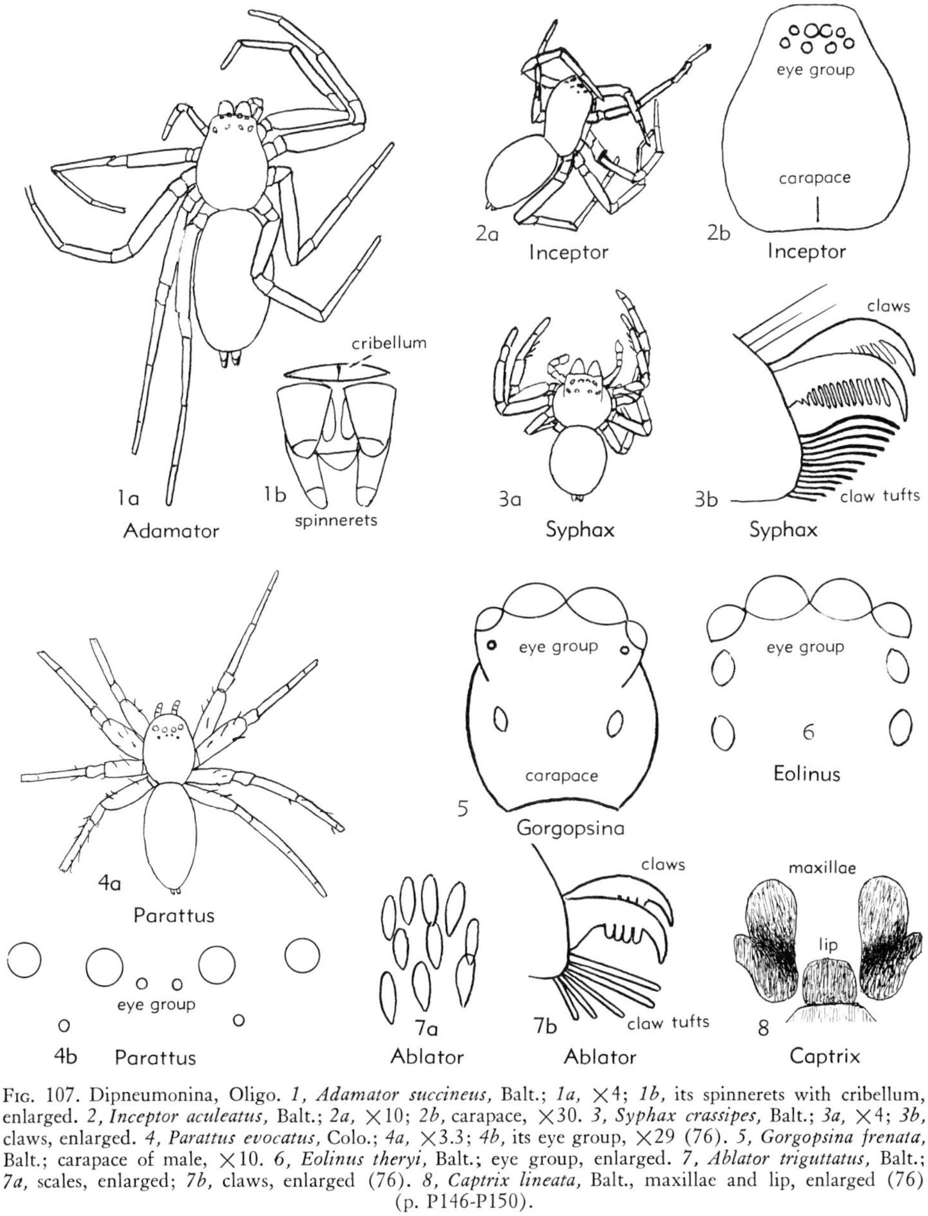

FIG. 107. Dipneumonina, Oligo. *1, Adamator succineus,* Balt.; *1a,* ×4; *1b,* its spinnerets with cribellum, enlarged. *2, Inceptor aculeatus,* Balt.; *2a,* ×10; *2b,* carapace, ×30. *3, Syphax crassipes,* Balt.; *3a,* ×4; *3b,* claws, enlarged. *4, Parattus evocatus,* Colo.; *4a,* ×3.3; *4b,* its eye group, ×29 (76). *5, Gorgopsina frenata,* Balt.; carapace of male, ×10. *6, Eolinus theryi,* Balt.; eye group, enlarged. *7, Ablator triguttatus,* Balt.; *7a,* scales, enlarged; *7b,* claws, enlarged (76). *8, Captrix lineata,* Balt., maxillae and lip, enlarged (76) (p. P146-P150).

Carapace narrowed in front. Eye group as wide as head (37). *Oligo.*, Balt.

Memoratrix PETRUNKEVITCH, 1942 [**M. rydei*]. Carapace narrowed in front. First leg 9 times as long as carapace (37). *Oligo.*, Balt.

Subfamily THERIDIOSOMATINAE Simon, 1892

Sternum broadly truncated behind. Legs without spines. Third claw at least as long as upper claws. *Oligo.-Rec.*

Elucus PETRUNKEVITCH, 1942 [**E. inermis*]. Head much wider than eye group. Clypeus as high as quadrangle. Sternum convex (37). *Oligo.*, Balt.

Superfamily ARCHAEOMETOIDEA Petrunkevitch, nov.

[*nom. transl.* PETRUNKEVITCH, herein (*ex* Archaeometidae PETRUNKEVITCH, 1949)]

Presumptive Trionychi with segmented abdomen. *Carb.*

Family ARCHAEOMETIDAE Petrunkevitch, 1949

Legs prograde. *Carb.*

Archaeometa POCOCK, 1911 [**A. nephilina*]. Abdomen long, cylindrical. Carapace with almost parallel sides (45). *Carb.*, Eu.—FIG. 110,*2*. **A. nephilina*, Eng.; ×4.1 (76).

Arachnometa PETRUNKEVITCH, 1949 [**A. tuberculata*]. Abdomen subconical. Carapace with strongly converging sides (45). *Carb.*, Eu.—FIG. 110,*1*. **A. tuberculata*, Eng.; ×4.1 (76).

Eopholcus FRITSCH, 1904 [**E. pedatus*]. Abdomen almost globular. Carapace with pair of eyes (45). *Carb.*, Eu.—FIG. 110,*6*. **E. pedatus*, Czech., ×8.7 (76).

Division DIONYCHI Petrunkevitch, 1933

Dipneumone spiders with 2 claws and claw tufts. Three pairs of cardiac ostia. *Oligo.-Rec.*

Superfamily ZOROPSIDIOIDEA Simon, 1892

[*nom. correct.* PETRUNKEVITCH, herein (*pro* Zoropidiformia CAPORIACCO, 1938, *nom. transl. ex* Zoropsidae SIMON, 1892)]

Dionychi with cribellum and calamistrum. *Oligo.-Rec.*

Family ZOROPSIDAE Simon, 1892

Eyes homogeneous, in 2 rows. First and 2nd tarsi and metatarsi scopulate. Cribellum divided. *Oligo.-Rec.*

Adamator PETRUNKEVITCH, 1942 [**A. succineus*]. Legs in order 4123. Both rows of eyes recurved, 2nd row much wider than 1st row (37). *Oligo.*, Eu.—FIG. 107,*1*. **A. succineus*, Balt.; *1a*. female, holotype, ×4; *1b*, spinnerets with divided cribellum, enlarged (76).

Superfamily THOMISOIDEA Sundevall, 1833

[*nom. correct.* PETRUNKEVITCH, herein (*pro* Thomisiformia CAPORIACCO, 1938, *nom. transl. ex* Thomisidae SUNDEVALL, 1833)]

Dionychi with laterigrade legs. Abdomen not segmented. *Oligo.-Rec.*

Family EUSPARASSIDAE Petrunkevitch, 1949

[=Sparassidae BERTKAU, 1878]

Tarsi usually scopulate. Trichobothria numerous. Claws similar. *Oligo.-Rec.*

Subfamily EUSPARASSINAE Petrunkevitch, 1949

[=Sparassinae SIMON, 1897]

Spinnerets normal. Ventral spines on anterior tibiae not unusually long. *Oligo.-Rec.*

Collacteus PETRUNKEVITCH, 1942 [**C. captivus*]. Legs in order 4231. First and 2nd tarsi and metatarsi scopulate. *Oligo.*, Balt.

Caduceator PETRUNKEVITCH, 1942 [**C. minutus*]. Legs in order 4213, without spines. Tarsi and metatarsi not scopulate (37). *Oligo.*, Eu.—FIG. 108,*1*. *C. quadrimaculatus* PETR., Balt.; *1a*, dorsal side showing coloration pattern, ×7.5; *1b*, eye group, ×45 (76).

Adulatrix PETRUNKEVITCH, 1942 [**A. fusca*]. Legs in order 4213, with spines. First and 2nd tarsi and metatarsi scopulate (37). *Oligo.*, Balt.—FIG. 31,*5*. **A. fusca;* photo of ventral side of female, ×8.5 (76).

Zachria L.KOCH, 1875 [**Z. flavicoma*]. Legs in order 2143. All tarsi scopulate. *Oligo.-Rec.* [*Z. peculiata* PETRUNKEVITCH, Oligo., Balt.].

Subfamily SPARIANTHIDINAE Simon, 1897

Spinnerets on pedestal. Ventral spines on anterior tibiae and metatarsi very long and slender. *Oligo.-Rec.*

Eostasina PETRUNKEVITCH, 1942 [**E. aculeata*]. Anterior metatarsi with 3 pairs of ventral spines (37). *Oligo.*, Balt.

Eostaianus PETRUNKEVITCH, 1942 [**E. succini*]. Anterior metatarsi with 2 pairs of ventral spines (37). *Oligo.*, Balt.

Family SELENOPIDAE F.P.Cambridge, 1900

Body very flat. Eyes in 2 rows, 1st row with 6 eyes, 2nd row with 2 eyes. Eyes heterogeneous, posterior medians alone nocturnal. *Rec.*—FIG. 35. *Selenops* sp., a typical laterigrade spider from tropical Africa, ×6 (75).

Family THOMISIDAE Sundevall, 1833

Tarsi without scopula. Margins of chelicerae smooth, or with 2 or 3 teeth. Eyes homogeneous. Maxillae converging. *Oligo.-Rec.*

Subfamily STEPHANOPSINAE Simon, 1892

Anterior legs considerably longer than posterior ones; 1st leg longest. Hair simple. Claw tufts well developed. *Oligo.-Rec.*

Syphax KOCH & BERENDT, 1854 [**S. megacephalus*]. Head rectangular. Eyes on common tubercle. Quadrangle wider behind than in front (37). *Oligo.*, Eu.—FIGS. 31,*1;* 107,*3*. *S. crassipes* PETRUNKEVITCH, Balt.; 31,*1*, photo of male, ×6.5; 107,*3a,* dorsal side, ×4; 107,*3b,* claws and claw tufts, enlarged (76).

Facundia PETRUNKEVITCH, 1942 [**F. clara*]. First leg longest. Anterior median eyes much smaller than lateral eyes. Sternum suborbicular (37). *Oligo.*, Balt.

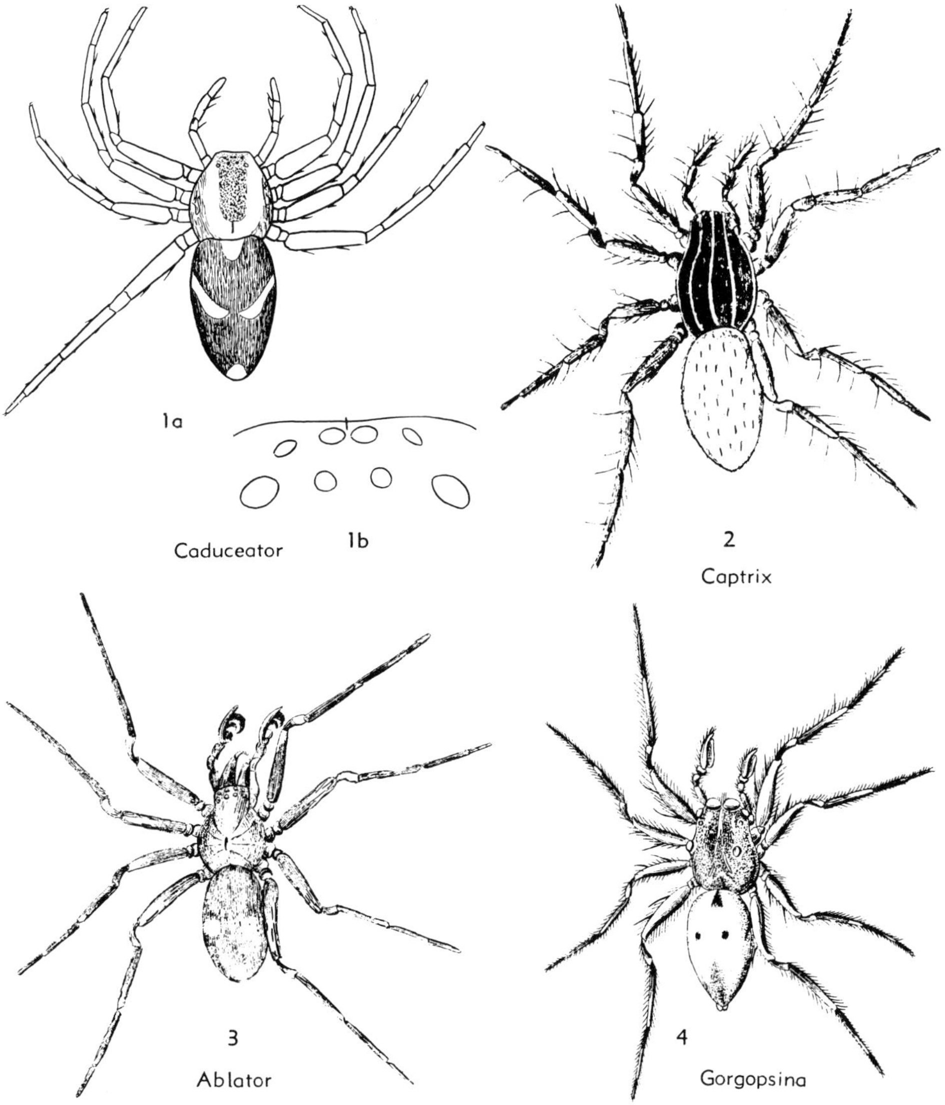

FIG. 108. Dipneumonina, Oligo., Balt. *1, Caduceator quadrimaculatus; 1a,* female, ×7.5; *1b,* its eye group, ×45 (76). *2, Captrix lineata,* ×7 (71). *3, Ablator triguttatus,* ×19 (71). *4, Gorgopsina melanocephala,* ×6 (71) (p. P146-P150).

Fiducia PETRUNKEVITCH, 1942 [*F. tenuipes*]. First leg longest. Eyes of 1st row equal. Sternum triangular (37). *Oligo.*, Balt.

Subfamily THOMISINAE Sundevall, 1833

[*nom. transl.* PETRUNKEVITCH, herein (*ex* Thomisidae SUNDEVALL, 1833)] [=Misumeninae SIMON *et auctt.*]

Second legs longest. *Oligo.-Rec.*

Thomisus WALCKENAER, 1805 [**Aranea alba* GMELIN, 1788]. Face angular. Clypeus high. *Oligo.-Rec.* [*T. defossus* SCUDDER, Oligo., N.Am.(Colo.); *T. oeningensis* HEER, Mio., Switz.].

Misumena LATREILLE, 1804 [**Araneus vatius* CLERCK, 1757]. Face obtusely truncate. Legs without spines (37). *Oligo.-Rec.* [*M. samlandica* PETRUNKEVITCH, Oligo., Balt.].

Subfamily PHILODROMINAE Simon, 1892

Fourth legs longest. Integument with plumose or squamose hair. *Oligo.-Rec.*

Eothanatus PETRUNKEVITCH, 1950 [**E. diritatis*]. Claw tufts present. Eyes sessile. *Oligo.*, Balt.

Medela PETRUNKEVITCH, 1942 [**M. baltica*]. Claw tufts lacking. Eyes of 1st row smallest. Patella strongly angular. Hair simple. *Oligo.*, Balt.

Filiolella PETRUNKEVITCH, *herein* [*nom. nov. pro Filiola* PETR., 1942 (ref. 37, p. 379) (*non* BARRANDE, 1881)] [**Filiola argentata* PETR., 1942]. Eyes subequal, on tubercle. Claw tufts lacking. Hair simple. *Oligo.*, Balt.

Superfamily PYRITARANEOIDEA Petrunkevitch, nov.

[*nom. transl.* PETRUNKEVITCH, herein (*ex* Pyritaraneidae PETRUNKEVITCH, 1953)]

Presumptive Dionychi with laterigrade legs and segmented abdomen. *Carb.*

Family PYRITARANEIDAE Petrunkevitch, 1953

Characters of superfamily. *Carb.*

Pyritaranea FRITSCH, 1899 [**P. tubifera*]. Carapace longer than wide. Legs slender, long, in order 2134 (45). *Carb.*, Eu.—FIG. 110,*3*. **P. tubifera*, Czech.; dorsal side, ×22.5 (76).

Dinopilio FRITSCH, 1904 [**D. gigas*]. Legs long, stout. Carapace not known (45). *Carb.*, Eu.—FIG. 110,*4*. **D. gigas;* holotype, ×1.5 (76).—FIG. 110, *5*. *D. parvus* PETRUNKEVITCH, Eng.; ×2.5 (76).

Superfamily CLUBIONOIDEA Simon, 1895

[*nom. correct.* PETRUNKEVITCH, herein (*pro* Clubioniformia CAPORIACCO, 1938, *nom. transl. ex* Clubionidae SIMON, 1895] [=Clubioniformia (*partim*)+Gnaphosaeformia (*partim*) CAPORIACCO, 1938]

Eyes in 2 rows. Legs prograde. Tarsi scopulate. Trichobothria numerous, on tibiae, metatarsi and tarsi. *Oligo.-Rec.*

Family DRASSODIDAE Petrunkevitch, 1942

[=Gnaphosidae *auctt.*] [Name of family derived from *Drassodes* WESTRING, 1851, *nom. correct. pro Drassus* WALCKENAER, 1805, of which it is a synonym; genotype of *Drassus* in doubt, changed twice, but of *Drassodes* undisputed. Family name Drassides SUNDEVALL, 1833, used as Drassidae until 1949, has priority over Gnaphosidae]

Anterior spinnerets wide apart. Maxillae with oblique depression. Eyes heterogeneous. *Oligo.-Rec.*

Captrix PETRUNKEVITCH, 1942 [**Textrix lineata* KOCH & BERENDT, 1854]. First row of eyes recurved, 2nd procurved. Quadrangle wider in front. *Oligo.*, Eu.—FIGS. 107,*9;* 108,*2*. **C. lineata* (KOCH & BERENDT), Balt.; 107,*9,* lip and maxillae, enlarged (76); 108,*2,* female, ×7 (71).

Palaeodrassus PETRUNKEVITCH, 1922 [**Titanoeca ingenua* SCUDDER, 1890]. Legs spinose, 4th leg longest. *Oligo.*, N.Am.(Colo.).

Family INCEPTORIDAE Petrunkevitch, 1942

Claw tufts wanting. Legs spinose, in order 4123. Cheliceral margins without teeth. *Oligo.*

Inceptor PETRUNKEVITCH, 1942 [**I. aculeatus*]. Spines erect. Posterior median eyes half as large as anterior ones (37). *Oligo.*, Eu.—FIG. 107,*2*. **I. aculeatus,* Balt.; *2a,* holotype, ×10; *2b,* carapace with eyes, ×30 (76).

Family CLUBIONIDAE Simon, 1895

Anterior spinnerets contiguous. Eyes homogeneous, in 2 rows. Maxillae without oblique impression. *Oligo.-Rec.*

Subfamily CLUBIONINAE Simon, 1895

[*nom. transl.* SIMON, 1903 (*ex* Clubionidae SIMON, 1895)]

Terminal segment of posterior spinnerets conical. Maxillae constricted in middle. Integument with plumose hair. *Oligo.-Rec.*

Clubiona LATREILLE, 1804 [**Aranea holosericea* LINNÉ, 1758 (=*Araneus pallidulus* CLERCK, 1757)]. Maxillae parallel. Legs spinose, 4th leg longest. *Oligo.-Rec.* [*C. arcana* SCUDDER, Oligo., Colo.].

Eoversatrix PETRUNKEVITCH, 1922 [**Clubiona eversa* SCUDDER, 1890]. Legs in order 1243. Hair simple (25). *Oligo.*, N.Am.(Colo.).

Eobumbratrix PETRUNKEVITCH, 1922 [**Clubiona latebrosa* SCUDDER, 1890]. Legs in order 2431 (35). *Oligo.*, N.Am.(Colo.).

Eostentatrix PETRUNKEVITCH, 1922 [**Clubiona ostentata* SCUDDER, 1890]. Legs in order 1423 (35). *Oligo.*, N.Am.(Colo.).

Subfamily MICARIINAE Simon, 1903

Terminal segment of posterior spinnerets

rounded. Maxillae truncate in front. *Oligo.-Rec.*

Massula PETRUNKEVITCH, 1942 [**M. klebsi*]. Legs without scopulae. Maxillae with parallel sides. Hair simple. Order of legs 4123 (37). *Oligo.*, Balt.

Ablator PETRUNKEVITCH, 1942 [**Ocypete triguttata* KOCH & BERENDT, 1854]. Legs with scopulae. Maxillae with ventrolateral protuberance. Legs in order 4123 (37). *Oligo.*, Eu.—FIGS. 107,*7;* 108,*3.* **A. triguttatus* (KOCH & BERENDT), Balt.; 107,*7a,* scales from abdomen, enlarged; *7b,* claws and claw tufts enlarged (76); 108,*3,* male, ×19 (71).

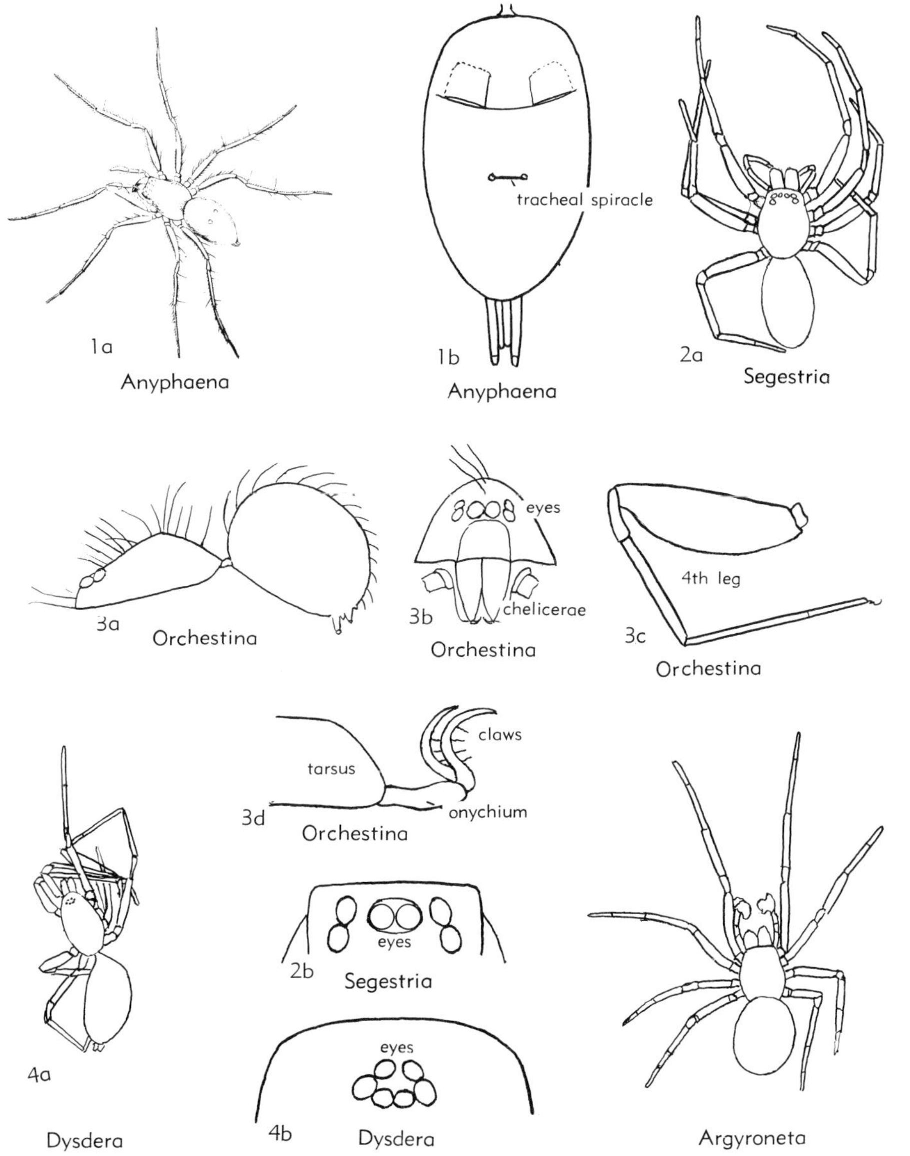

FIG. 109. Dipneumonina Quadrostiata. *1, Anyphaena fuscata,* Oligo., Balt.; *1a,* ×3 (71); *1b,* ventral side of abdomen showing position of spiracle, ×20 (76). *2, Segestris elongata,* Oligo., Balt.; *2a,* ×6; *2b,* eye group, enlarged. *3, Orchestina baltica,* Oligo., Balt.; *3a,* side view of body, ×20; *3b,* face, ×40; *3c,* 4th leg, ×90; *3d,* claws on onychium, ×3,000. *4, Dysdera scobiculata,* Oligo., Balt.; *4a,* ×4; *4b,* eye group, enlarged. *5, Argyroneta antiqua,* Mio. (Brown Coal), Ger.; ×2 (76) (p. P151-P152).

Desultor Petrunkevitch, 1942 [**D. depressus*]. Eye group much narrower than head. Chelicerae geniculated. Order of legs 4123 (37). *Oligo.*, Balt.

Family PARATTIDAE Petrunkevitch, 1922

Eyes of 2nd row much smaller than those of 1st row. Posterior row strongly recurved and much shorter than anterior row. *Oligo.*

Parattus Scudder, 1890 [**P. resurrectus*]. Anterior eyes equidistant, slightly procurved (35). *Oligo.*, N.Am.—Fig. 107,*4b*. *P. oculatus* Petrunkevitch, Colo.; eye group, ×29 (76).—Fig. 107,*4a*. *P. evocatus* Scudder, Colo.; female, ×3.3 (76).

Superfamily SALTICOIDEA F.P.Cambridge, 1900

[*nom. correct.* Petrunkevitch, herein (*pro* Salticiformia Caporiacco, 1938, *nom. transl. ex* Salticidae F.P. Cambridge, 1900] [=Attidae *auctt.*]

Eyes in 3 or 4 rows, homogeneous. Legs adapted for jumping. Claw tufts well developed. Integument with simple hair and scales. *Oligo.-Rec.*

Family SALTICIDAE F.P. Cambridge, 1900

Eight eyes in 3 rows, 1st row with 4 large eyes. Legs spinose. *Oligo.-Rec.*

Subfamily HELIOPHANINAE Simon, 1903

Retromargin of chelicerae with single, pointed tooth. Sternum broadly truncate in front. Eyes of 2nd row almost equidistant from anterior and posterior lateral eyes. Fourth leg longer than 3rd. *Oligo.-Rec.*

Parevophrys Petrunkevitch, 1942 [**P. succini*]. Eyes of 2nd row small, with tuft of 4 stout bristles lateral to them. Anterior tibiae with lateral spines (37). *Oligo.*, Balt.

Subfamily GORGOPSININAE Petrunkevitch, 1942

[*nom. correct.* Petrunkevitch, herein (*pro* Gorgopidinae Petrunkevitch, 1942)]

Both margins of chelicerae smooth. Carapace transversely depressed between 2nd and 3rd row of eyes. Claw tufts dense. *Oligo.*

Gorgopsina Petrunkevitch, *nom. nov.* [*pro Gorgopis* Menge, 1854 (ref. 22, p. 93) (*non* Huebner, 1820)] [**Phidippus frenatus* Koch & Berendt, 1854]. Eyes of 2nd row small, situated on swollen base of anterior lateral eyes. Hair simple (37). *Oligo.*, Eu.—Fig. 107,*5*. **G. frenata* (Koch & Berendt), Balt.; carapace of male, ×10 (76).—Fig. 108,*4*. *G. melanocephala* (Koch & Berendt), Balt.; male, ×6 (71).

Subfamily BOETHINAE Simon, 1903

Retromargin of chelicerae with several teeth. Eyes of 2nd row subequal in size to those of 3rd row. *Oligo.-Rec.*

Eolinus Petrunkevitch, 1942 [**E. succineus*]. Legs in order 4132. First tibia without dorsal spines (37). *Oligo.*, Eu.—Fig. 31,*2*; 107,*6*. *E. theryi* Petr., Balt.; 31,*2*, photo of immature female, ×11; 107,*6*, eye group, enlarged (76).

Paralinus Petrunkevitch, 1942 [**P. crosbyi*]. Legs in order 4312. Tibiae with at least 2 dorsal spines (37). *Oligo.*, Balt.

Cenattus Petrunkevitch, 1942 [**C. exophthalmicus*]. Legs in order 4312. All eyes on pedestals. Third row of eyes wider than head (37). *Oligo.*, Balt.

SALTICIDAE Incertae sedis

Eoattopsis Petrunkevitch, *nom. nov.* [*pro Attopsis* Gourret, 1886 (Rec. Zool. Suisse, v. 4, p. 441) (*non* Heer, 1849)] [**A. hirsutus* Gourret, 1886]. Thorax higher than head. Second leg longest, 3rd leg shortest. *Oligo.*, Fr.

Attoides Brongniart, 1901 [**A. eresiformis*]. Carapace low and long. Eye group narrower behind than in front. Legs in order 1243. Tibia of male palp with long and straight, prolateral apophysis. *Oligo.*, Fr.

Steneattus Bronn, 1856, *nom. subst.* [*pro Leda* Koch & Berendt, 1854 (*non* Schumacher, 1817)] [**Leda promissa* Koch & Berendt, 1854]. Head with parallel sides, almost as wide as thorax. Sternum heart-shaped. Legs thin. *Oligo.*, Balt.

Propetes Menge, 1854 [*P. felinus* (first species of 5 listed)]. Only character mentioned: eyes of 2nd row almost as large as those of 3rd row. *Oligo.*, Balt.

Division QUADROSTIATI Petrunkevitch, 1933

Dipneumone spiders with 2 pairs of cardiac ostia. *Oligo.-Rec.*

Superfamily DYSDEROIDEA C.L.Koch, 1837

[*nom. correct.* Petrunkevitch, herein (*pro* Dysderaeformia Caporiacco, 1938, *nom. transl. ex* Dysderidae C.L Koch, 1837)]

A pair of tracheal spiracles close behind openings of book lungs. *Oligo.-Rec.*

Family SEGESTRIIDAE Petrunkevitch, 1933

Third pair of legs directed forward. Three claws. Six eyes, homogeneous, in 2 rows, 1st row of 4 eyes, the 2 eyes of posterior row wide apart. *Oligo.-Rec.*

Segestria LATREILLE, 1804 [**Aranea florentina* ROSSI, 1790]. Retromargin of chelicerae with 2 teeth (37). *Oligo.-Rec.*—FIG. 109,*2*. *S. elongata* KOCH & BERENDT, Oligo., Balt.; *2a*, female, ×6; *2b*, eye group, enlarged (76).

Family DYSDERIDAE C.L.Koch, 1837

Third pair of legs directed backward; 2 or 3 claws. Six eyes, homogeneous, arranged in a more or less closed, transverse ellipse. *Oligo.-Rec.*

Dysdera LATREILLE, 1804 [**D. erythrina*]. Two claws. Terminal segment of posterior spinnerets with several spinning tubes (37). *Oligo.-Rec.*—FIG. 109,*4*. *D. scobiculata* MENGE, Oligo., Balt.; *4a*, female, ×4; *4b*, eye group, enlarged (76).

Harpactes TEMPLETON, 1834 [**Aranea hombergi* SCOPOLI, 1763]. Three claws. Terminal segment of posterior spinnerets with a single spinning tube. *Oligo.-Rec.* [*H. extinctus* PETRUNKEVITCH, Oligo., Balt.].

Thereola PETRUNKEVITCH, *nom. nov.* [*pro Thereo*

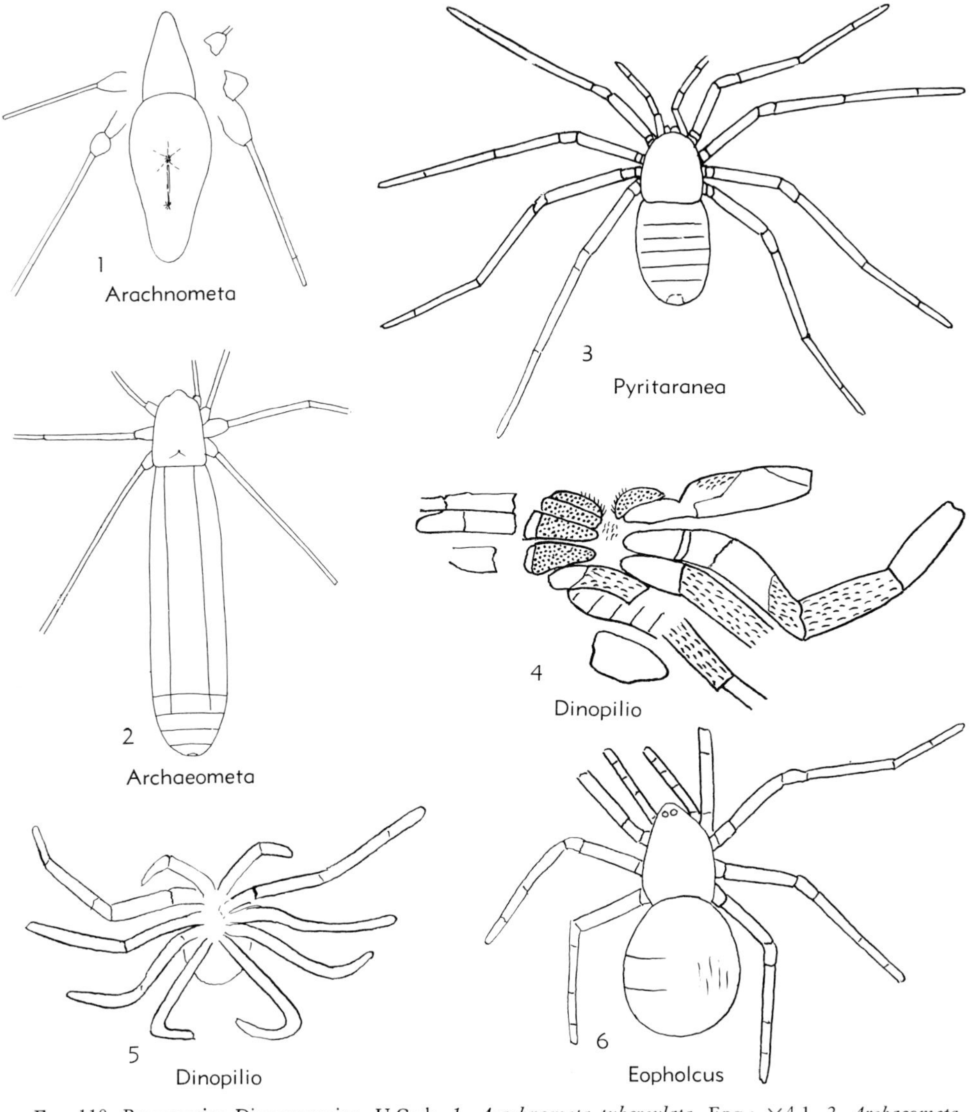

FIG. 110. Presumptive Dipneumonina, U.Carb. *1, Arachnometa tuberculata*, Eng.; ×4.1. *2, Archaeometa nephilina*, Eng.; ×4.1. *3, Pyritaranea tubifera*, Czech.; ×22.5. *4, Dinopilio gigas*, Czech.; ×1.5. *5, Dinopilio parvus*, Eng.; ×2.5. *6, Eopholcus pedatus*, Czech.; ×8.7 (76) (p. P146-P148).

KOCH & BERENDT, 1854 (ref. 22, p. 75) (*non* BILBERG, 1820)] [*T. *petiolata*]. Anterior pair of eyes on a transversely elliptic tubercle. Claws not known. *Oligo.*, Balt.

Family OONOPIDAE Simon, 1892

Six eyes, homogeneous, nocturnal. Two claws on onychium. *Oligo.-Rec.*

Orchestina SIMON, 1892 [**O. pavesii*]. Abdomen with soft integument. Fourth pair of legs longest, adapted for jumping, with distended femur (37). *Oligo.-Rec.*—FIG. 109,*3*. *O. baltica* PETRUNKEVITCH, Oligo., Balt.; *3a*, side view of body, ×20; *3b*, face, ×40; *3c*, 4th leg, ×90; *3d*, claws, ×3,000 (76).

Superfamily ANYPHAENOIDEA Bertkau, 1878

[*nom. transl.* PETRUNKEVITCH, herein (*ex* Anyphaenidae BERTKAU, 1878)]

Tracheal spiracle single. Two claws. *Oligo.-Rec.*

Family ANYPHAENIDAE Bertkau, 1878

Eight eyes, homogeneous, sessile, in 2 rows. Tarsi scopulate. *Oligo.-Rec.*

Anyphaena SUNDEVALL, 1833 [**Aranea accentuata* WALCKENAER, 1802]. Retromargin of chelicerae with several teeth. Eyes of 1st row subequal (37). *Oligo.-Rec.*—FIG. 109,*1*. *A. fuscata* KOCH & BERENDT, Oligo., Balt.; *1a*, female, ×3 (71); *1b*, ventral side of abdomen, showing position of tracheal spiracle, ×20 (76).

Superfamily ARGYRONETOIDEA Menge, 1869

[*nom. transl.* PETRUNKEVITCH, herein (*ex* Argyronetidae MENGE, 1869)]

Tracheal spiracle single. Three claws. *Oligo.-Rec.*

Family ARGYRONETIDAE Menge, 1869

Aquatic spiders building a bell-shaped web under water. Eight eyes, heterogeneous, in 2 rows. *Mio.-Rec.*

Argyroneta LATREILLE, 1804 [**Aranea aquatica* LINNÉ, 1758]. Legs in order 4123. Retromargin of chelicerae with 2 teeth (37). *Mio.-Rec.*—FIG. 109,*5*. *A. antiqua* VAN HEYDEN, Mio., Ger.; male, ×2 (76).

Suborder APNEUMONINA Dahl, 1913

[*nom. correct.* PETRUNKEVITCH, herein (*pro* Apneumones DAHL, 1913)] [Type: *Caponia* SIMON, 1887]

Araneida with diaxial chelicerae, 1 or 2 pairs of tracheal tubes in place of book lungs which are always lacking, and 2 pairs of cardiac ostia (36). No fossil spiders of this suborder have ever been found. *Rec.*

ARANEIDA INCERTAE SEDIS

Although all genera listed in this section have been placed by their respective authors in one or another family, none of them can now be placed in any known family, because their figures and descriptions are quite inadequate for correct placement. Even C. L. KOCH and MENGE, both very distinguished arachnologists of their time, are not exempt from this failure, because our knowledge has greatly progressed since their time. To list these genera in the families to which they were assigned by their authors, would be only misleading and contrary to scientific principles. For the same reason I have abstained from giving new names to such generic names as are preoccupied but cannot be placed. In absence of their type specimens which are reported to be lost or destroyed, a new name would be futile and contrary to all sense.

Amphiclotho GOURRET, 1886 [**A. breviuscula*]. *Oligo.*

Amphithomisus GOURRET, 1886 [**A. barbatus*]. *Oligo.*

Anandrus MENGE, 1856 (*nom. nud.*). *Oligo.*

Anatone MENGE, 1854 [**A. spinipes*]. *Oligo.*

Androgeus KOCH & BERENDT, 1854 [**A. militaris*]. *Oligo.*

Antopia MENGE, 1854 [**A. punctulata*]. *Oligo.*

Athera MENGE, 1856 [**A. exilis*]. *Oligo.*

Cercidiella GOURRET, 1886 [**C. aequisextana*]. *Oligo.*

Clubionella GOURRET, 1886 [**C. antiqua*]. *Oligo.*

Clya KOCH & BERENDT, 1854 [**C. lugubris*]. *Oligo.*

Clythia KOCH & BERENDT, 1854 [**C. alma*]. *Oligo.*

Corynitis MENGE, 1854 [**C. spinosa*]. *Oligo.*

Dielacata MENGE, 1856 (*nom. nud.*). *Oligo.*

Entomocephalus HOLL, 1829 [**E. formicoides*]. *Oligo.*

Eresioides GOURRET, 1886 [**E. orbicularis*]. *Oligo.*

Erithus MENGE, 1854 [**E. applanatus*]. *Oligo.*

Hersilioides GOURRET, 1886 [*H. thanatiformis*]. *Oligo.*

Heteromma MENGE, 1854 [**H. intersecta*]. *Oligo.*

Idmonia MENGE, 1854 [**I. virginea*]. *Oligo.*

Linoptes MENGE, 1854 [**L. oculeus*]. *Oligo.*

Lycosoides GOURRET, 1886 [**L. hersiliiformis*]. See *Trimeropus.*

Mastigusa MENGE, 1854 [**M. acuminata*]. *Oligo.*

Mizalia KOCH & BERENDT, 1854 [**M. punctulata*]. *Oligo.*

Ocia MENGE, 1854 (*nom. nud.*). *Oligo.*

Onca MENGE, 1854 [**O. pumila*]. *Oligo.*

Opisthophylax MENGE, 1856 [**O. exarata*]. *Oligo.*

Phalangopus MENGE, 1854 [*P. *subtilis*]. *Oligo.*

Prodysdera GOURRET, 1886 [*P. *intermedia*]. *Oligo.*

Protochersis GOURRET, 1886 [*P. *spinosus*]. *Oligo.*

Protolachesis GOURRET, 1886 [*P. *annulata*]. *Oligo.*

Protolycosa GOURRET, 1886 [*non.* RÖMER, 1866] [*P. *attiformis*]. *Oligo.*

Pseudothomisus GOURRET, 1886 [*P. *articulatus*]. *Oligo.*

Pytonyssa C.L. KOCH, 1837 [=*Gnaphosa* LATREILLE, 1804]. *Rec.*

Schellenbergia HEER, 1865 [*S. *rotundata*]. *Mio.*

Sphaeconia MENGE, 1854 [*S. *brevipes*]. *Oligo.*

Sosybius KOCH & BERENDT, 1854 [*S. *minor*]. *Oligo.*

Thyelia KOCH & BERENDT, 1854 [*T. *tristis*]. *Oligo.*

Trimeropus THORELL, 1891 [*nom. subst. pro Lycosoides* GOURRET, 1886 [*non* LUCAS, 1846)] [*L. *hersiliiformis*]. *Oligo.*

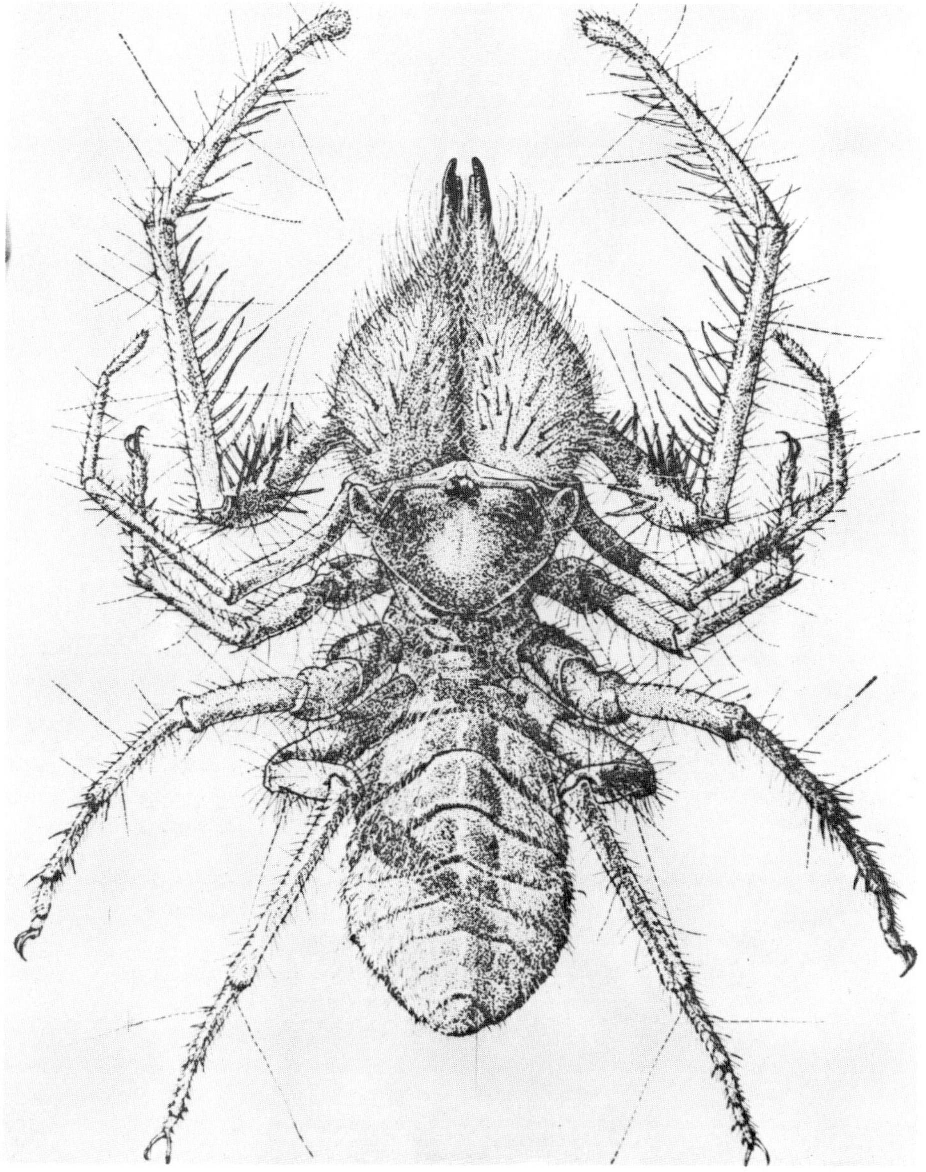

FIG. 111. *1, Galeodes arabs,* a common solpugid from Arabia, ×1.5 (by permission from *Traité de Zoologie,* tome 6, sous la direction du PR. GRASSÉ, Masson & Cie., Editeurs, Paris).

Superorder ROSTRATA Petrunkevitch, 1949

[Type: *Galeodes* OLIVIER, 1791] [=Mycetophorae LANKESTER, 1910]

Caulogastra with secondary mouth opening at end of beak formed by fusion of upper lip with deuterosternite. *Carb.-Rec.*

Order SOLPUGIDA Leach, 1815

[*nom. correct.*, PEARSE, 1936 (*pro* Solpugidea CAMBRIDGE, 1872, *pro* Solpugae C.L. KOCH, 1842, *pro* Fam. I. Solpugides LEACH, 1815)] [=Solifugae SUNDEVALL, 1833]

Carapace segmented, composed of propeltidium and 3 thoracic tergites, with a pair of median eyes and 1 or 2 pairs of rudimentary lateral eyes on propeltidium. Abdomen composed of 11 segments (Fig. 112,*2a*). Chelicerae 2-jointed, powerful, chelate. Pedipalpi pediform, 6-jointed. First pair of legs 7-jointed, tactile; the other legs 8-jointed, with a pair of long claws on an onychium. Ventral surface of 4th legs always with racket organs (malleoli) (Fig. 112,*2b*, showing anterior sternites of abdomen and 3 proximal joints of 4th legs with malleoli; Fig. 112,*2d*, malleolus enlarged). Anus on 11th (last) segment, slit-like, without operculum, terminal in some, ventral in others.

Respiration by means of tracheae opening to the outside by 7 spiracles, as follows: 1st pair between 2nd and 3rd coxae; 2nd pair on 3rd abdominal sternite; 3rd pair on 4th sternite; 7th, single opening on 5th sternite. No book lungs. Fore-gut with pumping pharynx, esophagus and rudimentary gizzard. Anterior end of rostrum (beak) with a brush of bristles serving as filtering apparatus. Cavity of rostrum serves as secondary mouth cavity with primary mouth opening into it at its base. Mid-gut with thoracenteron and chylenteron. Hind-gut short. Pair of coxal glands in cephalothorax, with openings on dorsal surface of pedipalpal coxae. Heart with 8 pairs of ostia, 2 pairs of which are in cephalothorax, 6 in abdomen. Nervous system with 5 abdominal neuromeres incorporated in thoracic ganglionic mass, remaining 5 pairs forming abdominal ganglionic mass in 6th abdominal segment. Sexes separate. In both sexes a single median genital opening on 2nd abdominal sternite. Oviparous. *Carb.-Rec.*

DISCUSSION

Solpugida are tropical and subtropical arachnids with a few species in moderate zones (Fig. 111, *Galeodes arabs*, ×1.4, from N.Africa and Arabia). They are swift runners and on account of their chelicerae were always feared and supposed to be poisonous. But none of them have poison glands. Like the majority of arachnids, they are carnivorous, feeding on insects. In mating the male uses the fingers of his chelicerae for the introduction of the spermatophores into the female genital opening, after having picked them up from the ground on which they were ejected by him.

The most characteristic external features are: (1) the enormous chelicerae, always directed forward, with movable finger ventral in position, the structure of the chelicerae furnishing secondary sexual characters by which males can be distinguished from females; (2) the shape and size of the propeltidium which, in conjunction with the chelicerae, gives the solpugids the appearance of an arthropod with a distinct head as in insects; (3) the shape of the 3 thoracic tergites which are independent of each other; (4) the stout pedipalpi with a blunt end bearing an adhesive organ; (5) the slender 1st legs, tactile in function and held upward; (6) the configuration of the coxosternal region (Fig. 112,*1*) with complete absence of externally visible thoracic sternites, 2nd and 3rd coxae meeting in median line; (7) the construction of the trochanters of 2nd, 3rd and 4th legs of 2 joints; (8) the clear segmentation of the abdomen; and (9) the possession of racket organs, not found in any other order of arachnids. In size Solpugida vary from 9 to 70 mm. Recent Solpugida are divided into 10 families, 134 genera and about 600 species. The only known fossil species is placed here provisionally in a Recent family. It is probable that it belongs to an extinct family, but its preservation is not good enough to permit the erection of a new family. Its placement in the family Galeodidae is entirely arbitrary and was motivated by the fact that this family has a wide Old-World distribution.

Family GALEODIDAE Pocock, 1897

Anus terminal. A pair of small claws at end of 1st tarsus. *Carb.-Rec.*

Protosolpuga PETRUNKEVITCH, 1913 [**P. carbonaria*]. Eyes lacking. Pedipalpi stout and long. *Penn.*, N.Am.—FIG. 112,*3*. **P. carbonaria,* Ill.; ×2 (drawing corrected from original after re-examination of holotype) (76).

Superorder PODOGONA Cook, 1899

[Type: *Ricinoides* EWING, 1929 (=*Cryptostemma* GUÉRIN, 1838)] [=Cucullifera PETRUNKEVITCH, 1945]

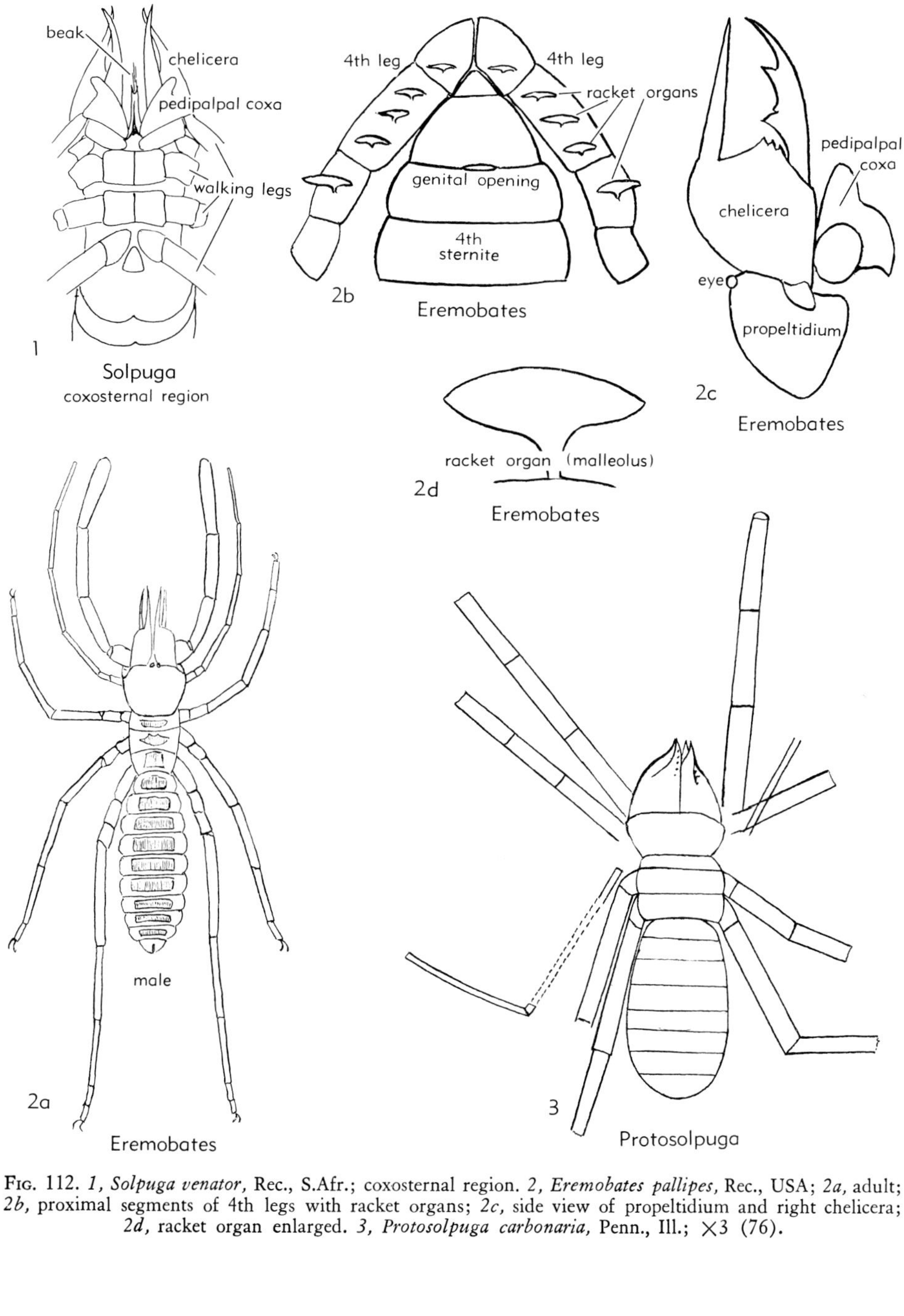

FIG. 112. *1, Solpuga venator,* Rec., S.Afr.; coxosternal region. *2, Eremobates pallipes,* Rec., USA; *2a,* adult; *2b,* proximal segments of 4th legs with racket organs; *2c,* side view of propeltidium and right chelicera; *2d,* racket organ enlarged. *3, Protosolpuga carbonaria,* Penn., Ill.; ×3 (76).

Caulogastra with movable plate (cucullus) in front of the carapace. *Carb.-Rec.*

Order RICINULEIDA Thorell, 1892

[*nom. correct.* PETRUNKEVITCH, herein (*pro* Ricinulei THORELL, 1892)] [Type: *Ricinoides* EWING, 1929] [=Rhinogastra COOK, 1899)]

Carapace entire, with a cucullus articulated to its anterior edge and capable of covering chelicerae from below (Fig. 113, *Cryptocellus foedus* female, dorsal side, ×10). Eyes wanting. First abdominal embryonic segment completely lost; abdomen of adult composed of 9 segments, the 1st of which (embryonic 2nd) forms petiolus and is concealed under dorsal anterior edge of following segment. Last 3 segments reduced to a pygidium. Chelicerae 2-jointed, chelate (Fig. 114,*1,* coxosternal region of *Cryprocellus dorotheae* GERTSCH, enlarged), their movable finger lateral and longer than immobile one. Pedipalpi 6-jointed, chelate, their coxae fused in median line, forming with the upper lip a camarostome similar to that in Thelyphonida and Schizomida. Pedipalpal tibia longest of the 6 joints, chela small, with movable finger longer than immobile one. First pair of legs 7-jointed, with single-jointed trochanter and tarsus; 2nd pair also with single-jointed trochanter, but with 5-jointed tarsus; 3rd pair with 2-jointed trochanter and 4-jointed tarsus; 4th pair also with 2-jointed trochanter, but 5-jointed tarsus. Only 4th coxae movable. Metatarsus and tarsus of 3rd pair of legs in males modified as organs of copulation and have the shape of a scoop, presumably to hold sperm. At end of all tarsi a pair of curved, smooth claws concealed in a depression of the tip. Body and legs heavily sclerotized. Sexes separate. Oviparous. Tropical and subtropical arachnids from 4 to 10 mm. in size. *Carb.-Rec.*

DISCUSSION

The juncture between the cephalothorax and the abdomen is accomplished by the 1st (embryonic 2nd) abdominal segment, and the abdomen is held in position by a

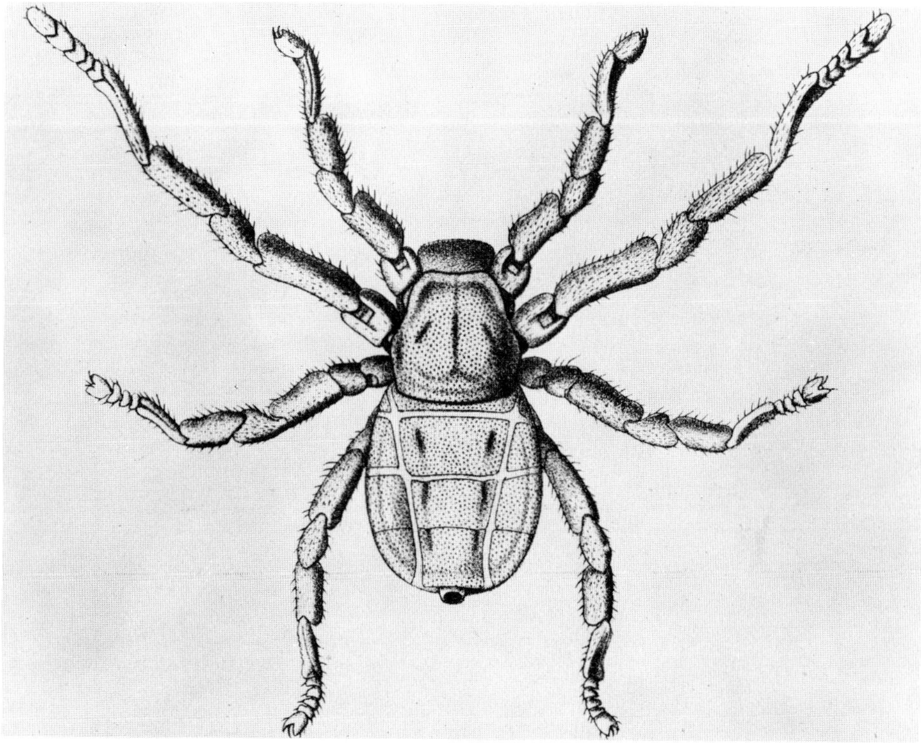

FIG. 113. Ricinuleida. *Cryptocellus foedus,* Rec., Brazil; female, ×10 (68).

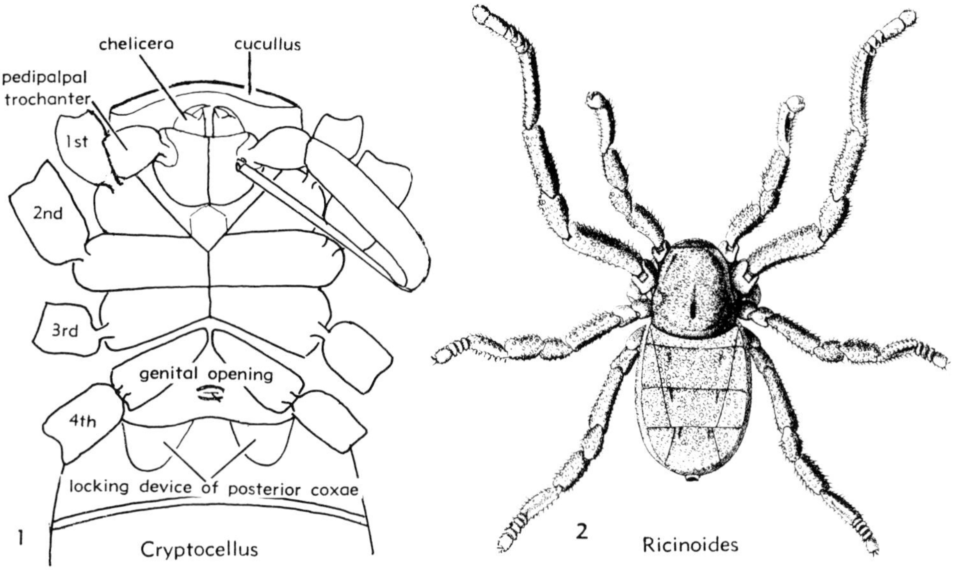

FIG. 114. Ricinuleida. *1, Cryptocellus dorotheae,* Rec., Tex.; coxosternal region, enlarged (76). *2, Ricinoides karschii,* Rec., Afr.; female, ×5 (68).

ridge of the posterior edge of the carapace fitting into a transverse groove on the 2nd (1st visible) abdominal tergite, which is in reality the tergite of the 3rd embryonic segment. The coupling is aided by the 4th coxae fitting into a pair of pouches of the postgenital sternite (Fig. 114,*1*). In consequence of such coupling, the animal appears to have a broad juncture between the carapace and the abdomen, and the 1st visible tergite looks like the true 1st tergite, whereas it is the tergite of the 2nd (embryonic 3rd) segment. Thus in all Recent Ricinuleida, and in the Carboniferous ones of the family Poliocheridae, only 4 tergites are visible, longitudinally divided by a pair of grooves into a median and a pair of lateral plates. The sternites are entire in Recent Ricinuleida and fossil Poliocheridae. On the other hand, in fossil Curculioididae the back of the abdomen, with exception of the pygidium, is covered by a single shield which is longitudinally divided by a median line, thus giving the arachnid the appearance of an insect with a pair of elytra (Fig. 115,*2b*). At the same time their 3rd, 4th and 5th abdominal sternites are subdivided by a pair of longitudinal lines into a median and a pair of lateral plates (Fig. 115,*2a*). The pygidium is always composed of 3 cylindrical segments which are telescoped in Recent species, but fully exposed in fossil ones.

As in Camarostomata, the mouth opens into the camarostome which serves as filter. The pharynx is quite rudimentary, the esophagus short. It opens into a true, pumping gizzard (according to my own, not yet published observations). The diverticles of the mid-gut are almost entirely restricted to the abdomen where they represent a true chylenteron, although arranged more or less longitudinally. Respiration is accomplished by tracheae which open by a pair of spiracles above 3rd coxae. Nervous system entirely concentrated in the cephalothorax. Nothing is known of the life and development of Ricinuleida, except that they pass through a 6-legged larval stage.

Ricinuleida are a small order represented by a single Recent family with 2 genera and 16 species and 2 fossil families with 2 genera and 9 species. Both Carboniferous families are extinct. The differences between the Carboniferous family Curculioididae and the Recent family Ricinoididae are great and it seems certain that the former represent the end of a special evolutionary line. The Carboniferous Poliocheridae are in many respects closely related to the Recent Ricinoididae. Their abdomen has the

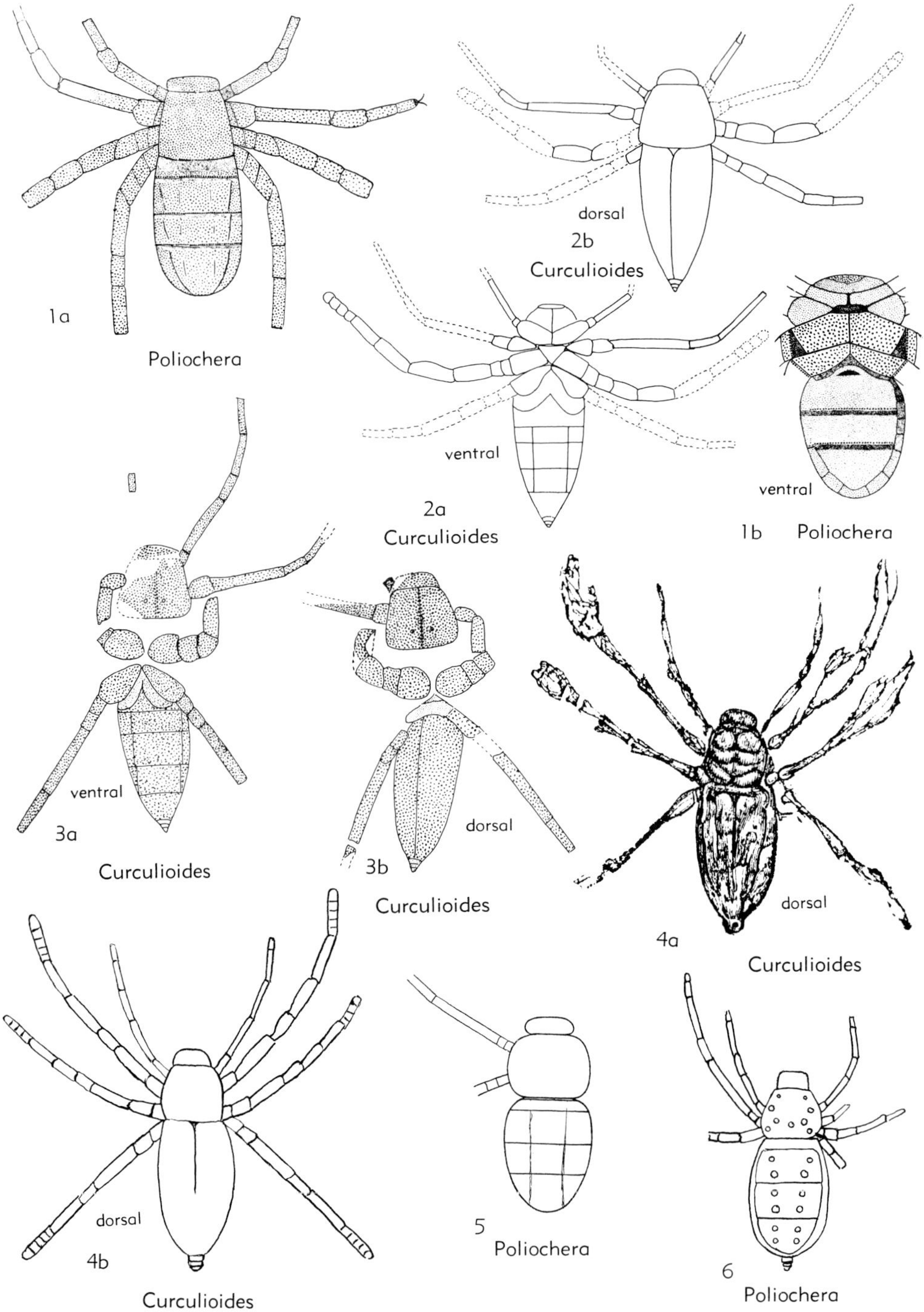

FIG. 115. Ricinuleida. *1, Poliochera punctulata,* Penn., Ill.; *1a,* dorsal, ×2; *1b,* ventral, ×2 (76). *2, Curculioides granulatus,* Penn., Ill.; *2a,b,* ventral and dorsal, ×1.5 (76). *3, Curculioides eltringhami,* Carb., Eng.; *3a,b,* ventral and dorsal, ×1.5 (76). *4, Curculioides ansticii,* Carb., Eng.; *4a,* BUCKLAND's original figure, ×1.9; *4b,* dorsal surface, restored, ×1.9 (76). *5, Poliochera glabra,* Penn., Ill.; ×2. *6, Poliochera alticeps,* Carb., Eng.; ×2.8 (76) (p. P159).

same type of segmentation with similar dorsal grooves dividing the tergites and similarly undivided sternites. The coupling between the cephalothorax and the abdomen is the same in all Ricinuleida. The chief visible character separating the Recent from the fossil Ricinuleida is furnished by the configuration of their coxosternal region. In Ricinoididae the 1st pair of coxae is clearly separated (Fig. 114,*1*), the 2nd and 3rd coxae are in contact with each other and all 3 pairs immobile. The 4th coxae are free and mobile. In both fossil families the 2nd coxae are clearly separated (Fig. 115,*1b*,*2a*), while the 1st coxae are usually in contact with each other in the median line. The 2 fossil families are easily distinguished by the structure of their abdomen.

Family POLIOCHERIDAE Scudder, 1884

Abdominal tergites subdivided by a pair of longitudinal grooves or lines into a median and a pair of lateral fields. Sternites entire. *Carb.*

Poliochera SCUDDER, 1884 [**P. punctulata*]. *Carb.*, Eu.-N.Am.—FIG. 115,*1*. **P. punctulata*, Ill.; *1a*, dorsal side, ×2; *1b*, ventral side, ×2 (76).—FIG. 115,*5*. *P. glabra* PETRUNKEVITCH, Ill.; dorsal side, ×2.—FIG. 115,*6*. *P. alticeps* POCOCK, Eng.; dorsal side, ×2.8 (76).

Family CURCULIOIDIDAE Cockerell, 1916

[=Holotergidae PETRUNKEVITCH, 1913]

Abdomen with a single dorsal shield longitudinally divided by a median line. Sternites divided by a pair of lines into a median and a pair of lateral fields. *Carb.*

Curculioides BUCKLAND, 1837 [**C. ansticii*] (45). *Carb.*, Eu.-N.Am.—FIG. 115,*4*. **C. ansticii*, Eng.; *4a*, dorsal side, ×1.9 (BUCKLAND's original figure) (65); *4b*, same (reconstr.), ×1.9 (76).—FIG. 115,*3*. *C. eltringhami* PETRUNKEVITCH, Eng.;

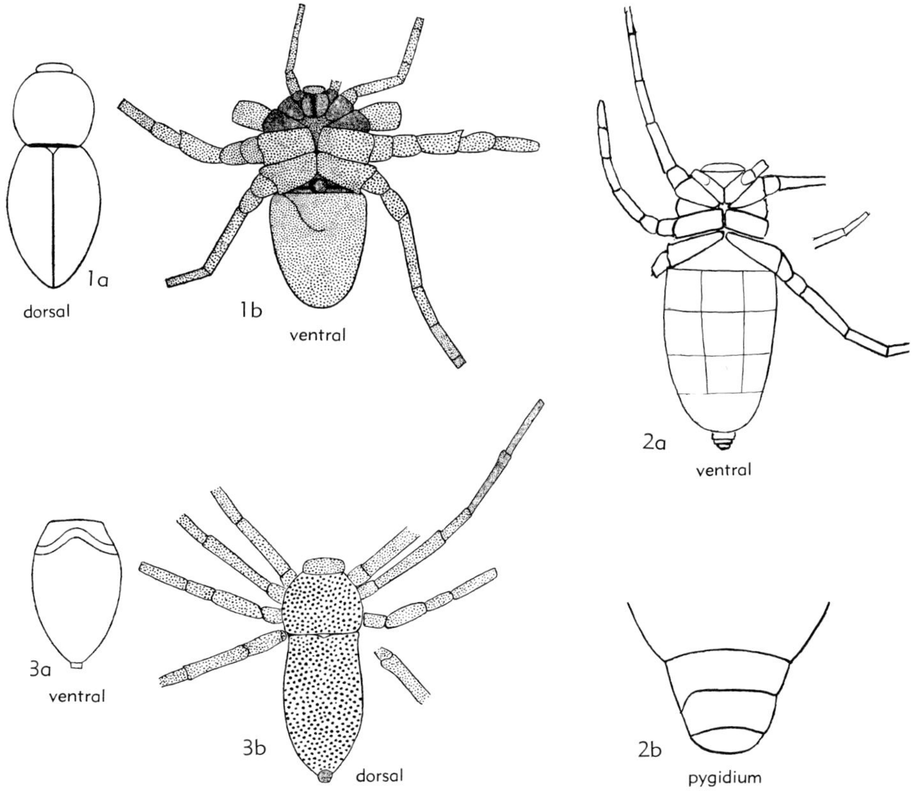

FIG. 116. Ricinuleida. *1*, *Curculioides scaber*, Penn., Ill.; *1a,b*, dorsal and ventral, ×2. *2*, *Curculioides gracilis*, Penn., Ill.; *2a*, ventral, ×2.2; *2b*, pygidium, ×14. *3*, *Curculioides sulcatus*, Penn., Ill.; *3a,b*, ventral and dorsal, ×1.7 (76) (p. P160).

3a, ventral side, ×1.5; *3b,* dorsal side, ×1.5 (76). —FIG. 115,*2. C. granulatus* PETR., Eng.; *2a,* ventral side, ×1.5; *2b,* dorsal side, ×1.5 (76).—FIG. 116,*1. C. scaber* (SCUDDER), Ill.; *1a,* dorsal side of body, ×2; *1b,* ventral side with legs, ×2 (76). —FIG. 116,*2. C. gracilis* PETR., Ill.; *2a,* ventral side, ×2.2; *2b,* pygidium, ×14 (76).—FIG. 116, *3. C. sulcatus* (MELANDER), Ill.; *3a,* ventral side of abdomen, ×1.7; *3b,* dorsal side of holotype, ×1.7 (76).

Family RICINOIDIDAE Ewing, 1929

[Type: *Ricinoides* EWING, 1929]
[=Cryptostemmidae WESTWOOD, 1874]

First coxae separated, 2nd and 3rd coxae contiguous. Abdomen as in Poliocheridae. *Rec.*

Ricinoides EWING, 1929 [**Cryptostemma westermannii* GUÉRIN, 1838]. *Rec.*—FIG. 114,*2. R. karschii* (HANSEN & SÖRENSEN), Afr.; female, ×5 (68).

GENERIC NAMES OF ARACHNIDA PLACED IN WRONG ORDERS AND INADEQUATELY DESCRIBED

Amphitrogulus GOURRET, 1886 [**A. sternalis*], placed in Phalangiida, is an inadequately described spider. *Oligo.*, Eu.

Phalangillum GOURRET, 1886 [**P. hirsutum*], placed in Phalangiida, is an inadequately described spider. *Oligo.*, Eu.

GENERIC NAMES WRONGLY PLACED IN ARACHNIDA

Hasseltides WEYENBERGH, 1869 [*pro Hasseltia* WEYENBERGH, 1869] [**Hasseltia primigenia* WEYENBERGH, 1869], placed in Phalangiida, is a crinoid (*Saccocoma*). *Jur.*, Eu.

Phalangites MÜNSTER, 1839 [**P. priscus*] [=*Palpipes* ROTH, 1851 (obj.)], placed in Phalangiida, is a crustacean larva.

Pleurolycosa FRITSCH, 1904 [**Arthrolycosa prolifera* FRITSCH, 1899], placed in Araneida, is quite unidentifiable.

REFERENCES

The following list of publications relating to Arachnida includes only references that are judged to be most helpful in furnishing additional information concerning this group and in offering a guide to other literature, which is not cited.

The index numbers enclosed in parentheses in the column at left are employed in the text for identification of the publications.

André, Marc

(1) 1949, *Ordre des Acariens:* in GRASSÉ, P.-P., Traité de Zoologie, Masson (Paris), v. 6, p. 794-893, 2 pl., fig. 794-893.

Baker, E. W., & Wharton, G. W.

(2) 1952, *An introduction to acarology:* Macmillan (New York), xiii+465 p., 377 fig.

Beier, Max

(3) 1937, *Pseudoscorpione aus dem baltischen Bernstein:* Gestschr. Embrik Strand, v. 2, p. 302-316, fig. 1-19.

Berland, Lucien

(4) 1949, *Ordre des Opilions:* in GRASSÉ, P.-P., Traité de Zoologie, Masson (Paris), v. 6, p. 761-793, fig. 553-597.

Bertkau, P.

(5) 1878, *Einige Spinnen und eine Myriapode aus der Braunkohle von Rott:* Verh. naturwiss. Ver. preuss. Rheinl. und Westf., ser. 4, v. 35, p. 346-360.

Brongniard, Ch.

(6) 1877, *Note sur une Aranéide fossile des terrains tertiaires d'Aix (Provence):* Ann. Soc. entomol. Fr., ser. 5, tome 7, p. 221-224.

Chamberlin, J. C.

(7) 1931, *The arachnid order Chelonethida:* Stanford Univ. Press (Calif.), 284 p., 71 fig.

Dawydoff, Constantin

(8) 1949, *Developpement embryonaire des Arachnides:* in GRASSÉ, P.-P., Traité de Zoologie, Masson (Paris), v. 6, p. 320-385, fig. 97-157.

Fritsch, Anton

(9) 1904, *Paläozoische Arachniden:* (Praha), 55 p., 15 pl.

Gerhardt, U., & Kästner, Alfred

(10) 1938, *Araneae, echte Spinnen:* in KÜKENTHAL, W., Handbuch der Zoologie, de Gruyter (Berlin, Leipzig), v. 3, p. 394-656, fig. 484-854.

Gourett, Paul

(11) 1886, *Recherches sur les Arachnides tertiaires d'Aix en Provence:* Rec. zool. suisse, tome 4, p. 431-496.

Guthörl, P.

(12) 1938, *Eophrynus waecherti n. sp. (Arachn. Anthracom.) aus der Tiefbohrung Stangenmühle,* Saar-Karbon: Senckenbergiana

(Stuttgart), v. 20, p. 462-470, fig. 1-3.
(13) 1940, *Zur Arthropoden-Fauna des Karbons und Perms, 10. Trigonomartus pustulatus (Scudder) (Arachn. Anthracom.) aus der Grube Nordstern Aachener Karbon:* Paläont. Zeitschr. (Stuttgart), v. 22, p. 63-74.

Haase, E.
(14) 1890, *Beiträge zur Kenntnis der fossilen Arachniden:* Zeitschr. deutsch. geol. Gesell., v. 42, p. 629-657, pl. 30-31.

Hansen, J. J., & Sørensen, W.
(15) 1904, *On two orders of Arachnida. II. The order Ricinulei:* Cambridge Univ. Press (Eng.), p. 114-157, pl. 7-9.

Heer, O.
(16) 1876, *The primeval world of Switzerland:* Longmans, Green (London), ed. J. Heywood, v. 1, 393 p., 10 pl., 196 fig., v. 2, 324 p., pl. 11, 368 fig.

Hirst, S.
(17) 1923, *On some arachnid remains from the Old Red Sandstone (Rhynie Chert bed, Aberdeenshire):* Annals and Mag. Nat. Hist., ser. 9, v. 12, p. 455-474, fig. 1-13.

Karsch, F.
(18) 1884, *Neue Milben in Bernstein:* Entomol. Zeitschr. (Berlin), v. 28, p. 175-176.

Kästner, Alfred
(19) 1932, *Pedipalpi Latreille—Geissel-Scorpione; Palpigradi Thorell; Ricinulei Thorell:* in Kükenthal, W., Handbuch der Zoologie, de Gruyter (Berlin, Leipzig), v. 3, p. 1-116, fig. 1-156.
(20) 1933, *Solifugae Sundevall = Walzenspinnen:* Same, v. 3, p. 193-300, fig. 268-362.
(21) 1935, *Opiliones:* Same, v. 3, p. 300-393, fig. 363-483.

Koch, C. L., & Berendt, G. C.
(22) 1854, *Die im Bernstein befindlichen Crustaceen, Myriapoden, Arachniden und Apteren:* (Berlin), 124 p., 17 pl.

Lehmann, W. M.
(23) 1944, *Palaeoscorpius devonicus, n.g., n.sp., ein Skorpion aus dem rheinischen Unterdevon:* Neues Jahrbuch für Mineralogie, etc., p. 177-185, fig. 1-4.

Menge, A.
(24) 1854, *Lebenszeichen vorweltlicher im Bernstein eingeschlossener Thiere:* Programm der Petrischule (Danzig), 32 p.

Millot, Jacques
(25) 1949, *Morphologie générale et anatomie interne [Arachnida]:* in Grassé, P.-P., Traité de Zoologie, v. 6, Masson (Paris), p. 263-319, fig. 47-96.
(26) 1949a, *Ordre des Palpigrades:* Same, p. 520-532, fig. 274-286.
(27) 1949b, *Ordre des Uropyges:* Same, p. 533-562, fig. 287-317.
(28) 1949c, *Ordre des Amblypyges:* Same, p. 562-588, fig. 318-348.
(29) 1949d, *Ordre des Aranéides:* Same, p. 589-743, fig. 349-528.
(30) 1949e, *Ordre des Ricinuléides:* Same, p. 744-760, fig. 529-551.

——— & Vachon, Max
(31) 1949, *Ordre des Scorpions:* Same, p. 386-436, fig. 158-195.
(32) 1949a, *Ordre des Solifuges:* Same, p. 482-519, fig. 242-273.

Pampaloni, L.
(33) 1902, *Microflora e microfauna nel disodile di Melilli in Sicilia:* Rend. Accad. Lincei, v. 11, p. 252.

Petrunkevitch, Alexander
(34) 1913, *A monograph of the terrestrial Palaeozoic Arachnida of North America:* Trans. Conn. Acad. Arts Sci. (New Haven), v. 18, p. 1-137, pl. 1-13, fig. 1-88.
(35) 1922, *Tertiary spiders and opilionids of North America:* Same, v. 25, p. 211-279, fig. 1-39.
(36) 1933, *An inquiry into the natural classification of spiders, based on a study of their internal anatomy:* Same, v. 31, p. 299-389, pl. 1-13.
(37) 1942, *A study of amber spiders:* Same, v. 34, p. 119-464, pl. 1-69.
(38) 1945, *Palaeozoic Arachnida of Illinois:* Ill. Sci. Papers (Springfield), v. 3, no. 2, p. 1-72, pl. 1-4, fig. 1-34.
(39) 1945a, *Calcitro fisheri, a new fossil arachnid:* Am. Jour. Sci., v. 243, p. 320-329, pl. 1, fig. 1-8.
(40) 1946, *Fossil spiders in the collection of the American Museum of Natural History:* Am. Mus. Novitates, no. 1328, p. 1-36, fig. 1-80.
(41) 1948, *The case of Phalangiotarbus subovalis (Woodward):* Am. Jour. Sci., v. 246, p. 353-362, pl. 1, fig. 1-4.
(42) 1948a, *Arachnida:* Encycl. Britannica, 10 p., 17 fig.
(43) 1949, *A study of Palaeozoic Arachnida:* Trans. Conn. Acad. Arts. Sci. (New Haven), v. 37, p. 60-315, pl. 1-83.
(44) 1950, *Baltic amber spiders in the Museum of Comparative Zoology:* Bull. Mus. Comp. Zool., Harvard, v. 103, p. 259-337, fig. 1-215.
(45) 1953, *Paleozoic and Mesozoic Arachnida of Europe:* Geol. Soc. America, Mem. 53, 128 p., 58 pl.

Pierce, W. D.

(46) 1950, *Fossil arthropods from onyx marble:* Bull. Southern Calif. Acad. Sci., v. 49, p. 101-104, fig. 1-1a.

(47) 1951, *Fossil arthropods from onyx marble:* Same, v. 50, p. 34-49, fig. 1-7.

Protescu, O.

(48) 1937, *Étude géologique et paléobiologique de l'ambre Roumain:* Bull. Soc. Roumaine Géol., v. 3, p. 65-110.

Röwer, C. F.

(49) 1934, *Solifugae:* in BRONN, H. G., Klassen und Ordnungen des Tierreichs, v. 5, p. 1-637, fig. 1-362.

(50) 1934a, *Palpigradi:* Same, p. 640-723.

(51) 1939, *Opilioniden im Bernstein:* Palaeobiologica, v. 7, p. 1-5.

(52) 1940, *Chelonethi oder Pseudoskorpione:* in BRONN, H. G., Klassen und Ordnungen des Tierreichs, v. 6, p. 1-354, fig. 1-258.

Scudder, S. H.

(53) 1890, *The Tertiary insects of North America (Arachnides Latreille):* U. S. Geol. Survey Terr. (Hayden), p. 45-90, pl. 11.

Sellnick, M.

(54) 1913, *Milben im Bernstein:* Berstein-Forschungen, p. 148-180.

(55) 1919, *Die Oribatidae der Bernsteinsammlung der Universität Königsberg:* Schr. phys.-ökonom. Gesell. Königsberg, v. 59, p. 21-42.

Vachon, Max

(56) 1949, *Ordre des Pseudoscorpions:* in GRASSÉ, P.-P., Traité de Zoologie, Masson (Paris), v. 6, p. 437-481, fig. 196-241.

Vitzthum, H.

(57) 1931, *Acari:* in KÜKENTHAL, W., Handbuch der Zoologie, de Gruyter (Berlin, Leipzig), v. 3, p. 1-160.

(58) 1940-43, *Acarina:* in BRONN, H. G., Klassen und Ordnungen des Tierreichs, v. 5, p. 1-1011, fig. 1-522.

Werner, Franz

(59) 1935, *Scorpiones:* Same, v. 5, p. 1-316, fig. 1-290.

(60) 1935a, *Pedipalpi:* Same, v. 5, p. 317-490.

Wills, L. J.

(61) 1947, *A monograph of British Triassic scorpions:* Palaeontogr. Soc. (London), 137 p., 12 pl., 53 fig.

SOURCES OF ILLUSTRATIONS

(62) **André, Marc**
(63) **Berland, Lucien**
(64) **Börner, Carl**
(65) **Buckland, W.**
(66) **Chamberlin, J. C.**
(66a) **Fritsch, Anton**
(67) **Haase, E.**
(68) **Hansen, H. J., & Sørensen, W.**
(69) **Heyden, C. van**
(70) **Hirst, S.**
(71) **Koch, C. L., & Berendt, G. C.**
(72) **McCook, H. C.**
(73) **Meek, F. B., & Worthen, A. H.**
(74) **Menge, A.**
(75) **Millot, Jacques**
(75a) **Vachon, Max**
(76) **Petrunkevitch, Alexander**
(77) **Pierce, W. D.**
(78) **Pocock, R. I.**
(79) **Pruvost, Pierre**
(80) **Scudder, S. H.**
(81) **Vitzthum, H.**
(82) **Wills, L. J.**

PYCNOGONIDA

By Joel W. Hedgpeth

CONTENTS

MORPHOLOGICAL FEATURES

The pycnogonids are exclusively marine arthropods, superficially resembling the Chelicerata in having the 1st pair of appendages chelate, but otherwise distinct from that subphylum, since they lack the well-developed abdomen of chelicerates, which houses the gonads and bears the **gonopores** on its 2nd segment. The 2nd pair of appendages is palpiform, but the 3rd is a highly modified pair of legs, the **ovigers** (Figs. 117; 120,*3*), adapted in the male to carrying the eggs during incubation. Generally there are 4 pairs of **walking legs** and 4 corresponding **trunk somites;** a few species have an extra trunk somite and pair of legs, and one 12-legged form is known (Fig. 120, *1b*). All appendages are uniramous. No well-formed body or prosoma occurs, the so-called trunk being merely a series of cylindrical somites bearing lateral outgrowths (**lateral processes**) from which the legs originate. In many species the trunk has been coalesced into a compact disc. The 1st segment of the trunk which bears the 3 pairs of anterior appendages is termed cephalic; a dorsal tubercle on this segment carries the eyes, although eyes are lacking in many deep-sea forms. The **abdomen** is a small papilliform or elongate tubercle (Figs. 117, 119-122). The most conspicuous external feature of a pycnogonid is the so-called **proboscis,** a large structure composed of a dorsal and 2 lateroventral antimeres, housing the specialized stomodeum. The integument is thin and fragile, unfavorable for fossilization.

The nervous system is of the basic arthropod type, consisting of paired ventral ganglia, a circumesophageal ring and a dorsal "brain." There is a simple dorsal heart with 2 or 3 pairs of lateral ostia, but no respiratory system. The digestive system is a simple tube without specialized glands (digestion being intracellular), with branches into the legs and 1st pair of appendages (**chelifores**). The reproductive system also has lateral diverticula extending into the legs, and eggs ripen in the swollen femurs of the female. The gonopores are typically on the ventral surface of the 2nd coxae of the last 2 pairs of legs in the males and they occur on all pairs in the females in some genera, thus placing them from the 4th to 7th pair of appendages. The males of many species have femoral glands which secrete a substance that aids in binding eggs together in a compact ball. The sexes are separate (although one hermaphroditic form is known); the male carries the eggs until they hatch.

The larval stage, termed **protonymphon**

(Fig. 118), has a proboscis, 2 simple eyes and 3 pairs of appendages, the first chelate. While this superficially resembles the nauplius, it does not have biramous appendages and resemblance is more apparent than real. It is not free-living.

MODE OF LIFE

Pycnogonida are stenohaline, benthic succivorous predators, although a few ascend to the surface to spawn and one is bathypelagic. They occur from the intertidal zone to a depth of nearly 7,000 m. and are especially abundant in arctic and antarctic waters, although a large variety of species occurs in tropical seas. They range in size from a few mm. in some intertidal species to a span of more than 40 cm. in the deep-sea *Colossendeis* (Fig. 120,*1*). The young are characteristically parasitic; many species encyst in hydroids soon after hatching, some form galls on hydroids or octocorals, others live in or on mollusks and some have been found in hydromedusae. In some species, the young stay with their fathers until able to shift for themselves. The adults feed

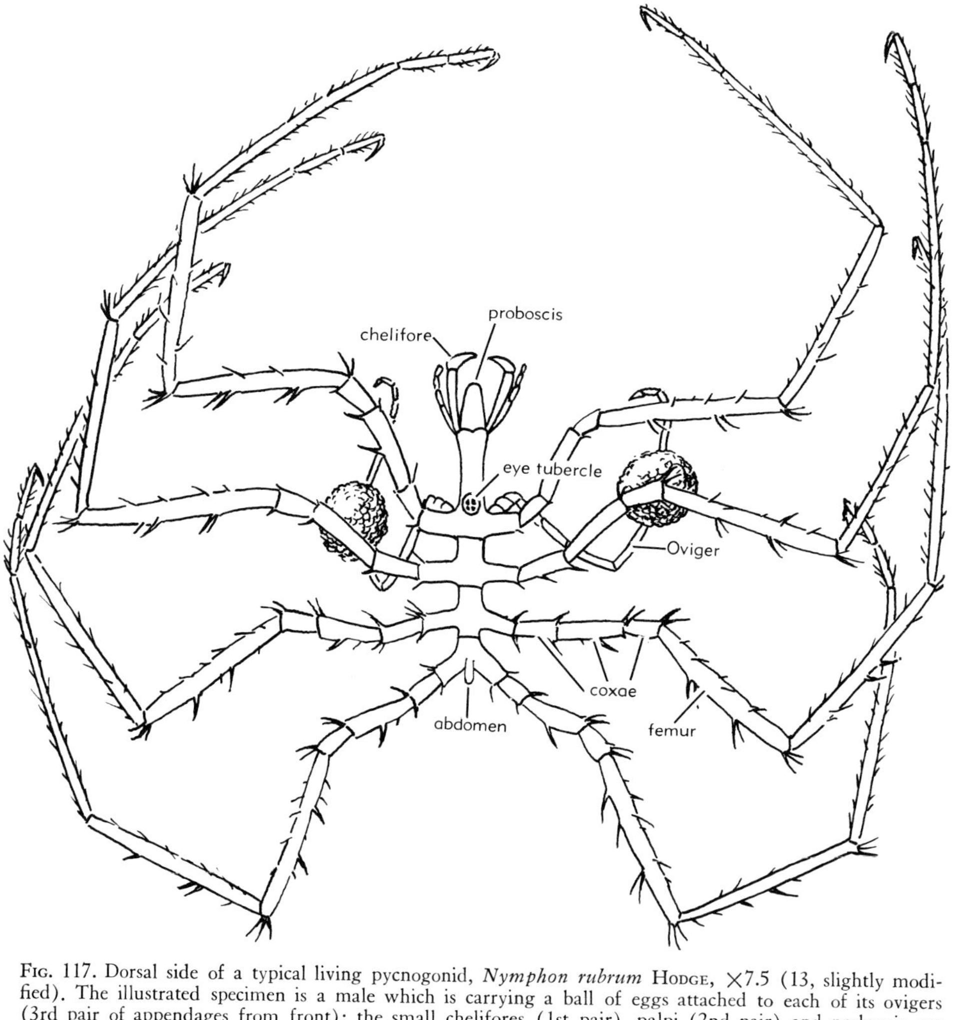

FIG. 117. Dorsal side of a typical living pycnogonid, *Nymphon rubrum* HODGE, ×7.5 (13, slightly modified). The illustrated specimen is a male which is carrying a ball of eggs attached to each of its ovigers (3rd pair of appendages from front); the small chelifores (1st pair), palpi (2nd pair) and proboscis are borne by the anterior prolongation of the 1st body somite (cephalic segment). This genus has 4 pairs of walking legs joined to lateral processes of the body somites. At the rear end is the diminutive tail-like abdomen.

usually on coelenterates, nudibranchs, and similar soft-bodied organisms.

CLASSIFICATION

About 500 species of pycnogonids are known, living forms being divided into 8 families which contain approximately 70 genera; more than 100 of the known species belong to the genus *Nymphon*. The wealth of transitional forms makes it impossible to recognize different suprafamilial groups even though some authors (Bouvier, 1913; Fage, 1949, ref. 3) have distinguished orders named Colossendeomorpha Cole (1903), Nymphonomorpha Pocock (1904), Pycnogonomorpha Pocock (1904) and Ascorhynchomorpha Pocock (1909). All living species are assigned here to a single order, Pantopoda. The Palaeopantopoda of Broili (1929) may be recognized provisionally as another order which includes the fossil *Palaeopantopus,* of Devonian age (Fig. 122). Only the lack of a proboscis and presence of a 2-segmented abdomen, instead of a nonsegmented abdominal tubercle at the rear end of the body, distinguish *Palaeopantopus* from Recent forms; the extra leg segments seen in *Palaeopantopus* may be a secondary development. All other anatomical characters, insofar as discerned, are not of a nature to prevent inclusion of this Devonian form with Recent species; hence, the classification adopted here is based more on absence of definite differing characters than on presence of clear-cut features indicating relationship with living forms.

In the past, the Pyconogonida have been classified most commonly with the Arachnida, or they have been treated somewhat vaguely as an appendix to this class. Some zoologists refer them to a meaningless assemblage called Arachnoidea. In recent years, the pycnogonids have been considered to be of chelicerate stock, constituting an aberrant group assignable to a separate class having rank equal to that of the Merostomata and Arachnida. Actually, the Pycnogonida differ from chelicerates in having the 3rd pair of appendages modified as ovigers, the anterior part of the body being produced as a proboscis and the abdomen reduced to almost vestigial proportions; in addition, the pycnogonids possess multiple genital openings (gonopores) located on preabdominal (body) somites and the legs have 3 coxal segments (Fig. 119). The development of intratarsal muscles in terminal segments of the ovigers is a feature found in no other arthropods. Effort to fit the Pycnogonida in a classificatory scheme that associates them closely with arachnids and other chelicerates should be abandoned. The group is recognized here as an independent subphylum.

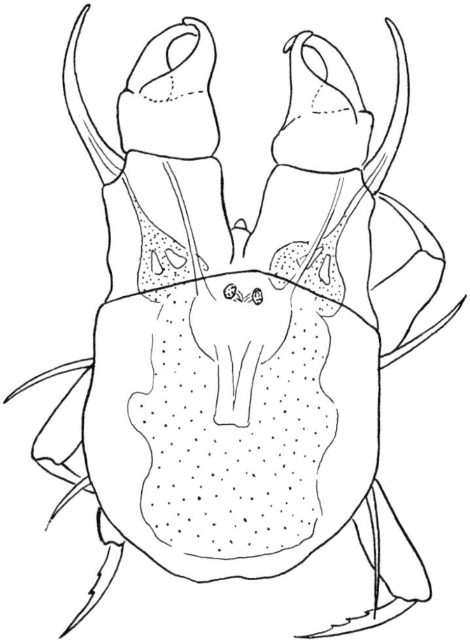

Fig. 118. Protonymphon larva of a pycnogonid, enlarged (12).

SYSTEMATIC DESCRIPTIONS

Subphylum PYCNOGONIDA Latreille, 1810

[*nom. transl.* Hedgpeth, herein (*ex* Pycnogonida Latreille, 1810, distinguished as an assemblage without explicitly designated taxonomic rank)]

Arthropoda superficially resembling Chelicerata in having 1st pair of appendages chelate, but differing from them in presence of gonopores on preabdominal or body somites and in lacking well-developed abdomen; 3rd pair of appendages modified as ovigers, which occur commonly in both

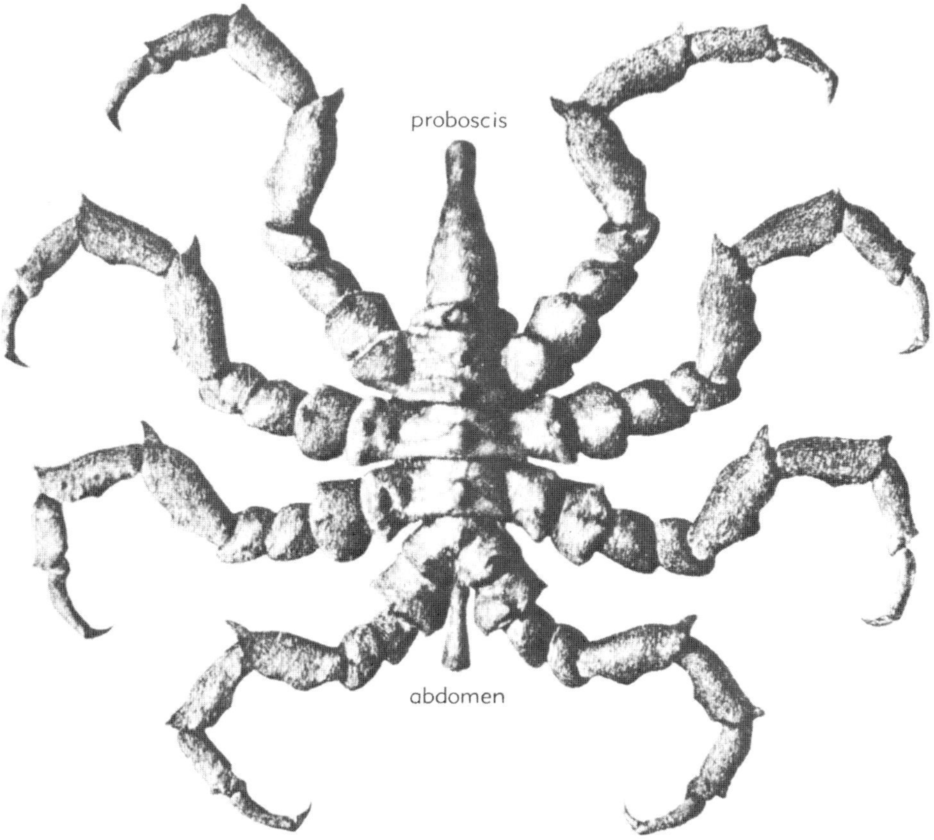

FIG. 119. *Pycnogonum littorale* (STRÖM), a modern pycnogonid, of the North Atlantic littoral zone, ×3 (14).

sexes but serve functionally only in males for carrying eggs; anterior part of body produced as a prominent proboscis. Trunk with 4 to 6 somites and legs with 3 coxal segments. Digestive system consisting of a simple tube with diverticles extending into legs but lacking specialized glands; respiratory system absent. Sexes typically separate but hermaphrodite forms may occur. Exclusively marine. *L.Dev.-Rec.*

Order PANTOPODA Gerstaecker, 1863

Proboscis well developed; chelifores, palpi, and ovigers present; postcoxal part of legs composed of 6 segments (from body outward, femur, 1st tibia, 2nd tibia, basitarsus, tarsus, dactyl); abdomen reduced to a tubercle. *Rec.*

Families of the Pantopoda are defined primarily by the presence of 3 pairs of appendages (chelifores, palpi, and ovigers) in front of the walking legs and the nature of these appendages, or alternatively, by absence of these appendages. Under each briefly diagnosed family, a few representative genera are cited but records of type species are omitted.

Family NYMPHONIDAE Wilson, 1878

Ovigers 10-segmented, in both sexes; chelifores 2-segmented, chelate; palpi 5-segmented. Includes a single decapodous genus. *Rec.*

Nymphon FABRICIUS, 1794.—FIG. 117. *N. rubrum* HODGE, ×7.5 (8).—FIG. 120,*2,3*. **N. grossipes* (FABRICIUS), ×6 (7). **Pentanymphon** HODGSON, 1904. **Heteronymphon** GORDON, 1932.

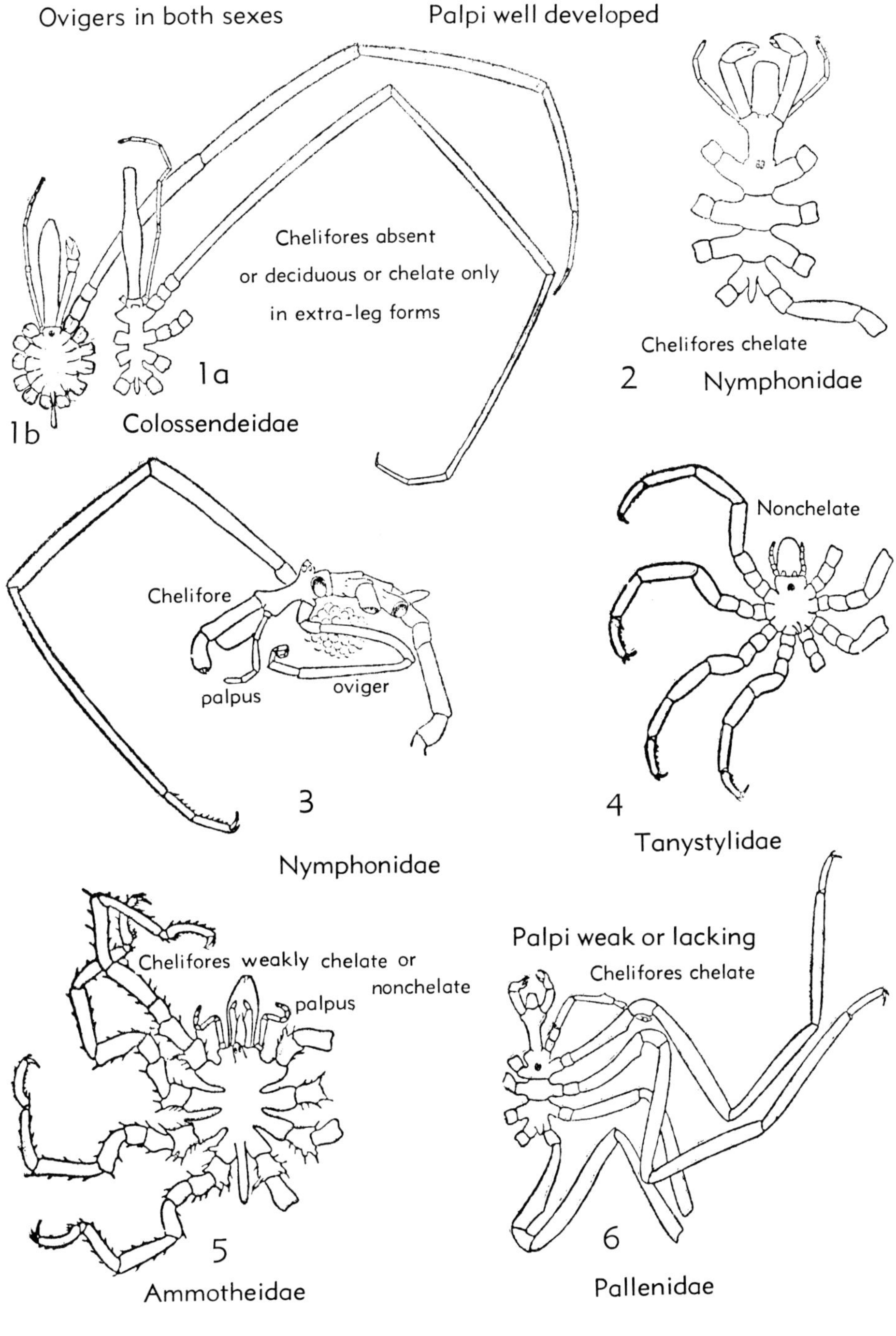

FIG. 120. Representative Pantopoda having ovigers in both sexes (p. P166-P168).

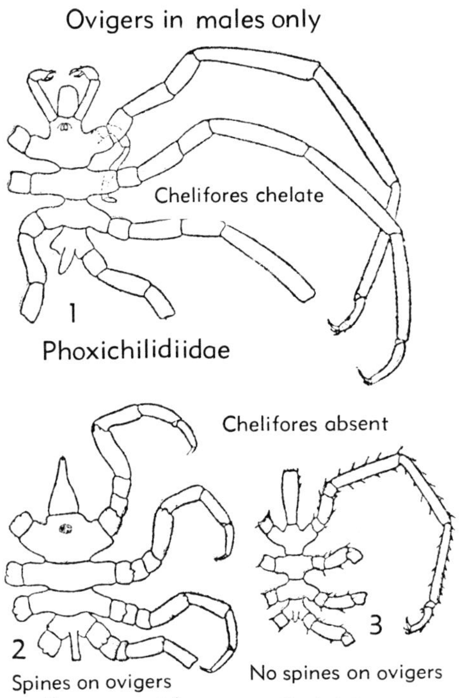

FIG. 121. Representative Pantopoda in which ovigers are carried by males only (p. P168).

Family PALLENIDAE Wilson, 1878

Ovigers 10-segmented, with or without terminal claw, in both sexes; chelifores usually chelate, proximal part (scape) with 1 or 2 segments; palpi represented by rudimentary knobs or lacking. *Rec.*

Callipallene FLYNN, 1929.—FIG. 120,6. **C. brevirostris* JOHNSTON, ×12 (7). **Pseudopallene** WILSON, 1878. **Pallenopsis** WILSON, 1881. **Hannonia** HOEK, 1881. **Propallene** SHIMKEVICH, 1909. **Austropallene** HODGSON, 1914. **Pycnothea** LOMAN, 1920. **Pigrogromitus** CALMAN, 1927. **Oropallene** SHIMKEVICH, 1930.

Family COLOSSENDEIDAE Hoek, 1881

Ovigers 10-segmented, with terminal claw, present in both sexes; chelifores generally lacking in adults but chelate when present; palpi 9- or 10-segmented. Trunk with 4 to 6 leg-bearing somites. *Rec.*

Colossendeis JARZYNSKY, 1870.—FIG. 120,1a, *C. colossea* WILSON, ×0.5 (6). **Dodecolopoda** CALMAN & GORDON, 1933.—FIG. 120,1b. **D. mawsoni* CALMAN & GORDON, ×0.5 (6). **Decolopoda** EIGHTS, 1837. **Rhopalorhynchus** WOOD-MASON, 1873. **Pentacolossendeis** HEDGPETH, 1943.

Family AMMOTHEIDAE Dohrn, 1881

Ovigers 9- or 10-segmented, present in both sexes; chelifores 2-segmented, usually subchelate; palpi 6- to 10-segmented but generally 8- or 9-segmented. *Rec.*

Achelia HODGE, 1864.—FIG. 120,5. *A. spinosa* (STIMPSON), ×12 (6). **Ammothea** LEACH, 1814. **Eurycyde** SCHIÖDTE, 1857. **Ascorhynchus** SARS, 1877. **Lecythorhynchus** BÖHM, 1879. **Boehmia** HOEK, 1881. **Nymphopsis** HASWELL, 1884. **Paranymphon** CAULLERY, 1896. **Ammothella** VERRILL, 1900. **Austroraptus** HODGSON, 1907. **Cilunculus** LOMAN, 1908. **Nymphonella** OHSHIMA, 1927. **Ephyrogymna** HEDGPETH, 1943. **Pycnofragilia** HEDGPETH, 1943. **Heterofragilia** HEDGPETH, 1943. **Calypsopycnon** HEDGPETH, 1948.

Family TANYSTYLIDAE Shimkevich, 1913

Ovigers 10-segmented, present in both sexes; chelifores 1- or 2-segmented, very small; palpi 4- to 6-segmented. *Rec.*

Tanystylum MIERS, 1879.—FIG. 120,4. *T. orbiculare* WILSON, ×12 (6). **Rhynchothorax** COSTA, 1861. **Discoarachne** HOEK, 1881. **Oorhynchus** HOEK, 1881. **Trygaeus** DOHRN, 1881. **Austrodecus** HODGSON, 1907. **Scipiolus** LOMAN, 1908.

Family PHOXICHILIDIIDAE Sars, 1891

Ovigers 5- or 6-segmented, present in males only; chelifores 2-segmented, chelate; palpi lacking. *Rec.*

Phoxichilidium EDW., 1840.—FIG. 121,1. **P. femoratum* (RATHKE), ×12 (6). **Anoplodactylus** WILSON, 1878. **Halosoma** COLE, 1904. **Hodgsonia** SHIMKEVICH, 1929.

Family ENDEIDAE Norman, 1908

Ovigers 7-segmented, present in males only; chelifores and palpi absent. *Rec.*

Endeis PHILIPPI, 1843.—FIG. 121,3. **E. spinosa* (MONTAGU), ×5 (6).

Family PYCNOGONIDAE Wilson, 1878

Ovigers 6- to 9-segmented, present in males only; chelifores and palpi absent. *Rec.*

Pycnogonum BRÜNNICH, 1764.—FIGS. 121,2; 119. **P. littorale* (STRÖM) ×9, ×25 (6). **Pentapycnon** BOUVIER, 1910. ?**Queubus** BERNARD, 1946 (females unknown).

Order PALAEOPANTOPODA Broili, 1930

Trunk with 4 somites, proboscis not evident, possibly absent; chelifores lacking; palpi (?ovigers) present; abdomen with 2 or 3 segments, with lateral processes that bear annular markings or swellings; legs with 3 coxal segments, ?5 longer segments, and several (?5) short terminal segments; gonopores doubtful. *L.Dev.*

This order is represented by a single known species described from 2 specimens discovered in the Hunsrück shale (L. Dev.) of western Germany (1, 2). The fossils do not resemble very closely any other arthropods, living or extinct, although they are enough like the Pantopoda to suggest the name chosen by BROILI (*Palaeopantopus*) who classed them without question as members of the Pycnogonida (Fig. 122). Such classification has been accepted with reservation by some students of modern pycnogonids (CALMAN & GORDON, 1933; HEDGPETH, 1947, ref. 4; FAGE, 1949, ref. 3), somewhat less skeptically by others (MARCUS, 1940, ref. 6), and unreservedly by still others (STØRMER, 1944; PETRUNKEVITCH, 1952, ref. 8).

Valid evidence for excluding *Palaeopantopus* from the arthropod assemblage which contains living pycnogonids cannot be cited, even though certain characteristic features of Pycnogonida are very obscure, to say the least, or are lacking. Neither of the described specimens shows any sign of a proboscis, but since only the dorsal surface is known, the possible existence of some sort of recurved, ventrally borne proboscideal structure must be allowed. The absence of eyes

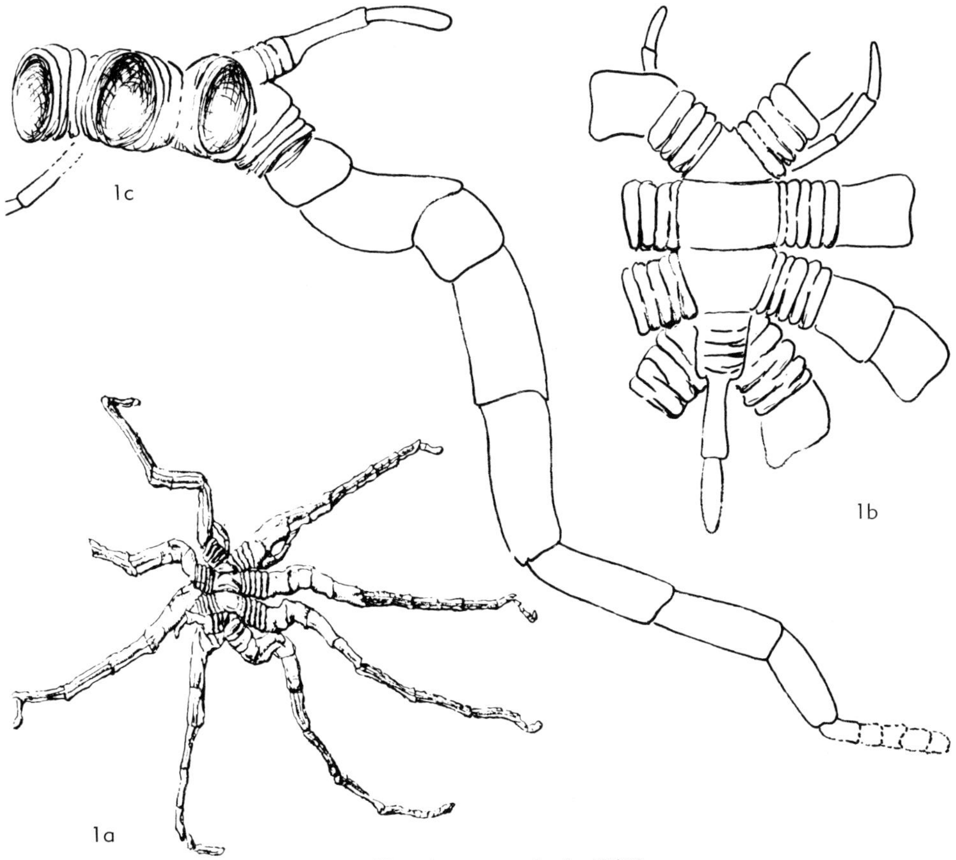

FIG. 122. Palaeopantopoda (p. P170).

and of various anterior appendages lacks significance because many living pycnogonids also are deficient similarly, and the numerous small distal leg joints suggest the condition in *Nymphonella.* The occurrence of 2 or 3 segments in the abdomen of *Palaeopantopus* is compatible with the range of variation observed in existing pycnogonids, for some of these possess 2 pairs of abdominal ganglia. The segmented abdomen might be cited as evidence of migration of the gonads into the prosoma during early history of the group along with gradual reduction of the abdomen from the most posterior somites, narrowed to little more than a jointed tube. PETRUNKEVITCH (1949, ref. 7) has demonstrated the evolutionary tendency of arachnids to lose segmentation of the abdomen and has presented evidence that the major chelicerate groups completed their differentiation by Devonian time. Accordingly, differentiation of a pycnogonid-like animal early in the Devonian, already well separated from any chelicerate stock, may be accepted with equanimity.

A logical disposition of *Palaeopantopus* is to recognize it as representative of an order (Palaeopantopoda) distinct from that (Pantopoda) containing the modern pycnogonids and to assign both to the Pycnogonida, designated as a subphylum. Definition of a class having scope identical to that of the subphylum seems to be unnecessary. By accepting *Palaeopantopus* as a pycnogonid, we are forced to conclude that within this peculiar group the trend toward reduction of the arthropod abdomen had been completed largely before mid-Devonian time. This signifies that instead of being a relatively new group of arthropods, as was my view at earlier date (4), the living Pycnogonida may be survivors of one of the oldest arthropod groups.

Family PALAEOPANTOPODIDAE Hedgpeth, nov.

Characters of the order. *L.Dev.*

Palaeopantopus BROILI, 1929 [**P. maucheri*]. *L. Dev.(Hunsrück.),* Ger.—FIG. 122,*1*. **P. maucheri; 1a,* dorsal side of nearly complete specimen, ×0.7 (6); *1b,* dorsal side of body and proximal part of appendages (reconstr.), enlarged (7); *1c,* side view of body showing walking leg attached (reconstr.), enlarged (7).

REFERENCES

Broili, F.

(1) 1929, *Beobachtungen an neuen Arthropodenfunden aus den Hunsrückschiefern; ein Pantopode aus dem rheinischen Unterdevon:* Sitzungsber. Bayer. Akad. Wiss., math.-naturhist. Abt., 1929, p. 272-280, pl. 5, fig. 1-6.

(2) 1930, *Über ein neues Exemplar von Palaeopantopus:* Same, 1930, p. 209-214, fig. 1.

Fage, Louis

(3) 1949, *Classe des Pycnogonides:* in GRASSÉ, P.-P., Traité de Zoologie, Masson (Paris), v. 6, p. 906-941, fig. 687-719.

Hedgpeth, J. W.

(4) 1947, *On the evolutionary significance of the Pycnogonida:* Smithson. Misc. Coll. (Washington), v. 106, no. 18, p. 1-53, pl. 1, fig. 1-16.

(5) 1954, *On the phylogeny of the Pycnogonida:* Acta Zoologica (Stockholm), Bd. 35, p. 193-213, fig. 1-9.

Marcus, E.

(6) 1940, *Os Pantopoda brasileiros e os demais sul-americanos:* Bol. Fac. Filos. Ciên. Letr. São Paulo, v. 19 (Zool. 4), p. 3-179, pl. 1-7.

Petrunkevitch, Alexander

(7) 1949, *A Study of Palaeozoic Arachnida:* Trans. Conn. Acad. Arts. & Sci., v. 37, p. 69-315, pl. 1-83.

(8) 1952, *Macroevolution and the fossil record of Arachnida:* Am. Scientist, v. 40, p. 99-122, fig. 1-12.

Sars, G. O.

(9) 1891, *Pycnogonidea:* Norweg. N. Atl. Exped., v. 6 (Zool. 20), p. 1-163, pl. 1-5, chart.

Shimkevich, V. M.

(10) 1929, *Mnogokolenchatyie (Pantopoda) fauna SSSR [Pantopodes (Pantopoda) faune de l'U.R.S.S.]:* (Leningrad), pt. 1, cxiv+224 p., 4 pl., 57 fig.

SOURCES OF ILLUSTRATIONS

(11) **Broili, F.**
(12) **Hedgpeth, J. W.**
(13) **Sars, G. O.**
(14) **Shimkevich, V. M.**

PALAEOISOPUS

By Joel W. Hedgpeth

INTRODUCTION

The large, dorsoventrally flattened arthropod named *Palaeoisopus problematicus*, from the Hunsrück shale (L.Dev.) of western Germany, when first discovered was attributed by Broili (1) to the Isopoda. Subsequently, when specimens of the form called *Palaeopantopus maucheri* were obtained from the same strata and region, their resemblance to pycnogonids (order Pantopoda) and to *Palaeoisopus* led Broili to classify both genera as fossil representatives of the Pycnogonida (2, 3). Despite obvious dissimilarities of *Palaeopantopus* and *Palaeoisopus,* both have been accepted as somewhat aberrant ancient pycnogonids by Størmer (8) and Moore *et al.* (5). As one who has made extensive studies of modern pycnogonids, it is obvious to me that whatever affinities the creature known as *Palaeoisopus* has with other arthropod groups, they are surely insufficient to warrant classification of this fossil with the Pycnogonida.

DESCRIPTION

Palaeoisopus is known from 4 specimens which have been described by Broili (1-3), who finally prepared a reconstruction (Fig. 123) showing the animal with outspread appendages as it might appear swimming gracefully amid a group of crinoids in the moderately deep quiet waters of the Hunsrück sea.

This arthropod was a relatively large creature, having a span of approximately 20 cm. across the last pair of legs. It possessed 4 pairs of appendages, the rearmost pair being largest and paddle-shaped, which suggests a swimming habit. In front of the body, composed of 4 somites bearing lateral processes, was an elongate proboscis which was apparently formed by at least 5 distinct segments. The bulbous abdomen was short and segmented; relatively it much exceeded the abdomen of any pycnogonid, differing both in shape and size also from the segmented abdomen of *Palaeopantopus.* The 1st 3 pairs of appendages were apparently 8-jointed, the last 7-jointed; all the legs terminated in hooklike claws. There was no well developed body, thus suggesting the usual form of a pycnogonid, and the cephalic segment seems to have been the smallest. A most curious feature, which *Palaeoisopus* shares with *Palaeopantopus,* is the formation of ringlike markings or structures on the lateral extensions of the body somites. These markings are apparent in only one leg of another nearly complete fossil (3). They are, however, very conspicuous in *Palaeopantopus.* At any rate, this ringlike structure is actually the only external feature, aside from the general reduction of the body to a pycnogonid-like aspect, that these 2 fossils share in common.

This arthropod *Palaeoisopus* is a strange and fascinating object, with its jointed anterior process and bulbous posterior. Broili discerned no less than 5 segments in this proboscis, behind which he counted the 1st trunk segment (Fig. 117). If we accept this interpretation, the last pair of legs falls to the 9th somite (counting from the anterior extremity of the proboscis). This, of course, proves nothing, and confronts us with the difficulty of homologizing several apparently cylindrical somites with those bearing the chelifores, palpi, and ovigers in living pycnogonids. If we are to take this at face value, the anterior nozzle-like structure of *Palaeoisopus* represents a high degree of specialization. The posterior end of the fossil poses some additional difficulties. As Petrunkevitch (7) has pointed out, the loss of abdominal segments is one of the most important evolutionary trends in the Arthropoda. In the living Pycnogonida this reduction has reached an end point, the abdomen being no more than a tubercle with a distal anus, although in a few species it may be rather long and spinose. The large abdomen of *Palaeoisopus* is evidently connected to the trunk by a constriction, suggesting a specialized condition, which if the animal belongs to chelicerate stock at all, seems to set it off on a divergent line from the basic arachnid type. A large abdomen suggests retention of the gonads in the structure, so we cannot say that the Pycnogonida are derived from *Palaeoisopus* by simply pinching off the abdomen at the constriction without somehow getting the gonads through it into the thorax or prosoma, which seems unlikely. In *Limulus,* in which the gonads have anterior branches into the prosoma, the abdomen is broadly joined to the prosoma; if the Pycnogonida have stemmed from some chelicerate stock, it is more reasonable to suppose that the gonads have attained their present anterior position by migrating through a similarly broad connection between the abdomen and the prosoma.

In those chelicerates which have been studied in detail, the germ primordia appear very early in development, although the position of the genital openings is remarkably constant (on the 2nd abdominal segment). It is difficult to see how new genital openings could be developed on all body somites in response to a forward migration of the gonad from the abdomen. However, the possibility of early development of the gonads in the Pycnogonida might have been associated with an inhibition of the original posterior migration. That is, the gonads might have matured before their migration into the posterior part of the body could take place.

It is possible that the ventral groove suggested in the abdomen by Broili in his reconstructed *Palaeoisopus* is the genital orifice (Fig. 123); if so, it seems to be well back of the 2nd abdominal somite, which Petrunkevitch (6) considers one of the few good reference points in the anatomy of fossil arachnids. On the basis of the available evidence, therefore, we cannot even place *Palaeoisopus* near the Chelicerata, and if anything is plain about this fossil, it is that it lacks 2 pairs of antennae and biramous appendages, so it cannot be a crustacean. Possibly it constitutes a larval form of some sort of marine arachnid, although larval forms usually possess anterior ap-

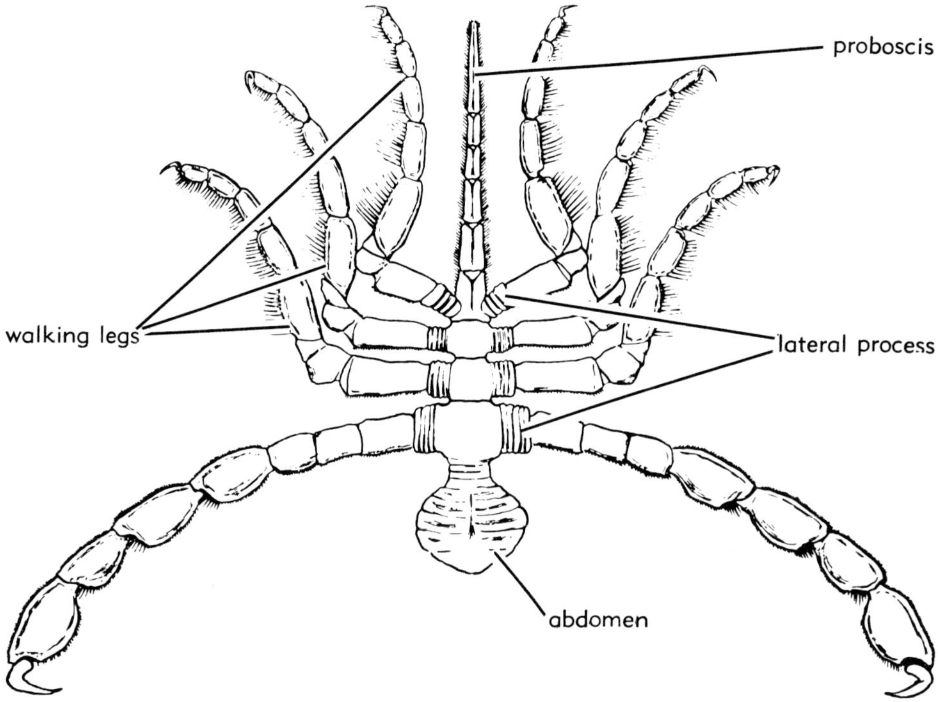

Fig. 123. *Palaeoisopus problematicus* Broili, ventral view (reconstr.), ×0.7 (after Broili)

pendages; the well-developed abdomen suggests maturity and it seems very unlikely that 4 larval specimens of similar form would be found without associated remains of adults. *Palaeoisopus* may be a neotenic form. If it is sexually mature, it seems reasonable to assume that, as in Recent arachnids, a large part of the abdomen was occupied by gonads. In the Pantopoda, the gonads open on the 1st to 4th pairs of legs (4th and succeeding pairs of appendages), a characteristic that distinguishes them from all living Chelicerata (4). No gonopores are discernible on the legs of *Palaeoisopus.*

In summary, all we can say of *Palaeoisopus* is that it belongs among the arthropods. It is classified here as Arthropoda *incertae sedis.*

Palaeoisopus BROILI, 1928 [**P. problematicus*]. Trunk reduced, with 4 segments, as in Recent Pantopoda; proboscis jointed, with 5 separate segments; 4 pairs of legs, the last pair pleopod-like, adapted for swimming; abdomen large, oval-shaped, with at least 5 segments. There are no indications of other appendages or of eyes. *L.Dev.,* Ger.—FIG. 123. **P. problematicus;* ventral side (reconstr.), ×0.7 (after BROILI).

REFERENCES

Broili, F.

(1) 1928, *Crustaceenfunde aus dem rheinischen Unterdevon. I. Über Extremitätenreste:* Sitzungsb. Bayer. Akad. Wiss. (Math.-Naturh. abt.), 1928, p. 197-201, pl. 1.

(2) 1932, *Palaeoisopus ist ein Pantopode:* Same, 1932, p. 45-60, pl. 1-3.

(3) 1938, *Weitere Beobachtungen an Palaeoisopus:* Same, 1938, p. 33-47, fig. 1-5, pl. 1-5.

Hedgpeth, J. W.

(4) 1954, *On the phylogeny of the Pycnogonida:* Acta Zoologica (Stockholm), Bd. 35, p. 193-213, fig. 1-9.

Moore, R. C., Lalicker, C. G., & Fischer, A. G.

(5) 1952, *Invertebrate Fossils:* McGraw-Hill (New York), xiii+766 p., illus.

Petrunkevitch, Alexander

(6) 1949, *A Study of Palaeozoic Arachnida:* Trans. Conn. Acad. Arts & Sci., v. 37, p. 69-315, pl. 1-83.

(7) 1952, *Principles of classification as illustrated by studies of Arachnida:* Systematic Zool., v. 1, p. 1-19, fig. 1-8.

Størmer, Leif

(8) 1944, *On the relationships and phylogeny of fossil and recent Arachnomorpha:* Skrifter Norske Vidensk.-Akad. Oslo (Mat.-Naturv. Klasse), no. 5, pp. 1-158, 30 figs.

INDEX

Names included in the following index are classified typographically as follows: (1) Roman capital letters are used for suprafamilial taxonomic units which are recognized as valid in classification; (2) italic capital letters are employed for suprafamilial categories which are considered to be junior synonyms of valid names; (3) generic and family names accepted as valid and morphological terms are printed in roman type; and (4) generic and family names classed as invalid, including junior homonyms and synonyms, are printed in italics.

Printed in the U.S.A. by Malloy Lithographing, Inc., Ann Arbor, Michigan.